T0182089

Pattern and Security Requirements

Kristian Beckers

Pattern and Security Requirements

Engineering-Based Establishment
of Security Standards

 Springer

Kristian Beckers
paluno—The Ruhr Institute for Software
   Technology
University of Duisburg-Essen
Duisburg
Germany

ISBN 978-3-319-36587-9   ISBN 978-3-319-16664-3   (eBook)
DOI 10.1007/978-3-319-16664-3

Springer Cham Heidelberg New York Dordrecht London
© Springer International Publishing Switzerland 2015
Softcover reprint of the hardcover 1st edition 2015

Printed on acid-free paper

Springer International Publishing AG Switzerland is part of Springer Science+Business Media
(www.springer.com)

To my wife with love and gratitude

Foreword

Information security concerns have to be addressed in most domains. In particular, governments, private and public organizations and other enterprises have to protect themselves against ongoing and evolving security threats. The Norwegian government has issued a Cyber Security Strategy[1] for Norway to address the rising security issues in information and communication technology (ICT). The report states that security has to be analyzed and treated at a system level and not exclusively focus on fixing individual vulnerabilities of technical devices. The goal should be to achieve an acceptable security level for the entire organization.

System level security can be achieved by establishing a so-called information security management system (ISMS). An ISMS is a comprehensive and systematic process that ensures an organization can identify and protect itself adequately against security threats. Creating an ISMS is by no means a simple task and the Cyber Security Strategy for Norway recommends making use of recognized standards to support this process. Security standards such as ISO 27001 are helpful descriptions of core concepts of security management. ISO 27001 is quite general, but most of the established security management methodologies can be understood as instantiations of the activities it prescribes. ISO 27001 explains how to deal with these changing security issues without restricting itself to a specific methodology. Therefore, leading to numerous ambiguities in the text, which has to be resolved when creating a precise and compliant method for security management.

My own experience in information security is based on the design and use of the CORAS method for model-driven risk analysis. CORAS offers a set of well-defined steps to conduct risk analysis in practice. Security risk analysis is an integrated part of security management. Not all security threats identified by risk analysis can be treated, this is due to limited budgets, with assets whose value is smaller than it costs to implement safeguards, and so on. Making the decision on which security threats to treat by a safeguard, requires a cost–benefit analysis.

---

[1]Norwegian Ministries, Cyber Security Strategy for Norway: https://www.regjeringen.no/globalassets/upload/fad/vedlegg/ikt-politikk/cyber_security_strategy_norway.pdf

CORAS is based on the ISO 31000 standard for risk management and its steps are refinements of the activities described in ISO 31000. We relied on ISO 31000 to ensure that we were accommodating the industry best practices. Understanding the standard and resolving its ambiguities was a daunting task. ISO 31000, similar to numerous security and risk management standards, focuses on the what, but explains little to nothing about the how. Therefore, I am painfully aware of the difficulty involved when creating a standard compliant methodology.

This book provides a framework for practitioners, as well as graduate and undergraduate students, particularly in the IT domain, to support them in working with security standards. The support includes showing activities contained in security standards, how supporting methods for security management based on well-known existing methodologies such as CORAS can be created, and how the methods in this book can be applied. These activities are all needed when we want to successfully establish an acceptable security level throughout an entire ICT infrastructure.

An important objective of this book is to support the certification efforts of security standards. To have a certification body to analyze the security management efforts of an organization, and issue a certificate if these efforts are adequate, helps to build trust for the organization. It is based on documenting the actions taken to ensure an acceptable security level is secured. Therefore, security standard-compliant methods not only have to show step-by-step how to conduct a security analysis, but also how to document all activities according to the demands of the standard. Hence, it is of utmost importance to provide descriptions of how to document the results of each step and the methods used.

Methods can rely on formal or graphical models, as well as natural language descriptions. While formal or graphical models help to express circumstances precisely, they must also be understandable for the certification bodies that should certify the security issues based on the documentation. Hence, CORAS provides a mapping from graphical models to English prose. This feature makes CORAS an ideal method for supporting security standards. This book also exemplifies the use of other methods based on Si* and Problem Frames and shows how to create standard-compliant documentation from these, so in this respect, this book provides a generic framework too.

The approach taken in this book also includes the use of general patterns that allow an easy instantiation for a specific domain. Each of these context-patterns have clearly defined inputs and outputs and provides guidance on instantiation through defined instantiation rules. Context-patterns are accompanied by rules for security analysis of their instantiations. The security analysis is conducted systematically in manageable steps by instantiating patterns and analyzing results. The set of patterns can be expanded according to a pattern language outlined in this book. A pattern-based approach for security management has the potential to ease the effort demanded. The reduction of effort is an important goal, because we have to provide small and medium businesses with methods and tools to conduct security management within their budgets.

In contrast to other approaches, this book provides a systematic approach to security management compliant to standards such as ISO 27001 and Common Criteria. This includes methodological guidelines on how to maintain and update security-standard-compliant documentation for certification and to create your own standard-compliant methodologies, something that to a large extent is missing from other security management approaches.

Oslo, Norway, February 2015   Ketil Stølen

Preface

Information security management is a challenging topic, due to the difficulty of exhaustively modeling attackers for an entire system and the threats they cause to it. The idea of security standards and their respective certification schemes is an excellent one. Companies can use a security analysis process in a standard and establish a security product, e.g., a secure software or a process for information security. Security standards are based on best practices from industry and agreed upon in respective consortiums. After a security standard is established, a certification body checks the security product for compliance with the standard. The certification body shall either certify the successful efforts with an official document or provide guidance on how to improve their security product to achieve certification.

The alternative to a security certification process is to deal with security problems at random points in a system. These isolated efforts can prevent the exploitation of vulnerabilities of several parts of a system, but without a structured and systematic method the security level of the entire organization cannot be determined. Furthermore, inventing an effective process with regard to security not based on best practices or standards is extremely difficult. Not to mention the missing certification option, which includes an outside review of the security analysis. A security standard with a certification infrastructure has the potential to help with these problems effectively.

In 2013 the International Organization for Standardization (ISO) registered in total 22 293 ISO 27001 (ISO/IEC, 2005b) certifications[2] and 246 Common Criteria certifications.[3] These numbers show a low adoption rate of security standards in comparison to other standards such as ISO 9001 for which ISO registered 1 129 446 certifications[4] in 2013. These numbers suggest that security standards are still in the early stages of industrial adoption. Note also that the numbers published by ISO do

---

[2]ISO statistic: http://www.iso.org/iso/iso_survey_executive-summary.pdf

[3]Common Criteria Statistic: https://www.commoncriteriaportal.org/products/stats/

[4]ISO statistic: http://www.iso.org/iso/iso_survey_executive-summary.pdf

not reflect the size of the organization. It can be assumed that the numbers are even lower when only considering small and medium businesses, due to the required efforts to establish a security standard.

A scientific analysis of understanding why these standards are not largely adopted in industry and what we can do to improve the situation would be a timely research effort with relevance for our society. We can only encourage researchers to look into this problem in a more detailed analysis than we do in this book. We assume, based on discussions with practitioners, that the root cause for the low adoption of security standards is the ambiguity of their texts. Identifying and removing ambiguities is a timely research subject in the requirements engineering community,which is of particular relevance for information security standards. The reason is that these shall be applicable to all kinds of current and future hardware and software products. This requires the standards to be more ambiguous than other standards, due to the frequent changes in these information technology products. This situation should be addressed by providing structured methods, tools, and other support for resolving the ambiguity issues with information security standards. This book provides several such artifacts as examples for how to address this problem.

Who is This Book For

This book is for software engineers, security analysts, and other professionals that are tasked with establishing a security standard, as well as researchers who aim to investigate the problems with establishing security standards. Furthermore, this book can be used to teach undergraduate students about security standards, security requirements engineering methods, context-patterns, and the relations between these topics. The numerous examples and explanations in this book are meant to be understandable by all these readers.

What Is the Reward for Reading This Book

The book presents an analysis of the construction of security standards and the elementary concepts of security requirements engineering methods. We explain how to use this knowledge to build customized methods supporting security standard establishment. Note that this book does not focus on a single approach, but instead shows how to create step-by-step support using different security requirements engineering methods. The methods can be chosen by individual preferences such as familiarity with a particular method. Providing these multiple ways of achieving security standard support makes this book unique.

In addition, we acknowledge the fundamental importance of context elicitation and analysis for security standards. Numerous security issues in the past were possible due to incomplete context descriptions. Our context-patterns approach helps to close this gap and also to ease the reuse of analysis results. Once readers have understood how to use this approach including describing their own context-patterns, they are empowered to utilize this approach to ease their work with security standards and to improve the results of security standard establishments.

How to Use This Book

This book can be used in one or more of the activities explained below.

Understand and Compare Security Standards using our CAST methodology including available tool support for comparing standards (see Chap. 4). We show a way to describe standards in a high-level and comparable fashion. We described all the standards used in this book using CAST and present the results of a structured comparison. Moreover, the conceptual model CAST is based upon provides insights into the general construction of security standards. In addition, the reader can apply CAST to further standards with little effort.

Use our Methods for Security Standard Establishment presented in this book (see Chaps. 6–8). Each of these methods supports a particular standard and focuses on different aspects of security analysis. Hence, readers can choose the method that supports their demands best. The methods also serve as examples on how different the methodologies for security standards can be, due to their ambiguity. The numerous examples in this book even help readers to discover their preferences and identify the right type of method for their needs.

Create Customized Methods for Establishing Security Standards by using our relations between security standards and security requirements engineering methods (see Chap. 5). This enables engineers to tailor a given SRE method to support security standard establishment. The methods can be selected by any criteria, such as personal preferences or specific analysis demands.

Apply Pattern-Based Security Standard Establishment based on our context-pattern approach, which helps to reuse domain knowledge and support the identification of security analysis tasks that have the potential to be partially automized (see Chaps. 10–13). Our context-pattern language and example methodology show how to use this approach, which includes the option for the readers to describe their own context-patterns and to create methodologies based on these.

In summary, this book can be used to learn about security standards and security requirements engineering. In addition, it shows how to apply several step-by-step SRE-based methodologies for security standard establishment. Moreover, readers can develop the skills for understanding security standards and create methodologies for supporting their establishment. All this is possible by learning about the fundamental concepts of security standards and security requirements engineering methods presented in conceptual models in this book. The reader learns how existing standards and methods relate to these models and how to use these relations to create supporting approaches for security standard establishment. Furthermore, our context-pattern approach illustrates how to create supporting methods with a focus on the reuse of analysis results for particular domains and identifying analysis tasks that can be computer-aided.

Dortmund, Germany, February 2015   Kristian Beckers

Acknowledgments

First and foremost, I thank my wife Clarissa for her encouragement and patience during my countless struggles when writing this book and to give me the strength to follow a topic I believe in.

Moreover, I am grateful for the support of the numerous fellow researchers and practitioners that have discussed and contributed to my research. Maritta Heisel supported me to pursue this research topic at paluno—The Ruhr Institute of Software Technology—at the University Duisburg-Essen as a member of her group and provided valuable feedback. I also thank my colleague Stephan Faßbender for encouraging and insightful discussions about this work. I thank Rene Meis for our productive discourse regarding the formal expressions of this book. My thanks go also to Christina Menges for helping me improve my English writing.

Thanks are also in order for the practitioners of the ITESYS company that supported validating and improving my work from a security standard consultant viewpoint. I am thanking Isabelle Côté, Ludger Goeke, and Denis Hatebur for their help. Furthermore, I thank Thomas Frese from Ford Werke GmbH, for a great collaboration with ITESYS and myself, which gave me an opportunity to show that my research developed for security standard establishment can be transferred to support safety standard establishment.

I am thanking Bjørnar Solhaug and Ketil Stølen from the SINTEF research institute in Norway for helping me to start working in the field of risk management and in particular with their CORAS method. These insights provided the foundation for several of my subsequent research efforts in this field. In addition, my thanks go to Federica Paci and Le Minh Sang Tran from the University Trento for increasing my understanding of goal-based requirements engineering and in particular the application of the Si* notation. I am thanking Aljosa Pasic from ATOS and Jorge Cueller from SIEMENS for their valuable feedback. Furthermore, I am thanking Stefan Hofbauer from Amadeus, as well as, Stefan Fenz and Gerald Quirchmayr from the University Vienna for their support of this work. I am also thanking the Fraunhofer Institute for Software and Systems Engineering ISST for introducing me to the topic of security standard establishment for cloud computing systems.

In particular, my thanks go to Holger Schmidt and Jan-Christoph Küster for insightful discussions regarding this work.

This research was partially supported by the Ministry of Innovation, Science, Research and Technology of the German State of North Rhine-Westphalia and EFRE (Grant No. 300266902 and Grant No. 300267002) and the EU project Network of Excellence on Engineering Secure Future Internet Software Services and Systems (NESSoS, ICT-2009.1.4 Trustworthy ICT, Grant No. 256980). I am grateful for the many opportunities for conducting, validating, and improving my research these projects gave me.

Thanks to Beverley Ford and James Robinson from Springer for their professional help to make this book a reality.

Contents

Acronyms

ADIT	Analysis, Design, Implementation, and Test
AktG	Stock Corporation Act
AO	Tax Code
API	Application Programming Interface
ASIL	Automotive Safety Integrity Level
AURUM	Automated Risk and Utility Management
BDSG	German Federal Data Protection Act
BITKOM	German Federal Association for Information Technology, Telecommunications and New Media
BSI	German Federal Office for Information Security
CAN	Controller Area Network
CAP	Cloud System Analysis Pattern
CBK	Common Body of Knowledge
CC	Common Criteria
CF	Conceptual Framework
CHED	Consumer Home Energy Display
CLASP	Comprehensive, Lightweight Application Security Process
CLS	Controllable Local Systems
CO	Consumers
CPU	Central Processing Unit
CSA	Cloud Security Alliance
CSAP	Cloud System Analysis Pattern Tool
DoS	Denial of Service
EAL	Evaluation Assurance Level
EC	European Commission
EMF	Eclipse Modeling Framework
ENISA	European Union Agency for Network and Information Security
ES	Energy Supplier
ESCL	Electronic Steering Column Lock
EU	European Union
FI	Future Internet

FIPs	Fair Information Practice Principles
GEF	Graphical Editing Framework
GMF	Graphical Modeling Frame- work
GPU	Graphical Processing Unit
GUI	Graphical User Interface
HA	Home Agent
HAN	Home Area Network
HGB	Commercial Code
IaaS	Infrastructure as a Service
ICT	Information and Communication Technology
ICTG	ICT Gateway
IEC	International Electrotechnical Commission
IOI	Items of Interest
ISMS	Information Security Management System
ISO	International Organization for Standardization
ISSRM	Information System Security Risk Management
IT	Information Technology
KA	Knowledge Area
KO	Knowledge Object
KonTraG	Law on Monitoring and Transparency in Businesses
KWG	German Banking Act
LDSG	State Data Protection Acts
LMN	Local Metrological Network
MaRisk	Minimum Requirements for Risk Management
MS	Microsoft
MSDL	Microsoft's Security Development Lifecycle
NESSoS	Network of Excellence on Engineering Secure Future Internet Software Services and Systems
NIST	National Institute of Standards and Technology
NoE	Network of Excellence
OCL	Object Constraint Language
OECD	Organisation for Economic Co-operation and Development
OR	Objectives Rational
OS	Operating System
OSP	Organizational Security Policy
OWASP	Open Web Application Security Project
P2P	Peer-to-Peer
PaaS	Platform as a Service
PACTS	PAttern-Based Method for Establishing a Cloud-specific Information Security Management System
PDCA	Plan-Do-Check-Act
PI	Personal Information
PP	Protection Profile
ProPAn	Problem-Based Privacy Analysis
QoS	Quality of Service

SA	Smart Appliances
SaaS	Software as a Service
SAR	Security Assurance Requirement
SDL	Security Development Lifecycle
SDLC	Software Development LifeCycle
SEPP	Security Engineering Process using Patterns
SFR	Security Functional Requirement
SLA	Service Level Agreement
SM	Small and Medium Enterprises
SME	Smart Meter
SO	Security Objective
SOA	Service-oriented Architectures
SO-OE	Security Objective for the Environment
SPD	Security Problem Definition
SPIT	Spam over Internet Telephony
SRD	Software Requirements Document
SRE	Security Requirements Engineering
SREP	Security Requirements Engineering Process
SRR	Security Resources Repository
ST	Security Target
TMG	Telemedia Act
ToE	Target of Evaluation
UML	Unified Modeling Language
UML4PF	UML for Problem Frames
US	United States of America
VIP	Very Important Person
WAN	Wide Area Network
WpHG	Securities Trading Act

Chapter 1
Introduction

Abstract Security threats are a significant problem for information technology companies. We discuss the possibility to mitigate security threats using security standards and the problems that engineers face when doing so, which are caused by ambiguity in these standards. In particular, we are concerned with the ISO 27001 standard and the Common Criteria standard. We provide an overview of existing research approaches to address this problem and illustrate our research roadmap, which outlines the remainder of this book. The overall research question of this book is: How can patterns and existing security requirements engineering methods support the security analysis and documentation demands of security standards? We show how to refine this research question and provide a step-by-step overview of how each chapter addresses a particular refinement of the overall question. For example, we address how concepts of security requirements engineering methods can be used and improved such that they support the establishment of security standards and similar questions.

1.1 Motivation

Information and Communication Technology (ICT) has a significant impact on our society. An analysis[1] of the German Federal Association for Information Technology, Telecommunications and New Media (BITKOM) estimates the size of the ICT market in Germany of €153 billion and more than 900 000 jobs in 2013. ICT also evolves constantly, e.g., in Germany more than 21 million mobile ICT devices were sold in 2012. BITKOM concludes in a further analysis[2] that this leads to a connection between private and commercial ICT systems via initiatives such as "*Bring your own device.*" This initiative encourages employees to use their private mobile devices, e.g., smartphones at their place of employment for work-related purposes. The study is concerned with the increasing number of security issues, which might be caused

---

[1] BITKOM press notification: http://www.bitkom.org/files/documents/BITKOM_Presseinfo_CeBIT_Jahres-PK_04_03_2013.pdf (last visited on 18.10.2013).

[2] BITKOM IT compass (guidelines for security): http://www.bitkom.org/files/documents/Kompass_der_IT-Sicherheitsstandards_it-sa_Broschuere_Web.pdf (last visited on 18.10.2013).

© Springer International Publishing Switzerland 2015
K. Beckers, *Pattern and Security Requirements*,
DOI 10.1007/978-3-319-16664-3_1

by this and similar developments. For example, the German police reported[3] 63 959 criminal acts involving ICT in 2012 and an increase of ICT criminality of 8 % between 2011 and 2012. Symantec stated in their Norton Cybercrime report[4] for 2012 that the damages of cybercrime worldwide result in losses of $110 billion.

In order to improve the security of an ICT system, all relevant aspects of this system and its environment have to be considered. It is by no means sufficient just to focus on a single security mechanism or safeguard, e.g., firewalls. Several well-known experts in the area such as Schneier (2000), Bishop (2003), Pfleeger and Pfleeger (2007), and Anderson (2008) agree that security is a system property of ICT systems and has to be addressed via "technical, procedural, operational, and environmental safeguards against threats" (Pfleeger and Pfleeger 2007, p. xxi).

Security standards exist that provide procedures to manage ICT security on a system level. This is achieved via establishment of a security standard. We define the term *establishment*[5] of a security standard in this book as follows: "The establishment of a security standard is the effort to realize the procedure described in a security standard with respect to a particular ICT system." The ISO 27001 standard (ISO/IEC 2005) and the Common Criteria (ISO/IEC 2012) are two widely accepted security standards. In 2012 the International Organization for Standardization (ISO)[6] registered 19 577 ISO 27001 certifications, which represents an increase of 13 % as compared to 2011. The ISO also registered 300 Common Criteria certifications in 2012, a 14 % increase compared to 2011.[7] These numbers cover certifications world-wide and show the relevance of security standards.

However, the number of certifications is low in comparison to the overall number of companies in ICT. BITKOM has published a statistic[8] mentioning there are 85 080 ICT companies in Germany. 76 502 of these companies are small and medium enterprises (SMEs) with less than €1 million revenue per year. These are only the companies that make their main revenues in ICT. The overall number of companies that use ICT is much larger. Hence, we can assume that only a small fraction of all companies use security standards. In Germany, 488 ISO 27001 certifications were

---

[3] Bundeskriminalamt cybercrime statistic in Germany: http://www.bka.de/nn_224082/SharedDocs /Downloads/DE/Publikationen/JahresberichteUndLagebilder/Cybercrime/cybercrimeBundeslage bild2012,templateId=raw,property=publicationFile.pdf/cybercrimeBundeslagebild2012.pdf (last visited on 18.10.2013).

[4] Symantec Norton Cybercrime Report 2012 http://now-static.norton.com/now/en/pu/images/ Promotions/2012/cybercrimeReport/2012_Norton_Cybercrime_Report_Master_FINAL_050912. pdf (last visited on 18.10.2013).

[5] This definition is in alignment of the use of the term in ISO 27001 (ISO/IEC 2005, pp. 4 and 9).

[6] ISO statistic: http://www.iso.org/iso/iso_survey_executive-summary.pdf (last visited on 18.10.2013).

[7] Common Criteria statistic: http://www.commoncriteriaportal.org/products/stats/ (last visited on 18.10.2013).

[8] BITKOM statistic on the number of ICT companies in Germany: http://www.bitkom.org/files/ documents/Anzahl_ITK-Unternehmen_2011.pdf (last visited on 18.10.2013).

achieved in 2012[9] and 60 Common Criteria certifications.[10] The possibility exists that companies use security standards without aiming for a certification, but these are not possible to consider, because of a lack of reporting. To sum up, the presented numbers show an alarmingly low rate of security standard certification. This rate has to be improved, and supporting methods and tools are necessary to achieve this goal. Moreover, these methods should support reuse of analysis results and system descriptions in order to reduce the efforts in terms of time for applying them.

In addition, several government agencies such as the German Federal Office for Information Security (BSI)[11,12] and the European Union Agency for Network and Information Security (ENISA)[13,14] endorse the use of these security standards. Moreover, both organizations offer support to establish these particular standards. The offers for supporting security standard establishment from these governmental organizations suggest that the establishment of a security standard is not a simple task.

The BSIMM5 study from 2013[15] investigates the efforts spent by 67 leading ICT companies such as Microsoft, Google, and SAP to ensure software security. The study creates a maturity model for software security with numerous categories. The effort of each company in these categories is rated, but not published in the study for confidentiality reasons. The study published only the average ratings. The BSIMM model can be instantiated for any given company, and the results can be compared with the average ratings documented in the study. The maturity model of the study considers *standards and requirements* as one of its categories.[16] The relation between standards and requirements is fundamental for our work because without considering security requirements, the selection of proper security measures is not possible. We refer to the credo of security requirements engineering (Fabian et al. 2010) that states *How can I select the right measure, if I do not know what right is?* The ISO 27001 and the Common Criteria consider the importance of security requirements (Chaps. 2 and 5 of this book). Hence, if companies do not consider security requirements, the establishment of security standards is not likely to be effective.

---

[9]ISO statistics about ISO 27001 certifications: http://www.iso.org/iso/database_iso_27001_iso_survey.xls (last visited on 18.10.2013).

[10]Common Criteria Portal list of all certifications: http://www.commoncriteriaportal.org/products/certified_products.csv (last visited on 18.10.2013).

[11]German BSI Common Criteria certifications: https://www.bsi.bund.de/EN/Topics/Certification/TechnicalGuidelines/certtechguide.html (last visited on 18.10.2013).

[12]German BSI ISO 27001 usage: https://www.bsi.bund.de/DE/Themen/ITGrundschutz/ITGrundschutzZertifikat/ISO27001Zertifizierung/iso27001zertifizierung_node.html (last visited on 18.10.2013).

[13]ENISA ISO 27001 usage: http://www.enisa.europa.eu/activities/cert/support/guide2/external-relations/management/part-of (last visited on 18.10.2013).

[14]ENISA Common Criteria usage: http://www.enisa.europa.eu/activities/risk-management/current-risk/risk-management-inventory/roadmap/risk-management-integration (last visited on 18.10.2013).

[15]The fifth iteration of the BSIMM study: http://bsimm.com/download/ (last visited on 5.1.2014).

[16]Note that the term standard in the study also includes the creation of security standards for specific technologies or security controls such as authentication.

Furthermore, security standards are ambiguous on purpose, because these standards shall be usable for a large set of different scenarios. The establishment of a security standard requires to remove all ambiguities, elicit concrete security requirements, and select appropriate security measures. Security requirements engineering (SRE) methods exist to support the elicitation and refinement of stakeholders' goals and requirements. They help to remove ambiguities of requirements, but do not support the establishment of security standards. Although some SRE methods use the procedures contained in security standards as an inspiration (Mellado et al. 2006a, b; Schneider et al. 2012), these methods do not support the establishment of the standards. Initial work exists that propose to map the output of security requirements engineering methods to parts of security standards (Rottke et al. 2002; Schmidt 2010; Hatebur 2012). However, these works do not aim at establishing a security standard, either. They lack examples of their applications specifically for standards. These works just provide the idea to reuse generated artifacts for standards. In addition, they do not provide support for fulfilling the documentation demands of standards.

Current research in security engineering provides guidance on how to interpret parts of security standards (Calder 2009; Kersten et al. 2011; Klipper 2010) or to describe the concepts of standards in models, e.g., Cheremushkin and Lyubimov (2010), Lyubimov et al. (2011), Mayer et al. (2007). Several works also propose to add steps to the procedures described in security standards (Ardi and Shahmehri 2009; Yin and Qiu 2010) or to use ontologies to relate important aspects of security standards such as threats, assumptions, and security controls (Bialas 2009; Białas 2009; Chang and Fan 2010; Ekelhart et al. 2009). In summary, there is a lack of security requirements engineering methods, which support engineers in the effort of establishing security standards and satisfying the standards' documentation demands.

However, even if such methods existed, SRE methods have to be repeated entirely each time they are applied. While these methods have the potential to remove the ambiguity of the standards, they do not support reuse of information gathered for a specific scenario. Patterns for similar ICT contexts, which capture the most relevant elements of a scenario, are missing. Such patterns could in turn be used for security analysis for these elements. These patterns can also serve as input for SRE methods or even be integrated into these methods. Combining security requirements engineering methods with patterns allows to remove the ambiguity of security standards, and also to reuse the gained knowledge from removing the ambiguity in future applications of the method.

The work of Gamma et al. (1994) is widespread, which concerns patterns for the design phase of software engineering. The authors use their experience in software development to identify and describe a series of common design issues and solutions. The authors argue that "Graphical notations, while important and useful, are not sufficient. They simply capture the end product of the design process as relationships between classes and objects. To reuse the design, we must also record the decisions, alternatives, and tradeoffs that lead to it." (Gamma et al. 1994, p. 6). The authors state that they "describe design patterns using a consistent format. Each pattern is divided into sections according to the following template. The template lends a uniform structure to the information, making design patterns easier to learn, compare, and use." (Gamma et al. 1994, p. 6). The template includes a unique name, problem description, a solution, and consequences (Gamma et al. 1994, p. 3).

Schumacher et al. (2006) apply the pattern concept to designing secure software, and Hafiz (2006) applies the concept to designing privacy preserving software. Fowler (1996, 2002) proposes analysis patterns for software engineering. These analysis patterns describe organizational structures in models. Few further works exist for the analysis phase of software engineering, e.g., Fernandez et al. (2007) propose analysis patterns that describe Voice-over-IP systems in UML models. However, structured methods for describing how to create analysis patterns for software engineering are missing. Moreover, the application of analysis pattern to security is missing. Besides, some work exists to provide patterns for writing textual software requirements (Withall 2007). Nevertheless, the work does not consider specific patterns for security requirements.

The *Security Engineering Process using Patterns (SEPP)* (Schmidt 2010) decomposes security issues into isolated concerns, e.g., a specific threat. Each security concern is analyzed in relation to existing functional requirements and refined into security requirements. These security concerns are presented in a so-called *security problem frame*, which is a kind of pattern that can be reused for different situations. However, security problem frames capture detailed concerns in isolation and do not try to create patterns for entire kinds of systems such as clouds. A security analysis based on a pattern for, e.g., a cloud enables the reuse of the results for each time a security analysis concerns clouds. The patterns shall use graphical representation and also templates as argued by Gamma et al. (1994). This reuse of security concerns would help security engineers to focus on the individual characteristics of each cloud. Hence, allowing a more effective use of the time of security engineers.

1.2 Research Questions

The motivation of this book leads to the following main research question:

How can patterns and existing security requirements engineering methods support the security analysis and documentation demands of security standards? We have refined the main research question into several more detailed research questions, in order to reduce the complexity of our main question.

RQ 1 Which concepts of security requirements engineering methods can be used and improved such that they support the establishment of security standards?

RQ 2 Are the identified techniques to extend security requirements engineering methods for security standards isolated to the security knowledge area or can the techniques also be applied to other knowledge areas such as safety?

RQ 3 How can security analysis contexts be described in a uniform and reusable way in alignment with the documentation demands of security standards?

RQ 4 Can security standard establishment be based on context-patterns and security requirements patterns?

The contributions of this book are enclosed in the PEERESS (Pattern- and sEcurity-rEquirements-engineeRing-based Establishment of Security Standards) framework. The framework contains extensions to security requirements engineering methods to support ISO 27001 and Common Criteria establishment. In addition, our framework contains the context-pattern approach, a way to describe domain knowledge of systems in a reusable way, and to enable the reuse of security standard compliant security analysis results, as well. Section 1.3 introduces the PEERESS framework in more detail.

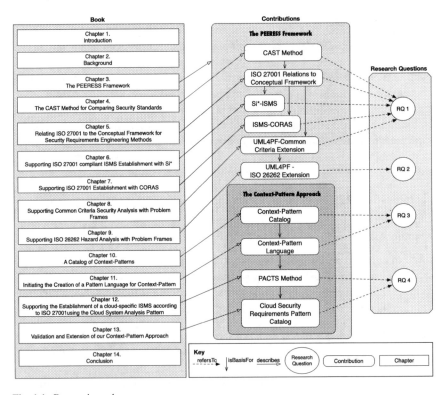

Fig. 1.1 Research roadmap

The principal claim of this book is as follows:

The PEERESS framework is a prime example for using patterns and existing security requirements engineering methods to support the security analysis and documentation demands of security standards.

We illustrate the relations between our research questions, contributions, and the chapters of this book in a research roadmap, depicted in Fig. 1.1. This figure is inspired by the work of Faily (2011). The relations are depicted as arrows, as shown in the key at the bottom of Fig. 1.1. Each contribution of this book *refersTo* at least one research question. A chapter *describes* a contribution and it is possible that one contribution *isBasisFor* another.

1.3 Overview

This book presents the PEERESS framework that contains extensions to security requirements engineering methods (SRE), which particularly support the establishment and documentation demands of security standards. PEERESS also contains another framework (Fabian et al. 2010), which is a conceptual framework for security requirements engineering methods that maps the overall concepts of SREs to several existing SRE methods. We use these mappings from SRE concepts to existing methods to identify suitable SRE methods for establishing security standards, and we introduce the identified methods and relevant standards in Chap. 2. Chapter 3 shows our PEERESS framework, while Chap. 4 explains the CAST method for comparing and analysing security standards, and we explain how we identified suitable SRE methods in Chap. 5. In particular, we contribute an extension to the high-level SRE notation Si* for ISO 27001 (Chap. 6). The resulting method called *Si*-ISMS* investigates the goals and views of all stakeholders of the system. Si*-ISMS considers threats based on structured goal models and considers also the attacker as a stakeholder with particular goals. Hence, the method considers all stakeholder goals and focuses only on the relevant software artifacts to these goals, which is aligned with the demands of ISO 27001. Moreover, since ISO 27001 has a strong emphasis on risk management, we contribute an extension to the risk-management-based SRE method CORAS, called *ISMS-CORAS* (Chap. 7). The SRE method CORAS is a natural fit for this demand, because CORAS focuses on model-driven and asset-based risk management. CORAS identifies assets and determines threats to these assets. CORAS considers only system artifacts that have a relation to an asset and does not represent the complete system. The SRE method Problem Frames uses an abstraction of the complete system and considers the environment of the system. Hence, it is our choice for satisfying the Common Criteria demands for system documentation. We contribute an extension to the ADIT software development process, which is based on UML4PF. UML4PF is an extension of the UML notation for problem frames. Chapter 8 describes our contribution called *UML4PF-CC*, which supports Common Criteria compliant system documentation and security analysis using ADIT. The

Common Criteria has strict demands for the type of documentation, e.g., assets have to be described with certain attributes, among them is the *need for protection* attribute. We extended UML4PF to account for these demands. We provide tool support for creating Common Criteria documents from UML4PF models including automatic consistency checks for UML4PF models, and support for security reasoning with automatic checks, as well. These checks are formalized as OCL (UML Revision Task Force 2010a) expressions, which are shown in detail in Annex A of this book.

Chapter 9 converts the research ideas from the previous chapter from the security to the safety knowledge area, considering the differences of these knowledge areas. We show a UML4PF-based method including tool support for our extension called *UMP4PF-ISO 26262*. This work presents a proof that our claim to use SRE methods to establish security standards also works for the safety knowledge area. Hence, our work is not restricted to the security knowledge area, but can serve further knowledge areas (in our case: safety).

The PEERESS framework contains our context-pattern approach, as well. This approach is founded on a catalogue of context-patterns (Chap. 10) that is based on the idea of creating reusable context descriptions. Many SRE methods begin with an initial context description such as problem frames or CORAS, but these are not reusable for different scenarios. The increasing complexity of systems, e.g., clouds, results also in an increase of the complexity of tasks such as the identification of stakeholders. Our context-patterns help with these and similar issues. We contribute relations between our context-patterns and support identifying further context-patterns in our pattern language for context-patterns (Chap. 11). Chapter 12 illustrates our PACTS method that shows how context-patterns are used to support the establishment of the ISO 27001 standard. Furthermore, we validate our context-pattern approach by discussions with practitioners and application of the work together with the researchers of the ClouDAT[17] project. This collaboration resulted in a catalog of security requirements patterns that have a relation to our cloud specific context-pattern (Chap. 13). The textual patterns can be instantiated with elements from the cloud pattern, and the instantiation can be validated. In addition, the security requirements patterns are mapped to ISO 27001 controls, which reduces the effort of selecting controls. Chapter 14 concludes this book, discusses key findings, provides answers for the research questions, and provides directions for future research.

References

Anderson, R. (2008). *Security Engineering* (2nd ed.). New York: Wiley.
Ardi, S., & Shahmehri, N. (2009). Introducing vulnerability awareness to common criteria's security targets. In *Proceedings of the Fourth International Conference on Software Engineering Advances, 2009. ICSEA* (pp. 419–424). IEEE Computer Society.
Bialas, A. (2009). Ontology-based security problem definition and solution for the common criteria compliant development process. In *Proceedings of the Fourth International Conference on*

---

[17]The ClouDAT project: http://ti.uni-due.de/ti/clouddat/en/.

Dependability of Computer Systems, 2009. DepCos-RELCOMEX (pp. 3–10). IEEE Computer Society.

Białas, A. (2009). Internet—Technical development and applications. In *Ontological Approach to the IT Security Development* (pp. 261–269). Springer.

Bishop, M. (2003). *Computer Security: Art and Science* (1st ed.). Pearson.

Calder, A. (2009). *Implementing Information Security based on ISO 27001/ISO 27002: A Management Guide*. Netherlands: Van Haren Publishing.

Chang, S.-C., & Fan, C.-F. (2010). Construction of an ontology-based common criteria review tool. In *Proceedings of the 2010 International Computer Symposium (ICS)* (pp. 907–912). IEEE Computer Society.

Cheremushkin, D. V., & Lyubimov, A. V. (2010). An application of integral engineering technique to information security standards analysis and refinement. In *Proceedings of the International Conference on Security of Information and Networks* (pp. 12–18). ACM.

Ekelhart, A., Fenz, S., & Neubauer, T. (2009). AURUM: A framework for information security risk management. In *Proceedings of the Hawaii International Conference on System Sciences (HICSS)* (pp. 1–10). IEEE Computer Society.

Fabian, B., Gürses, S., Heisel, M., Santen, T., & Schmidt, H. (2010). A comparison of security requirements engineering methods. *Requirements Engineering—Special Issue on Security Requirements Engineering, 15*(1), 7–40.

Faily, S. (2011). *A framework for usable and secure system design*. Unpublished doctoral dissertation, University of Oxford.

Fernandez, E. B., Pelaez, J. C., & Larrondo-Petrie, M. M. (2007). Security patterns for voice over ip networks. In *Proceedings of the International Multiconference on Computing in the Global Information Technology* (pp. 19–29). IEEE Computer Society.

Fowler, M. (1996). *Analysis patterns: Reusable object models*. Reading: Addison-Wesley.

Fowler, M. (2002). *Patterns of enterprise application architecture*. Reading: Addison-Wesley.

Gamma, E., Helm, R., Johnson, R., & Vlissides, J. (1994). *Design patterns: Elements of reusable object-oriented software*. Reading: Addison-Wesley.

Hafiz, M. (2006). A collection of privacy design patterns. In *Proceedings of the 2006 Conference on Pattern Languages of Programs* (pp. 1–13). ACM.

Hatebur, D. (2012). *Pattern and component-based development of dependable systems*. Baden-Baden: Deutscher Wissenschafts.

ISO/IEC. (2005). Information technology—Security techniques - Information security management systems—Requirements (ISO/IEC 27001). Geneva, Switzerland: International Organization for Standardization (ISO) and International Electrotechnical Commission (IEC).

ISO/IEC. (2012). Common Criteria for Information Technology Security Evaluation (ISO/IEC 15408). Geneva, Switzerland: International Organization for Standardization (ISO) and International Electrotechnical Commission (IEC).

Kersten, H., Reuter, J., & Schröder, K.-W. (2011). IT-Sicherheitsmanagement nach ISO 27001 und Grundschutz. Vieweg+Teubner.

Klipper, S. (2010). Information Security Risk Management mit ISO/IEC 27005: Risikomanagement mit ISO/IEC 27001, 27005 und 31010. Vieweg+Teubner.

Lyubimov, A., Cheremushkin, D., Andreeva, N., & Shustikov, S. (2011). Information security integral engineering technique and its application in isms design. In *Proceedings of the International Conference on Availability, Reliability And Security (ARES)* (pp. 585–590). IEEE Computer Society.

Mayer, N., Heymans, P., & Matulevicius, R. (2007). Design of a modelling language for information system security risk management. In *Proceedings of the International Conference on Research Challenges in Information Science (RCIS)* (pp. 121–132). IEEE Computer Society.

Mellado, D., Fernandez-Medina, E., & Piattini, M. (2006a). A comparison of the common criteria with proposals of information systems security requirements. In *The first International Conference on Availability, Reliability and Security, 2006. Ares.* (pp. 654–661). IEEE Computer Society.

Mellado, D., Fernández-Medina, E., & Piattini, M. (2006b). Applying a security requirements engineering process. In *Proceedings of Computer Security-ESORICS 2006* (pp. 192–206). Springer.

Pfleeger, C. P., & Pfleeger, S. L. (2007). *Security in computing* (4th ed.). Upper Saddle River: Prentice Hall PTR.

Rottke, T., Hatebur, D., Heisel, M., & Heiner, M. (2002). A problem-oriented approach to common criteria certification. In *Proceedings of the 21st International Conference on Computer Safety, Reliability and Security* (pp. 334–346). Springer.

Schmidt, H. (2010). *A pattern- and component-based method to develop secure software*. Baden-Baden: Deutscher Wissenschafts.

Schneider, K., Knauss, E., Houmb, S., Islam, S., & Jürjens, J. (2012). Enhancing security requirements engineering by organizational learning. *Requirements Engineering, 17*, 35–56.

Schneier, B. (2000). *Secrets & lies: Digital security in a networked world* (1st ed.). New York: Wiley.

Schumacher, M., Fernandez-Buglioni, E., Hybertson, D., Buschmann, F., & Sommerlad, P. (2006). *Security patterns: Integrating security and systems engineering*. New York: Wiley.

UML Revision Task Force. (2010a). OMG Object Constraint Language: Reference.

Withall, S. (2007). *Software requirement patterns*. Redmond: Microsoft Press.

Yin, L., & Qiu, F.-L. (2010). A novel method of security requirements development integrated common criteria. In *Proceedings of the 2010 International Conference on Computer Design and Applications (ICCDA)* (Vol. 5, pp. 531–535). IEEE Computer Society.

Chapter 2
Background

Abstract The background that is required to follow the remainder of this book is provided in this chapter. We provide descriptions of security standards, which are supported by various methods and approaches presented in this book, namely the ISO 27001 and the Common Criteria. In addition, we show how to apply our research to the safety, as well. In contrast to security standards that are concerned with protecting a system from attackers, safety standards aim to prevent harm to humans arising from hazards. The safety standard ISO 26262 focuses on the automotive domain and we introduce the standard in this chapter, as well. This book provides methods and techniques of how to apply requirements engineering methods to the establishment of security and safety standards. Hence, we introduce a conceptual framework for security requirements engineering and several of these methods in detail such as Si* and CORAS. Finally, we show the agenda approach, which is the underlying conceptual foundation of all methods contributed by this book.

2.1 Overview

We describe the required knowledge about security standards (Sect. 2.2), the ISO 26262 safety (Sect. 2.3) standard, a framework for security requirements engineering methods (Sect. 2.4), several security requirements engineering methods (Sect. 2.5), and the Agenda Concept (Sect. 2.6) in this chapter.

2.2 Security Standards

In this section we introduce the security standards that are the basis for our work, namely the ISO 27001 (Sect. 2.2.2) and the Common Criteria (Sect. 2.2.4). The ISO 27001 is the mandatory standard of the ISO 27000 series of standards (Sect. 2.2.1). The fundamental difference between these standards is that ISO 27001 describes a process for security management, while the Common Criteria describes how to document a software product.

© Springer International Publishing Switzerland 2015
K. Beckers, *Pattern and Security Requirements*,
DOI 10.1007/978-3-319-16664-3_2

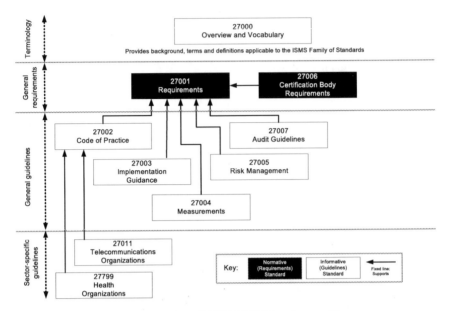

Fig. 2.1   ISO 27000 overview taken from (ISO/IEC 2009) (see footnote 1)

2.2.1 The ISO 27000 Series of Standards

The ISO 27000 series of standards addresses information security matters (ISO/IEC 2009) and the resulting Information Security Management System (ISMS). This is a system independent of vendors, technologies, or the size/type of organization that is part of the management system of an organization (Calder 2009).

The ISO 27000 series of standards is still in development. This means that numerous standards are not published yet. The standards in this family can be divided into several categories. The standards 27000–27005 concern the general description of an ISMS. The mandatory standard in the series is the ISO 27001 (ISO/IEC 2005) that defines how to built an ISMS. Certification of an implementation of the ISO 27001 process is possible. All the other standards of the series are specializations of this standard and describe parts or usage scenarios of the ISMS in detail (ISO/IEC 2009).

The ISO 27000 standard (ISO/IEC 2009) divides the standards of the ISO 27000 series of standards into four categories (see Fig. 2.1[1]).

The ISO 27000 standard itself defines the terminology of the series, the ISO 27001 states the general requirements for an ISMS. General guidelines specify parts of the ISMS, e.g., the ISO 27005 specifies risk management. Sector-specific guidelines

---

[1] Reproduced by permission of DIN Deutsches Institut fr Normung e.V. The definitive version for the implementation of this standard is the edition bearing the most recent date of issue, obtainable from Beuth Verlag GmbH, Burggrafenstraße 6, 10787 Berlin, Germany.

describe how an ISMS is to be implemented in a specific kind of organization, e.g., ISO 27011 concerns telecommunication organizations.

ISO 27000 provides a general overview and the vocabulary used in the ISO 27000 family. The following standards are also part of the ISO 27000 series of standards. ISO 27002 is a so-called *code of practice* and describes controls that can be used in an ISMS implementation.[2] ISO 27003 provides guidance in respect to project management when introducing an ISMS in an organization. ISO 27004 describes measurements for the effectiveness of an ISMS and ISO 27005 concerns risk management. In addition, the risk management in ISO 27005 uses the ISO Guide 73 that defines basics of risk management and the ISO 31000 family that exclusively addresses risk. ISO 31000 describes general principles and measurements and ISO 31010 presents risk assessment techniques (Klipper 2010). The ISO 27007 describes the auditing and certification of an ISMS developed in compliance with the ISO 27001 standard, while the ISO 27006 lists the certification body requirements. Organizations can get accreditation for certifying ISO 27001 realizations.

The remaining standards of the series describe a specific topic in relation to the ISMS. For instance, ISO 27010 describes how to combine different ISMS within one company, ISO 27031 describes business continuity management. The following standards are not published yet. ISO 27032 will provide information for an ISMS with specific types of software applications, e.g., Web 2.0, Software as a Service, and Blogging. ISO 27033 will provide information for the network security part of an ISMS, while ISO 27034's aim is application security. ISO 27035 will describe the incident management and ISO 27036 will provide guidelines for outsourcing (Klipper 2010).

However, even though numerous standards in the ISO 27000 series exist that specialize parts of the ISO 27001, it is not mandatory to use these specializations. The standard also allows to use different specifications, as long as they fulfill the requirements of the ISO 27001 (Klipper 2010). Hence, we focus in our work on the ISO 27001 standard.

The ISO considers also to publish a standard ISO 27017 to provide guidance for implementing an ISMS for clouds and the ISO 27018 to provide privacy guidelines for clouds. Both standards will be released in a draft status soon. However, neither of them will replace the ISO 27001 as the normative standard of the ISO 27000 series of standards.[3] Organizations can get accreditation for certifying an ISO 27001 implementation. However, the German Bundesamt für Sicherheit in der Informationstechnik (BSI) also provides a certification of ISO 27001 based upon their Grundschutz approach (BSI 2011).[4] In practical terms this means that the certification in this case is based on the BSI Grundschutz standards (BSI 2011). Nevertheless, the BSI certification is not accredited by the ISO (Klipper 2010).

---

[2] The standard was formerly known as ISO 17799 and later renamed to ISO 27002.

[3] Online Statement of the ISO 27000 series that ISO 27001 will remain the only mandatory standard of the series: http://www.iso27001security.com/html/27017.html.

[4] Note: This is the Bundesamt für Sicherheit in der Informationstechnik (BSI), a national government body that aims to increase IT security. This is not The British Standards Institution (BSI).

2.2.2 ISO 27001

The ISO 27001 defines the requirements for establishing and maintaining an ISMS (ISO/IEC 2005). In particular, the standard describes the process of creating a model of the entire business risks of a given organization and specific requirements for the implementation of security controls.

The ISO 27001 standard is structured according to the "Plan-Do-Check-Act" (PDCA) model, the so-called *ISO 27001 process* (ISO/IEC 2005). In the *Plan* phase an ISMS is established,[5] in the *Do* phase the ISMS is implemented and operated, in the *Check* phase the ISMS is monitored and reviewed, and in the *Act* phase the ISMS is maintained and improved. In the *Plan* phase, the *scope and boundaries* of the ISMS, its *interested parties, environment, assets*, and all the *technology* involved are defined. In this phase also the ISMS *policies, risk assessments, evaluations*, and *controls* are defined. Controls in the ISO 27001 are measures to *modify risk*. The ISO 27005 (ISO/IEC 2008) refines this process for risk management and extends it with a pre-phase for information gathering. The ISO 27001 standard demands a set of documents that describe how the requirements for the ISMS are fulfilled by a concrete implementation.

Changes in the organization or technology also have to comply with the documented ISMS. Furthermore, the standard demands periodic audits towards the effectiveness of an ISMS. These audits are also conducted using documented ISMS requirements. In addition, the ISO 27001 standard demands that management decisions, providing support for establishing and maintaining an ISMS, are documented as well.

The standard demands a set of documents for certification. In the following we list these documents, giving them names in order to simplify the reference to them later in the chapter.

1. The *Scope of the ISMS*;
2. the *ISMS Policy Statements* that contain general directions towards security and risk;
3. the *Procedures and Controls in Support of the ISMS*;
4. a description of the applied *Risk Assessment Methodology*;
5. a *Risk Assessment Report*;
6. a *Risk Treatment Plan*;
7. documented *Procedures to the effective planning, operation and control of the ISMS*;
8. *ISMS Records* that can provide evidence of compliance to the requirements of the ISMS;
9. the *Statement of Applicability* describing the control objectives and controls that are relevant and applicable to the organization's ISMS;
10. the *Management Decisions* that provide support for establishing and maintaining an ISMS.

---

[5] Note that we defined the term establishment with regard to standards in Chap. 1.

Note that ISO 27001 Sections 4.3.2 and 4.3.3 concern the control of documents and records that shall be specified in document (8). We focus on how to create the document and regard the protection of records as future work and do not consider it in this book.

2.2.3 ISO 27001:2013

In the end of 2013 ISO released a new version of the ISO 27001 standard (ISO/IEC 2013), hereafter called ISO 27001:2013. By the end of 2014 all certification bodies will support only the new version of the standard. Moreover, existing certifications of the old version of the standard, which we refer to if we only write ISO 27001, will have to be updated to ISO 27001:2013 within a period of 2–3 years after the release of ISO 27001:2013.[6] This book is being released during the transition phase of these two versions of the standards. Hence, we consider in our work both versions of the standard. We illustrate first how they support the old version and afterwards discuss the required changes to be compliant to the new version. Thus, we illustrate our research to readers that are familiar with the old version and show them how to transition to the new version. Moreover, we illustrate the origins and the evolution of the standard to readers not familiar with the approach. This way they can not only support the new version, but also help transitioning from one version to the next.

The new version of the standard does not explicitly consider the Plan-Do-Check-Act model anymore. Moreover, ISO 27001:2013 is compliant to a structure for management system standards defined by ISO in Annex SL of ISO/IEC Directives (ISO/IEC 2013, cf. p.1). The ISO 27001:2013 is also closely aligned with ISO 31000 (ISO 2009), a standard for risk management. This has led to changes in terminology, which we explain in Chap.5.

The structure of ISO 27001:2013 begins with a scope definition of the standard, states normative references to other standards, in this case to the new version of ISO 27000 from 2014 (ISO/IEC 2014). Followed by a section for terms and definition, which just contains a reference to the definitions in ISO 27000 (ISO/IEC 2014). The next section demands a context description of the organization aligned with ISO 31000 (ISO 2009). Afterwards the leadership commitments have to be defined. The following section concerns the risk management and security objectives of an ISMS. The remaining sections concern support, operation, evaluation, and improvement of an ISMS. The new version contains also the mandatory Annex A with security controls.

---

[6] http://www.bsigroup.com/en-GB/iso-27001-information-security/ISOIEC-27001-Revision/.

ISO 27001:2013 does not define document types that have to be written, but defines so-called *documentation information* in its sections. The following sections contain documentation information:

1. Section 4.3 demands documentation information about the *Scope of the ISMS*
2. Section 5.2 demands documentation information about the *Information Security Policy*
3. Section 6.1 demands documentation information about the *Risk Assessment* and the *Risk Treatment* including the *Statement of Applicability*
4. Section 6.2 demands documentation information about the *Information Security Objectives*
5. Section 7.2 demands documentation information about the *Competence Records*
6. Section 8.2 demands documentation information about the *Risk Assessment Results*
7. Section 8.3 demands documentation information about the *Risk Treatment Results*
8. Section 9.1 demands documentation information about the *Monitoring and Measuring Results*
9. Section 9.2 demands documentation information about the *Audit Programme and Results*
10. Section 9.3 demands documentation information about the *Management Review Results*
11. Section 10.1 demands documentation information about the *Evidence of Corrective Actions.*

The documentation can contain further documents if some topics have to be elaborated further, as suggested by e.g., Sections 7.5 and 8.1.

2.2.4 Common Criteria

The ISO/IEC 15408—Common Criteria for Information Technology Security Evaluation—(ISO/IEC 2012) is a security standard that can achieve comparability between the results of independent security evaluations from different security analysts of IT products. These are so-called *targets of evaluation (ToEs).*

The Common Criteria is based on a general security model. The model considers ToE owners that value their assets and wish to minimize risk to these assets via imposing countermeasures. These reduce the risk to assets. Threat agents want to abuse assets and give rise to threats for assets. The threats increase the risk to assets.

The concepts of the Common Criteria consider that potential ToE owners infer their security needs for specific types of ToE, e.g., a specific database. The resulting documents are called Security Targets (ST). Protection profiles (PP) state security needs for an entire class of ToEs, e.g., client VPN application. The evaluators check if a ToE meets its ST. PPs state the security requirements of ToE owners. ToE developers or vendors publish their security claims in an ST. A CC evaluation determines if the

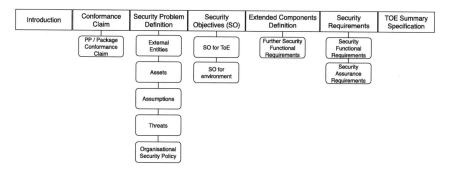

Fig. 2.2 The structure of a CC security target/protection profile (inspired by Fabian et al. 2010)

ST is compliant to a specific PP. The standard relies upon documents for certification, which state information about security analysis and taken measures.

The structure of a CC security target, depicted in Fig. 2.2 inspired by Fabian et al. (2010), starts with an ST *Introduction* that contains the description of the ToE and its environment. The *Conformance Claims* describe to which PPs the ST is compliant. The *Security Problem Definition* refines the external entities, e.g., stakeholders in the environment and lists all assets, assumptions about the environment and the ToE, threats to assets and organizational security policies. The *Security Objectives* have to be described for the ToE and for the operational environment of the ToE. The *Extended Component Definitions* describe extensions to security components described in the CCs part 2. The *Security Requirements* contain two kinds of requirements. The security functional requirements (SFR) are descriptions of security functions specific to the ToE. The security assurance requirements (SAR) describe the measures taken in development of the ToE. These are evaluated against the security functionality specified in the SFR. The Evaluation Assurance Level (EAL) is a numerical rating ranging from 1 to 7, which states the depth of the evaluation. Each EAL corresponds to an SAR package. EAL 1 is the most basic level and EAL 7 the most stringent.

2.3 Safety Standard ISO 26262

We propose a hazard analysis method based on problem frames (Sect. 2.5.3) compliant to ISO 26262 (Chap. 9). This section introduces this standard. ISO 26262 (ISO 2011) is a risk-based functional safety standard concerning safety-related systems that include one or more Electric and Electronic (E/E) systems, which are installed in passenger cars with a max gross weight up to 3500 kg. It addresses possible hazards caused by malfunctions of E/E safety-related systems, including the interaction of these systems and their subsystems.

ISO 26262 is derived from the generic functional safety standard ISO/IEC 61508 (International Organization for Standardization (ISO) and International Electrotechnical Commission (IEC) ISO/IEC 2000). It is aligned with the automotive safety lifecycle including specification, design, implementation, integration, verification, validation, configuration, production, operation, service, decommissioning, and management. ISO 26262 provides an automotive-specific risk-based approach for determining risk classes, which describe the necessary risk reduction for achieving an acceptable residual risk, the so-called *automotive safety integrity level (ASIL)*. The defined ASILs are *QM, ASIL A, ASIL B, ASIL C*, and *ASIL D*. ASIL D requires the highest risk reduction, while for functions with ASIL A, ASIL B, or ASIL C, fewer requirements on the development processes, safety mechanisms, and evidences are necessary. ISO 26262 demands just the normal quality measures applied in the automotive industry for a QM rating.

2.4 A Conceptual Framework for Security Requirements Engineering

Notions and terminology differ in different SRE methods (Fabian et al. 2010). In order to be able to compare different SRE methods, Fabian et al. (2010) developed a Conceptual Framework (CF) that explains and categorizes building blocks of SRE methods. In their survey the authors also use the CF to compare different SRE methods. Karpati et al. (2011) conclude in their survey that the only existing "uniform conceptual framework for translations"of security terms and notions for SRE methods is the work of Fabian et al. (2010). Therefore, we use this CF and base our relations between the ISO 27001 and SRE methods on it. We work on a subset of the CF (see Fig. 2.3 inspired by Fabian et al. 2010) that focuses on concerns relevant for our work. We derived this subset during the research done for this book.

The CF considers security as a system property using the terminology of Jackson (2001) as introduced in Sect. 2.5.3. The CF considers four main building blocks of SRE methods: *Stakeholder Views, System Requirements, Specification and Domain Knowledge*, and *Threat Analysis*, as depicted in Fig. 2.3. *Stakeholder Views* identify and describe the stakeholders and their functional and nonfunctional goals and resulting functional and nonfunctional requirements. Stakeholders express security concerns via security goals. These goals are described toward an asset of the stakeholder, and they are refined into security requirements. *System Requirements* result from a reconciliation of all functional, security, and other nonfunctional requirements, while the stakeholder view perspective focuses on the requirements of one stakeholder in isolation. Hence, the system requirements analysis includes the elimination of conflicts between requirements and their prioritization. The result is a coherent set of system requirements. Requirements are properties the system has after the machine is built. The *Specification and Domain Knowledge* building block consists of specifications, assumptions, and facts. The specification is the description of the interaction behavior of the machine with its environment. It is the basis for the

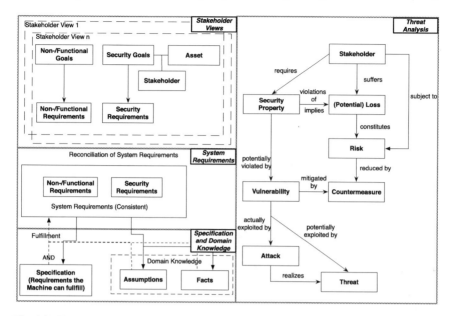

Fig. 2.3 The conceptual framework for security requirements engineering (inspired by Fabian et al. 2010)

construction of the machine. Assumptions and facts make up the domain knowledge. The domain knowledge describes the environment in which the machine will be integrated. In practical terms this means the security requirements have to be reviewed in context of the environment. The *Threat Analysis* focuses on security properties required by stakeholders. A violation of a security property is a potential loss for a stakeholder. This loss constitutes a risk for the stakeholder, which is reduced by countermeasures. A vulnerability may lead to a violation of a security property, and it is mitigated by a countermeasure. Attacks actually exploit vulnerabilities, while threats only potentially exploit vulnerabilities. Attacks realize threats.

Relations Between the Common Criteria and the Conceptual Framework for Security Requirements Engineering
The relations of the Common Criteria with the CF were analyzed by Fabian et al. (2010). We show the relation to the CF in Table 2.1. Fabian et al. (2010) described the relations in text and we show the relations in a table. We listed important terms of the conceptual framework in the left column of this table and stated the equivalent term(s) on the right. If both terms are identical, we listed a "~" in the table. The Common Criteria views security requirements as functional (or assurance) requirements. Other functional or nonfunctional requirements are not considered if they are not relevant to the security functionality, but the standard does not define a clear cut decision criteria for a relevant security functionality. In addition, requirements conflicts are not explicitly considered.

Table 2.1 Correspondence: terms of common criteria and the CF (inspired by Fabian et al. 2010)

CF Fabian et al.	Common criteria
System	~
Machine	The target of evaluation (ToE) is the software/hardware, which should be evaluated
Environment	~
Security goal	The common criteria considers high-level *security needs* of ToE owners
Security requirement	The *security objective* in a protection profile is a refined security need, which is similar to a security requirement being a refined security goal in the terminology of the CF
Specification	The *security objective* in the security target document describes security solutions, which are selected to address the threats. *Security Functional Requirements (SFRs)* are textual patterns in the common criteria, which have to be instantiated for a particular ToE. These refine the solutions of the elicited security objectives for the ToE
Stakeholder	The ToE owner of the common criteria's general security model and users in SFRs are stakeholders in the common criteria
Domain knowledge	The common criteria considers *assumption*s about the environment and also specific *security objectives* for the operational environment of the ToE, e.g., a securely configured firewall
Availability	~
Confidentiality	~
Integrity	~
Asset	~
Threat	~
Vulnerability	~
Risk	~

2.5 Security Requirements Engineering Methods

We introduce requirements engineering methods, which we enhance in our research (Chaps. 5–8). Note that the problem frame-based methods in Sect. 2.5.3 have security specific extensions, but are also useful for software engineering in general.

2.5.1 Si*

The Si* modeling language (Massacci et al. 2010; Asnar et al. 2011) has been proposed to capture security and functional requirements of socio-technical systems. Si* is founded on the concepts of *agent, role, goal, task, resource*, depicted in Fig. 2.4.

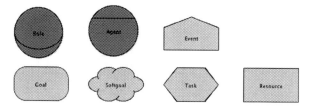

Fig. 2.4 Elements of the Si* notation

An agent is an active entity with concrete manifestations and is used to model humans as well as software agents and organizations. A role is the abstract characterization of the behavior of an active entity within some context. They are graphically represented as circles. Assignments of agents to roles are described by the *play* relation. The term actor refers to role and agent alike and is used in cases where these are not distinguished. A goal is a state of affairs whose realization is desired by some actor (objective), can be realized by some (possibly different) actor (capability), or should be authorized by some (possibly different) actor (entitlement). Soft goals are similar but have no clear criteria for stating if they are fulfilled or not. A task specifies the procedure used to achieve goals. A resource represents a physical or an informational entity without intentionality. A resource can be consumed or produced by a task. Events represent uncertain circumstances that have an impact on the fulfillment of goals or cause security concerns of a resource. In the graphical representation, goals, tasks, resources, and events are, respectively, represented as ovals, hexagons, rectangles, and pentagons. *Contribution* relations are used when the relation between goals is not the direct consequence of a deliberative planning but rather results from side effects. These relations have black arrowheads and a solid line. The impact can be positive or negative and is graphically represented as edges labeled with + and −, respectively. *Impact* relations describe the effect a task or an event has on a resource or a goal. These relations have white arrow heads and a dashed line. Negative impacts are denoted with the sign "−", and for significant negative impact with the sign "− −". Positive impacts are opportunities and denoted with the sign "+", and for significant positive impact with the sign "+ +". Finally, tasks or resources are linked to the goals that they intend to achieve using *means-end* relations. The relations have a solid line and a dart shaped arrowhead.

We provide an example of the Si* notation in Fig. 2.5. The example concerns the medical domain. *Dr. Smith* plays the role *Doctor*. *Dr. Smith* contributes to the soft goal *Quality Health Care* by his goal to *Treat Patients*. The resource *Stethoscope* is a means-end to this goal, as well as the task to *Inspect Patients Lung Functionality*. This task has a positive contribution on the event *Diagnose Lung Illness*. The event represents an uncertain circumstance that has a positive impact on the *Receive Treatment* goal of *Mrs. Jones*, who plays a *Patient*.

Goals and tasks of the same actor or of different actors are often related to one another in many ways. AND/OR decomposition combines AND and OR refinements

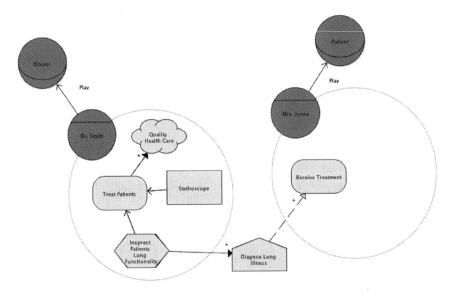

Fig. 2.5 Example healthcare in Si*

of a root goal into subgoals. However, neither are such goals possibly under the control of the actor, nor does the actor may have the capabilities to achieve them.

We illustrate a goal decomposition in Fig. 2.6. The role *Pharmacy* has the goal to *Sell Drugs*, which is decomposed into the goals *Provide Drugs* and *Manage Prescriptions*.

The relations between actors within the system are captured by the notions of *delegation* and *trust*. Assignment of responsibilities among actors can be expressed by *execution dependency*, when one actor depends on another actor for the achievement of a goal. Assignment of responsibilities among actors can also be expressed by *permission delegation*, when an actor authorizes another actor to achieve the goal. Usually, an actor prefers to appoint actors that are expected to achieve assigned duties and not misuse granted permissions. In the graphical representation, permission delegations are represented with edges labeled by **Dp** and execution dependencies with edges labeled by **De**.

Entitlements and capabilities of actors with regard to resources are modeled using the so-called *ECO Model* in Si*. The *own* relationship indicates that an actor has the property rights over a resource; *provide* indicates that an actor has the capabilities to make this resource available to another role or actor.[7] Own and provide are represented with edges between an actor and a resource labeled by **O** and **P**, respectively.

Figure 2.7 shows the application trust and ECO Model relations. A *Patient* has the goal to *Provide Medical Services* and the *Patient* applies a delegation execution

---

[7] Note that in contrast to our work Massacci applies the ECO Model also to relations between actors and goals.

Fig. 2.6 Example Si* goal
decomposition

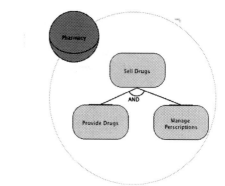

Fig. 2.7 Example trust
relations and ECO model
relations in Si*

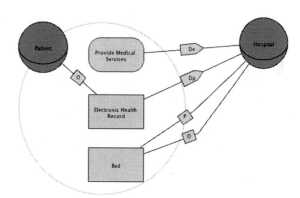

to the *Hospital* of this goal. The *Patient* owns an *Electronic Health Record* that
contains his/her medical information. The *Patient* gives a permission delegation to
the *Hospital* to access and use the *Electronic Health Record*. In addition, the *Hospital*
owns a *Bed*, which is provided to the *Patient* by the *Hospital*.

2.5.2 CORAS

The CORAS method (Lund et al. 2010) is a model-driven approach to risk analysis
and comes with a method, a language to support all steps of the method, as well
as a tool that is used throughout the process to conduct the tasks and document
the results. CORAS is classified as a risk-based security requirements engineering
method, because it works on a similar abstraction level (see the survey from Fabian
et al. 2010).

The method follows the five steps of ISO 31000 (ISO 2009), which are *context
establishment*, *risk identification*, *risk estimation*, *risk evaluation* and *risk treatment*.
The three activities in the middle are referred to as *risk assessment*.

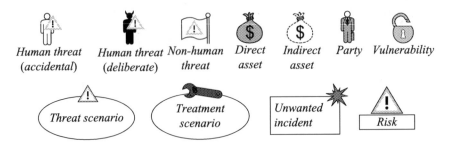

Fig. 2.8 CORAS symbols taken from Lund et al. (2010)

The symbols used in CORAS diagrams are depicted in Fig. 2.8. The elements asset, vulnerability, and threat are essential parts of any risk analysis. Assets are the things of value to a party that a threat can harm by exploiting a vulnerability. A threat scenario states how a threat exploits a certain vulnerability and leads to an unwanted incident. This incident causes harm to an asset. The likelihood or consequence of an unwanted incident result in a risk. If a risk is not acceptable, a treatment scenario has to be implemented that reduces the likelihood or consequence of the unwanted incident that causes the risk. Threats can arise from humans in an accidental or deliberate way. Threats can arise from a nonhuman entity as well. Note that we focus on security analysis in this work and on deliberate human threats.

Step 1: Context Establishment

Context establishment includes defining the scope and focus of the analysis, modeling the target of analysis at an adequate level of abstraction, identifying stakeholders and assets, and defining the risk evaluation criteria. The target is modeled using a (semi-) formal language, such as UML (UML Revision Task Force 2010), and assets are documented using CORAS *asset diagrams*. Figure 2.9 illustrates an example target description. This example concerns an eHealth scenario. A *Patient* of a hospital stores his medical information, e.g., X-rays, previous illnesses, current medication, and further information in an Electronic Health Record (EHR). The EHR is accessed using the *Hospital Information System* (HIS). The HIS and EHR are used by a *Nurse* and a *Doctor*. In addition, the HIS communicates with a *Local Physician*, who is out of scope of this analysis.

We show an example of an asset diagram in Fig. 2.10. The customer that conducts the risk analysis is the *Hospital* that evaluates the use of electronic health records. The risk analysis concerns two assets. The *Electronic Health Records* and the *Public's Trust in Electronic Health Records*. The arrow between the assets represents a *harm to* relationship, meaning harm to *Electronic Health Records* causes also harm to the *Public's Trust in Electronic Health Records*. CORAS considers two types of assets. The harm to a direct asset can be measured, while the harm to an indirect asset cannot. In addition, we can implement treatments for threats for direct assets. The reason for the harm to relation is to identify the direct assets that cause harm to the indirect ones. Treatments to the direct asset shall support risk reduction for indirect

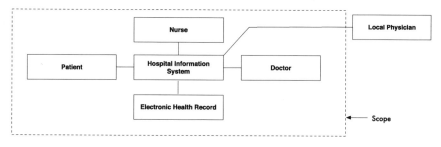

Fig. 2.9   The semi-formal target description of an eHealth scenario

Table 2.2   Example
high-level risk table

Who or what causes it?	How? What is the incident? What does it harm?	What makes it possible?
Hacker	System break-in and theft of electronic health record	Insufficient security

assets, as well. For example, if the confidentiality of the Electronic Health Records is breached and numerous incidents have been reported, the asset is harmed. This would lead to a reduction in the Public's Trust in Electronic Health Records.

   The context establishment also involves a high-level risk identification as part of defining the scope and focus. This is done using so-called *High-level Risk Tables*. We show an example in Table 2.2. The table states the cause of a threat in the first column, the incident and harm in the second column, and the issue that makes the incident possible in the third. The information for the table is the result of expert interviews and other sources of knowledge.

Step 2: Risk Identification
Risk identification is conducted by the identification, modeling, and documentation of threats, threat scenarios, vulnerabilities, and unwanted incidents with respect to the target of analysis and the identified assets. The modeling is conducted using CORAS *threat diagrams*. Figure 2.11 illustrates an example of a threat diagram. In this diagram a *Hacker* exploits the vulnerability *Insufficient Security* and causes the threat scenario *System Break-in and eavesdropping on Electronic Health Records*, which

Fig. 2.10   CORAS asset
diagram

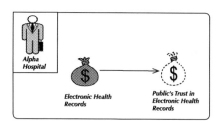

Table 2.3 Example
qualitative likelihood scale

Likelihood value	Description
Likely	A significant number of similar incidents have been recorded
Possible	Several similar incidents on record

Table 2.4 Example
qualitative consequence scale

Consequence	Generic interpretation
Major	Large-scale thefts have happened; can cause significant monetary reparations for the hospital
Moderate	Several thefts on a small scale have happened; can cause monetary reparations for the hospital

leads to the unwanted incident *Theft of Electronic Health Records*. This incident harms the asset *Electronic Health Record*.

Risk assessment can be conducted either quantitative or qualitative. Quantitative risk assessment demands that the likelihood and consequences scales contain numeric values. These have to express in which time frame a risk is likely and what the consequences are in, e.g., number of affected assets. Should these numbers not be available, because the system has not been built yet, likelihood and consequences tables can contain a qualitative scale that does not contain numbers. This is a starting point for risk assessment, and should the numeric values become available, a quantitative risk assessment should be done. We present likelihood (see Table 2.3) and consequence scales (see Table 2.4) for our example.

Step 3: Estimate Risk

The risk estimation is also conducted using threat diagrams, and involves the estimation of likelihoods and consequences for the identified incidents and scenarios. Figure 2.12 illustrates an annotated threat diagram with likelihoods and consequence. The likelihood of the threat scenario is *likely* and the likelihood of the unwanted incident is *possible*. These values are picked at random for this example. The consequence of the unwanted incident is *major* for the Electronic Health Records, because these

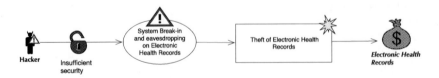

Fig. 2.11 CORAS threat diagram

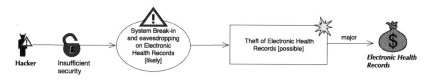

Fig. 2.12 CORAS threat diagram with likelihoods and consequence

Fig. 2.13 CORAS risk
diagram

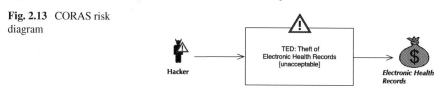

contain the illnesses of a person that shall stay confidential in order to avoid possible blackmail attempts.

We present a risk evaluation matrix for our example (see Table 2.5). The white fields in the matrix represent fields that are acceptable combinations of likelihood and consequences. Gray fields on the other hand are unacceptable risks that have to be treated.

Step 4: Evaluate Risk
Risk evaluation is also conducted using CORAS *risk diagrams*, and includes determining which risks need to be considered for possible treatment by comparing the risks against the evaluation criteria. Figure 2.13 shows a risk diagram for our example. The digram shows that a *Hacker* causes the risk *TED: Theft of Electronic Health Records*. The risk is classified as unacceptable and concerns the asset *Electronic Health Records*.

Step 5: Risk Treatment
Finally, risk treatment is conducted to identify means to mitigate unacceptable risks, and is conducted using CORAS *treatment diagrams*. Our example in Fig. 2.14 shows a treatment diagram. The treatment is to *Implement new Network Security Controls* and it shall reduce the likelihood of a Hacker exploiting the vulnerability *Insufficient security*. Treatments can also be shown in treatment overview diagrams, which show only the threat, risk, asset, and treatments. They do not show vulnerabilities or threat scenarios.

Legal CORAS
Legal CORAS (Lund et al. 2010) is an extension of CORAS specifically for considering legal aspects and legal risk. The approach is based on existing work on legal risk

Table 2.5 Example risk
evaluation matrix

		Consequence	
		Minor	Major
Likelihood	Likely		
	Possible		TED

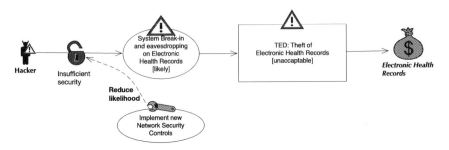

Fig. 2.14 CORAS treatment diagram

Fig. 2.15 The legal aspects symbol in CORAS

management (Mahler 2010). The initial target description in Legal CORAS contains a statement about whether and to what extent legal aspects should be considered in the risk analysis. The method elicits relevant legal aspects based upon the final target description.

The source of legal risks are legal norms, which are norms that stem from legal sources such as laws, regulations, contracts, and legally binding agreements. When assessing legal risks, there are two kinds of uncertainties that must be estimated. First, the legal uncertainty is the uncertainty of whether a specific norm actually applies to circumstances that may arise. Second, the factual uncertainty is the uncertainty of whether these circumstances will actually occur, and thereby potentially trigger the legal norm. It is by combining the estimates for these two notions of uncertainty that we can estimate the significance of a legal norm and its impact on the risk picture. Legal CORAS comes with the necessary analysis techniques and modeling support, but the involvement of a lawyer or other legal experts is usually required.

Legal CORAS introduces a new symbol to represent legal aspects, depicted in Fig. 2.15. The symbol states first the legal source, e.g., a law. Second, the legal norm, e.g., a part of the law and finally the legal consequence.

We show how legal aspects are integrated into a threat diagram in Fig. 2.16. The legal aspects are modeled as an additional information before the unwanted incident, which is a result of the legal aspects. In our example, we name the German Federal Data Protection Act and the incident is that the Hospital is fined, because a Hacker stole Electronic Health Records.

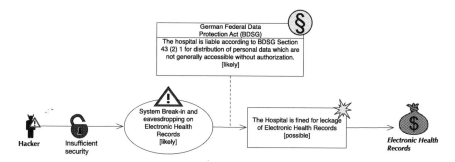

Fig. 2.16   CORAS threat diagram with legal aspects

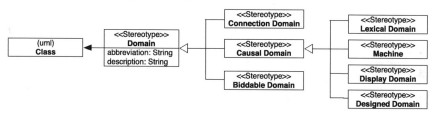

Fig. 2.17   Different domain types (inspired by Côté 2012)

2.5.3 Problem Frame-Based Methods

This section introduces a UML-based version of the problem frame notation, which we are using in this book. This section shows the analysis phase of the ADIT development process, which is based on problem frames, as well. We introduce tool support for ADIT in this section.

Problem Frames

Problem frames are a means to describe software development problems. They were proposed by Jackson (2001), who describes them as follows: "A problem frame is a kind of pattern. It defines an intuitively identifiable problem class in terms of its context and the characteristics of its domains, interfaces and requirement." It is described by a *frame diagram*, which consists of domains, interfaces between them, and a requirement. We describe problem frames using class diagrams extended by stereotypes as proposed by Hatebur and Heisel (2010) (see Fig. 2.17). All elements of a problem frame diagram act as placeholders, which must be instantiated to represent concrete problems. In doing so, one obtains a problem description that belongs to a specific kind of problem. Problem frames are an appropriate means to analyze not only functional requirements, but also dependability and other quality requirements (Hatebur and Heisel 2009; Alebrahim et al. 2011).

The UML profile for Jackson's problem frame notation is called UML4PF. The class with the stereotype machine represents the thing to be developed (e.g., the software). The other classes with some domain stereotypes, e.g., *CausalDomain* or *BiddableDomain* represent *problem domains* that already exist in the application

Fig. 2.18 An example
problem frame (taken from
Côté 2012)

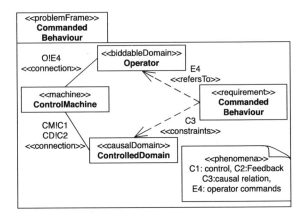

environment. Domains are connected by interfaces consisting of shared phenomena. Shared phenomena may be events, operation calls, messages, and the like. They are observable by at least two domains, but controlled by only one domain, as indicated by an exclamation mark. For example, in Fig. 2.18 the notation *O!E4* means that the phenomena in the set *E4* are controlled by the domain Operator. These interfaces are represented as associations, and the name of the associations contain the phenomena and the domains controlling the phenomena.

Jackson distinguishes the domain types CausalDomains that comply with some physical laws, LexicalDomains that are data representations, and BiddableDomains that are usually people. According to Jackson, domains are either designed, given, or machine domains. The domain types are modeled by the subclasses *BiddableDomain*, *CausalDomain*, and *LexicalDomain* of the class *Domain*. A lexical domain is a special case of a causal domain. This kind of modeling allows one to add further domain types, such as *DisplayDomain*s as introduced in Côté et al. (2008) (see Fig. 2.17).

Problem frames support developers in analyzing problems to be solved. They show what domains have to be considered, and what knowledge must be described and reasoned about when analyzing the problem in-depth.

Software development with problem frames proceeds as follows: first, the environment in which the machine will operate is represented by a *context diagram*. Like a frame diagram, a context diagram consists of domains and interfaces. However, a context diagram contains no requirements. Then, the problem is decomposed into subproblems. If ever possible, the decomposition is done in such a way that the subproblems fit to given problem frames. To fit a subproblem to a problem frame, one must instantiate its frame diagram, i.e., provide instances for its domains, phenomena, and interfaces. The instantiated frame diagram is called a *problem diagram*.

Since the requirements refer to the *environment* in which the machine must operate, the next step consists in deriving a *specification* for the machine (see Jackson and Zave 1995 for details). The specification describes the machine and is the starting point for its construction.

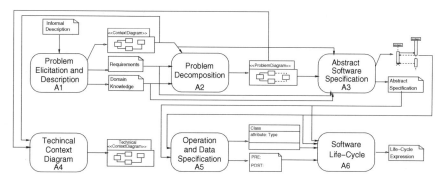

Fig. 2.19 ADIT analysis phase overview (taken from Côté 2012)

The Analysis Phase of the ADIT Software Development Process

In the following, we describe the development process called ADIT (**A**nalysis, **D**esign, **I**mplemen-tation, and **T**est). ADIT is a model-driven, pattern-based development process also making use of components. The process is almost based entirely on the UML notation (UML Revision Task Force 2010) and defines relations between different kinds of models. The ADIT process defines consistency checks of these models including tool support (Côté et al. 2011). The ADIT process also supports the traceability between development phases. For example, the process can answer the question which requirements refer to what design artifacts. The process is described in detail in Hatebur (2012) and Côté (2012). We focus on the analysis phase in this work and present this phase in the following. An overview of the different steps for the analysis phase is given in Fig. 2.19 taken from Côté (2012).

A1—Problem Elicitation and Description   The process begins with a description of the desired functionality of the software to be built. This description is refined into requirements and domain knowledge, which consists of facts and assumptions. We use a *context diagram* using our UML profile and tool support (Côté et al. 2011) that is based on Jackson (2001).

A2—Problem Decomposition   The second step decomposes the *context diagram* into *problem diagrams* (also according to Jackson 2001). These focus each on one or more requirements and the problem that the requirement expresses.

A3—Abstract Software Specification   Problem diagrams do not state the order in which the actions, events, or operations occur. They also concern requirements, which refer to problem domains, but not to the machine. Therefore, it is necessary to transform requirements into specifications (see Jackson and Zave 1995 for more details). We use UML *sequence diagrams* (UML Revision Task Force 2010) as models for our specifications. Sequence diagrams describe the interaction of the machine with its environment. Messages from the environment to the machine correspond to *operations* that have to be implemented. The resulting operations will be specified in detail in Step A5.

A4—Technical Context Diagram   This step describes the technical environment of the machine. For example, a web application machine may use the Apache web server. We use again the UML-based notation mentioned in Steps A1 and

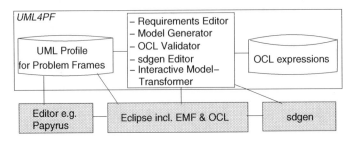

Fig. 2.20 UML4PF tool realization overview (taken from Côté 2012)

A2. We use a specific kind of context diagram, the technical context diagram. The technical context diagram describe technical means, e.g., on the API of the Apache web server.

A5—Operation and Data Specification   The purpose of this step is to set up the necessary internal data structures represented as analysis class diagrams. Furthermore, we specify the operations identified in Step Abstract Software Specification by providing pre- and postconditions for each relevant operation. We use OCL (UML Revision Task Force 2010) to express these operation specifications.

A6—Software Life-Cycle   We use life-cycle expressions proposed by Coleman et al. (1994) to describe the overall behavior of the machine. We express in particular the relations between the sequence diagrams and problem diagrams. This step does not rely on UML, but the notation proposed by Coleman et al. (1994).

UML4PF Tool Support

UML4PF (Côté et al. 2011) is the UML profile and support tool of the problem frame notation, which is used in the ADIT process (Sect. 2.5.3). The UML[8] *profile* contains formal validation conditions expressed in OCL[9] and an *Eclipse*[10] *plugin*.

Figure 2.20 shows UMP4PF's components. White boxes denote components particularly created for UML4PF and gray boxes denote reused components. In the following, we list the UML4PF tool functionalities and items:

- The *UML Profile for Problem Frames* defines the relevant stereotypes for the ADIT method, e.g., ≪ProblemDiagram≫.
- The *Requirements Editor* provides the functionality to add new requirements.
- The *Model Generator* automatically generates model elements, e.g., observed and controlled interfaces from association names.

---

[8] http://www.uml.org/.

[9] http://www.omg.org/spec/OCL/2.0/.

[10] http://www.eclipse.org/.

- The *OCL Validator* checks if a model is valid and consistent by evaluating UML4PF's *OCL expressions*. It also returns the names of invalid parts of the model. All in all, UML4PF contains more than 70 OCL validation conditions for the analysis phase.
- The *sdgen Editor* provides the means to edit sequence diagrams.
- The *Interactive ModelTransformer* serves to create software architectures via interactive model transformations.

2.6 The Agenda Concept

The *Agenda Concept* (Heisel 1998) concerns process knowledge in the context of software engineering. An agenda is a sequence of steps within a process. These steps produce one or more artifacts, e.g., textual documentation, models, formal expressions, or software. Each step provides methodological support via well-defined *activities*, *inputs*, *outputs*, and *validation conditions*. **Inputs** are descriptions of the essential "ingredients"to perform the activities of a step. **Activities** are descriptions of the processing of all inputs into outputs. A step can contain multiple activities. **Outputs** are the desired results of the activities of this step. **Validation conditions** support the detection of mistakes in outputs. These conditions check the results for, e.g., inconsistencies or mistakes made during the activities. A step can only be completed after the application of all validation conditions with positive results.

Agendas aim at making software engineering knowledge explicit and help engineers to learn and repeat efficient and effective methods for software engineering.

References

Alebrahim, A., Hatebur, D., & Heisel, M. (2011). A method to derive software architectures from quality requirements. In *Proceedings of the 18th Asia-Pacific Software Engineering Conference (APSEC)* (pp. 322–330). IEEE Computer Society.

Asnar, Y., Giorgini, P., & Mylopoulos, J. (2011). Goal-driven risk assessment in requirements engineering. *Requirements Engineering, 16*(2), 101–116.

BSI. (2011). BSI Grundschutz Homepage. Bonn, Germany: Federal Office for Information Security (BSI). (https://www.bsi.bund.de/DE/Themen/ITGrundschutz/itgrundschutz_node.html).

Calder, A. (2009). *Implementing information security based on iso 27001/iso 27002: A management guide*. Zaltbommel: Van Haren Publishing.

Coleman, D., Arnold, P., Bodoff, S., Dollin, C., Gilchrist, H., Hayes, F., et al. (1994). *Object-oriented development: The fusion method*. Englewood Cliffs: Prentice Hall.

Côté, I. (2012). *A systematic approach to software evolution*. Baden-Baden: Deutscher Wissenschafts-Verlag.

Côté, I., Hatebur, D., Heisel, M., Schmidt, H., & Wentzlaff, I. (2008). A systematic account of problem frames. In *Proceedings of the European Conference on Pattern Languages of Programs (EuroPLoP)*. Universitätsverlag Konstanz.

Côté, I., Hatebur, D., Heisel, M., & Schmidt, H. (2011). UML4PF—A tool for problem-oriented requirements analysis. In *Proceedings of the International Conference On Requirements Engineering (RE)* (pp. 349–350). IEEE Computer Society.

Fabian, B., Gürses, S., Heisel, M., Santen, T., & Schmidt, H., (2010). A comparison of security requirements engineering methods. *Requirements Engineering—Special Issue on Security Requirements Engineering, 15*(1), 7–40.

Hatebur, D. (2012). *Pattern and component-based development of dependable systems.* Deutscher Wissenschafts-Verlag (DWV) Baden-Baden.

Hatebur, D., & Heisel, M. (2009). A foundation for requirements analysis of dependable software. In *Proceedings of the International Conference on Computer Safety, Reliability and Security (SAFECOMP)* (pp. 311–325). Springer.

Hatebur, D., & Heisel, M. (2010). A UML profile for requirements analysis of dependable software. In *Proceedings of the International Conference on Computer Safety, Reliability and Security (SAFECOMP)* (pp. 317–331). Springer.

Heisel, M. (1998). Agendas—A concept to guide software development activities. In *Proceedings of the IFIP TC2 WG2.4 Working Conference on Systems Implementation: Languages, Methods and Tools* (pp. 19–32). Chapman & Hall London.

ISO. (2011). ISO 26262—Road Vehicles—Functional Safety. Geneva, Switzerland: International Organization for Standardization (ISO).

ISO/IEC. (2000). ISO/IEC 61508 Functional safety of electrical/electronic/programmable electronic safety-relevant systems. Geneva, Switzerland: International Organization for Standardization (ISO) and International Electrotechnical Commission (IEC).

ISO/IEC. (2005). Information technology—Security techniques—Information security management systems—Requirements (ISO/IEC 27001). Geneva, Switzerland: International Organization for Standardization (ISO) and International Electrotechnical Commission (IEC).

ISO/IEC. (2008). Information technology—Security techniques—Information security risk management (ISO/IEC 27005). Geneva, Switzerland: International Organization for Standardization (ISO) and International Electrotechnical Commission (IEC).

ISO/IEC. (2009). Information technology–Security techniques—Information security management systems—Overview and Vocabulary (ISO/IEC 27000). Geneva, Switzerland: International Organization for Standardization (ISO) and International Electrotechnical Commission (IEC).

ISO/IEC. (2012). Common Criteria for Information Technology Security Evaluation (ISO/IEC 15408). Geneva, Switzerland: International Organization for Standardization (ISO) and International Electrotechnical Commission (IEC).

ISO/IEC. (2013). Information technology—Security techniques—Information security management systems—Requirements (ISO/IEC 27001). Geneva, Switzerland: International Organization for Standardization (ISO) and International Electrotechnical Commission (IEC).

ISO/IEC. (2014). Information technology–Security techniques—Information security management systems—Overview and Vocabulary (ISO/IEC 27000). Geneva, Switzerland: International Organization for Standardization (ISO) and International Electrotechnical Commission (IEC).

ISO. (2009). ISO 31000 risk management—Principles and guidelines Geneva. International Organization for Standardization (ISO): Switzerland.

Jackson, M. (2001). *Problem frames: Analyzing and structuring software development problems.* Boston: Addison-Wesley.

Jackson, M., & Zave, P. (1995). Deriving specifications from requirements: An example. In *Proceedings of the 17th International Conference on Software Engineering* (pp. 15–24). ACM.

Karpati, P., Sindre, G., & Opdahl, A. L. (2011). Characterising and analysing security requirements modelling initiatives. In *Proceedings of the International Conference on Availability, Reliability and Security (ARES)* (pp. 710–715). IEEE Computer Society.

Klipper, S. (2010). Information Security Risk Management MIT ISO/IEC 27005: Risikomanagement MIT ISO/IEC 27001, 27005 und 31010. Vieweg+Teubner.

Lund, M. S., Solhaug, B., & Stølen, K. (2010). *Model-driven risk analysis: The CORAS approach* (1st ed.). London: Springer.

Mahler, T. (2010). *Legal risk management*. Unpublished doctoral dissertation, University of Oslo.

Massacci, F., Mylopoulos, J., & Zannone, N. (2010). Security requirements engineering: The SI* modeling language and the secure tropos methodology. *Advances in Intelligent Information Systems*, *265*, 147–174.

UML Revision Task Force. (2010). OMG unified modeling language: Superstructure.

Chapter 3
The PEERESS Framework

Abstract Establishing security standards is a challenging task, because the activities demanded by the standard have to be understood, executed, and well documented. We propose to ease the task by using security requirements engineering methods and patterns. Our Pattern- and sEcurity-rEquirements-engineeRing-based Establishment of Security Standards (PEERESS) framework provides support for using or creating support methodologies for the establishment of security standards. We provide several approaches outlined in the chapters of this book. Moreover, the framework helps in identifying requirements engineering methods or patterns that are useful as a basis for creating a support methodology for security standards. The framework contains exemplary methods for goal-, problem-, and risk management-based requirements engineering, as well as a method based on our context-patterns approach, which bases security analysis on reusable descriptions of systems and their environments. Furthermore, the PEERESS framework can help to create supporting methodologies for other domains as well. In particular, the framework contains an example of such a method for a safety standard.

3.1 Introduction

Requirements engineering methods aim to change reality with respect to a vision of the stakeholders. An example for such a vision is the goal of a company to produce a new car and sell it in significant numbers. Hence, a vision can be expressed using one or more goals that describe a needed change to reality. Goals do not talk about how that change shall be achieved (Pohl 2010, p. 42). In "security requirements engineering (SRE), a requirements engineering process must support engineers in identifying security goals of the security stakeholders" (Fabian et al. 2010, p. 8), meaning the vision focuses on the protection of a system and its assets against attackers.

Security standards support IT security evaluations by documenting and analyzing the overall security level of a system.[1] This analysis considers system descriptions, security goals and their refinement, as well Fabian et al. (2010). We propose in

---

[1] Note that depending on the standard a system can be an organization, an IT product (software or hardware) in its environment, etc.

© Springer International Publishing Switzerland 2015

K. Beckers, *Pattern and Security Requirements*,

DOI 10.1007/978-3-319-16664-3_3

this book to analyze and exploit the relations between SRE and security standards, because SRE methods provide structured techniques for eliciting and refining security goals and requirements. These techniques are vital for removing the ambiguities in security standards (cf. Chap. 1). Moreover, the reuse of information and analysis results is not well explored in research or practice. This statement is based on our experience in the field and on discussions with practitioners (cf. Chaps. 12 and 13). We aim to improve this situation as well.

The PEERESS (Pattern- and sEcurity-rEquirements-engineeRing-based Establishment of Security Standards) framework is our contribution that improves the discussed issues above. We describe the knowledge areas that the elements of our PEERESS framework address in Sect. 3.2 and provide an overview of the framework in Sect. 3.3. Section 3.4 shows how to apply our framework, and Sect. 3.5 summarizes our work in this chapter, which is based on Beckers et al. (2014). In this publication, the author of this book contributed the overall conceptual model for security standards and example applications of the framework. Stefan Fenz supported the risk management aspects of the framework and the other authors helped with improving the example applications.

3.2 Coverage of Knowledge Areas

Security standards are generally created by industrial consortia and governmental organizations,[2] containing the best practices of practitioners. In contrast, security requirements engineering methods are primarily developed by academic researchers (cf., Fabian et al. 2010). For this reason, the consideration of security standards establishment as an integral part of security requirements engineering methods is essential to foster the collaboration between industry and academia in the field of security engineering. The work in this book focuses on the intersection of these knowledge areas, as depicted in Fig. 3.1.[3]

We focus in our work on the security standards *ISO 27001* and *Common Criteria*, due to their relatively large-scale application, as described in Chap. 1. In addition, we identify relations between the Conceptual Framework (CF) for security requirements engineering (Fabian et al. 2010) and the ISO 27001 standard (abbreviated *CF - ISO 27001 Relations* in Fig. 3.1). Note that the CF already contains relations to the Common Criteria (see Sect. 2.4). We analyzed these relations and developed the following contributions. We extended the problem frames method (Jackson 2001), in particular the UML4PF form of the notation (Côté et al. 2011; Côté 2012) including its dependability extension (Hatebur 2012), to support the Common Criteria. Our

---

[2] The technical committee ISO/IEC JTC 1/SC 27 is writing, e.g., the ISO 27001 standard: http://www.jtc1sc27.din.de/cmd?level=tpl-bereich&menuid=63159&languageid=en&cmsareaid =63159 (last visited 20-10-2013).

[3] Note that *Problem Frames with UML4PF* is a (general) requirements engineering method, which we included for brevity's sake in the Security Requirements Engineering Methods.

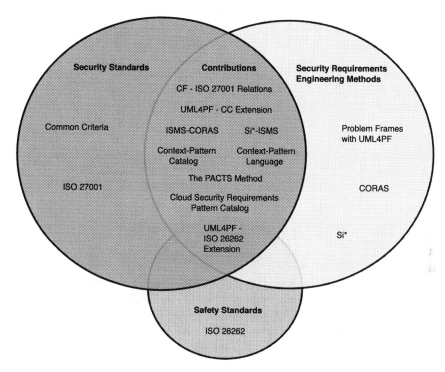

Fig. 3.1 Intersection of knowledge areas of the PEERESS framework (see footnote 3)

extension is called *UML4PF - CC Extension*. In addition, we extend the CORAS (Lund et al. 2010) and the Si* (Massacci et al. 2010) methods to support the ISO 27001 standard. These extensions are called *ISMS-CORAS* and *Si*-ISMS*.

Extended SRE methods for standards support the establishment of the standard, but their application is time consuming. One particular issue is that SRE methods demand their application from their first step on. Hence, every artifact an SRE method produces has to be created from scratch. We experienced a significant loss of time for creating artifacts for similar scenarios. For example, the first artifact that has to be created in UML4PF is a context diagram. When we apply the method twice for a cloud computing scenario we have to draw the context diagram twice. We aim to create specific patterns for SRE methods, which enhance reuse and are accompanied with structured analysis methods that use these patterns. Thus, some parts of a security analysis have to be conducted only once, e.g., the asset identification in a cloud scenario. The aim is to reduce the effort when applying SRE methods for standards via reuse of existing domain knowledge and previous analysis results. We describe a *Context-Pattern Catalog* and a corresponding *Context-Pattern Language* that shows the relations between our context-patterns. We illustrate how these context-patterns support the establishment of the ISO 27001 security standard in our PAttern-based method for establishing a Cloud-specific informaTion Security management system

(PACTS). *The PACTS Method* supports the analysis of the activities demanded by the standard to build an ISMS and presents patterns for these activities. In addition, the method incorporates existing security requirements methods where applicable. We describe a *Cloud Security Requirements Pattern Catalog*, which contains a set of security requirements patterns to enhance our structural cloud context-pattern with textual patterns for security requirements.

Finally, we show that SRE methods can support safety standards as well. In particular, we extend UML4PF to support the ISO 26262 (ISO 2011) standard for functional safety of road vehicles. This contribution is called *UML4PF - ISO 26262 Extension*.

3.3 An Overview of the PEERESS Framework

The PEERESS framework (see Fig. 3.2) supports the establishment of security standards and safety standards, which were introduced in Chap. 2. These are *building blocks* of our framework. Building blocks are existing works that our framework either supports or relies upon. Our framework further consists of *core parts*, which form our conceptual model for security standards. We show how several security standards relate to this conceptual model in Chap. 4.

Another element type of our framework are our *contributions*. Note that the core parts are contributions as well. These are just presented as a different element for illustrative purposes. Our framework contains *relates to* relations between the elements of the framework and activities inside the core parts that are *connected to* each other.

In this framework, we propose to extend requirements engineering and security requirements engineering methods to support security standard establishment. These extended SRE methods are building blocks of our framework as well, and were also introduced in Chap. 2. We also rely upon a conceptual model for security requirements engineering, which was introduced in Chap. 2. We contribute relations from the conceptual model to the ISO 27001 security standard (Chap. 5), while we also rely on already existing relations of the framework to the Common Criteria (Chap. 2). By these proof of concepts we are confident that the conceptual framework can be related to further security standards.

In addition, we use the relations between the mentioned security standards and the conceptual framework for security standards. In particular, we consider three different types of security requirements engineering methods. These types are introduced by Fabian et al. (2010) and are elaborated in Chap. 5. In short, *Goal-based Security Standard Establishment* focuses on identifying the involved stakeholders and eliciting their security goals toward the system-to-be. *Problem-based Security Standard Establishment* decomposes the problem of building a secure system into simple subproblems. *Risk Management Security Standard Establishment* applies methodologies that focus on analyzing and treating risk. We present an example extension of each method type in Chaps. 6–8. The context-pattern approach is based on the idea that

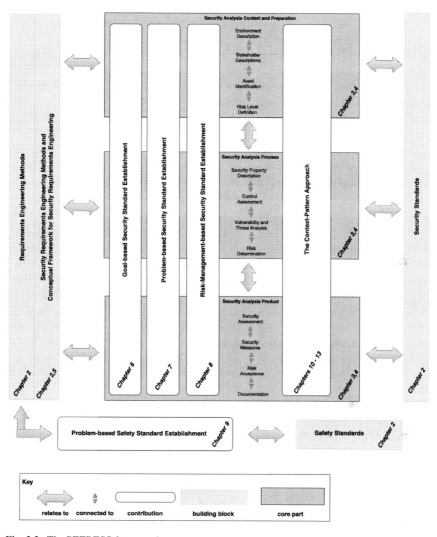

Fig. 3.2 The PEERESS framework

traditional requirements engineering methods have to be applied from scratch each time they are used. Context-patterns describe essential elements of a domain and can be used to support the establishment of security standards. Moreover, the security analysis results are based on a specific context-pattern, which means that some results can be reused for different instantiations of the context-pattern. We illustrate a number of different context-patterns, initiate a context-pattern language, and create an exemplary support method for cloud computing-specific ISO 27001 standard establishment in Chaps. 10–13. The context-pattern approach can be applied to other standards and domains, e.g., for Peer-to-Peer systems and the Common Criteria standard as well. This enables reuse of security analysis results and documentation artifacts for the domain Peer-to-Peer systems and Common Criteria standard establishment. Furthermore, our framework can be applied to support the establishment of safety standards. We provide a proof of concept example with an application and extension of a requirements engineering method to the ISO 26262 standard in Chap. 9. Hence, the PEERESS framework provides support beyond the security knowledge area. We will conduct further research into the applicability of the framework to additional knowledge areas. The core part of the PEERESS framework is a conceptual model for security standards published in Beckers et al. (2014) (cf. Chap. 4). The model consists of a sequence of *Standard Activities*. These are the activities that have to be conducted to establish a security standard. We structure our conceptual model into three parts: *Security Analysis Context and Preparation*, *Security Analysis Process*, and *Security Analysis Product* (see the gray boxes in the middle of Fig. 3.3). Note that inside each part of the framework we state the chapter of this book that elaborates on that particular part.

We explain the *Security Analysis Context and Preparation* part in the following. We split the *scope identification* of standards (cf. Sunyaev 2011) into an *environment description* and a *stakeholder description*. The reason is that security is about protection of assets, and harm to assets results in a loss to stakeholders. We have to understand the significance of the loss by describing the stakeholder. Moreover, stakeholders can cause threats to assets, and the identification of stakeholders in a scope is a research problem (Sharp et al. 1999; Pouloudi 1999; Ballejos and Montagna 2008). Moreover, we consider the activity *Risk Level Description* to include a mechanism to categorize assets already in the beginning of the security analysis. This is done to focus security analysis on assets with a high-risk level, as suggested by NIST SP 800-30 (Stoneburner et al. 2002) and IT Grundschutz (Bundesamt für Sicherheit in der Informationstechnik (BSI) 2008).

We describe the activities contained in the *Security Analysis Context and Preparation* part in the following and illustrate how it can be applied to support the understanding of security standards in Chap. 4.

Environment Description   The environment description states the scope of the standard. Hence, the environment in which the security system shall be integrated into, e.g., an organization or an information and communication technology (ICT)-based system or combinations of both.

Stakeholder Description   The stakeholder description describes all relevant persons, organizations, and government bodies that have a relation to the environment.

Asset Identification   The asset identification for the stakeholders collects all information or resources that have a value to the stakeholders. The assets shall be protected from harm caused by the environment.

Risk Level Description   For each asset a risk level description states the impact the loss of an asset has on a stakeholder. Hence, the risk level description classifies the assets into categories according to their significance for the environment.

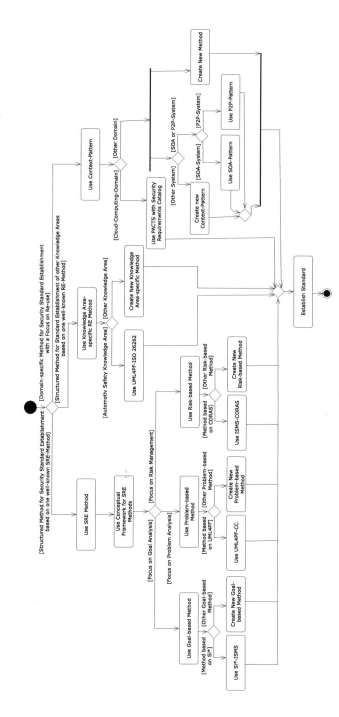

Fig. 3.3   Application of our PEERESS framework

In this part the risk level determination is based on the opinion of stakeholders and described on a high level of abstraction.

We explain the activities of the *Security Analysis Process* in the following. The initial activities are a security property definition for assets and an assessment of existing controls. The security properties provide an overview of high-level security goals, which should be separated from the *Control Assessment*, since it considers existing security solutions. Moreover, we combined the threat analysis and vulnerability identification, because threats exploit vulnerabilities (Fabian et al. 2010) and should be considered together in our view. We add also a *Risk Determination* activity to the *Security Analysis Process* that describes how likelihoods and consequences for the resulting threats are assessed.

Security Property Description   We initiate the *Security Analysis Process* with a high-level security property description, which determines security goals for assets. For example, the ISO 27001 standard uses high-level security objectives to "establish an overall sense of direction and principles for action with regard to information security" (ISO/IEC 2005, p. 4) as part of their ISMS policy. The ISMS policy includes all security policies described during the establishment of an ISO 27001 Information Security Management System.

Control Assessment   The control assessment determines which controls (either technical ones such as encryption mechanisms or nontechnical controls such as security policies) are already in place and their ability to ensure a security property of an asset.

Vulnerability and Threat Analysis   The threat analysis starts with assumptions about vulnerabilities of an asset. If these assumptions are validated by proving that the assumed vulnerabilities exist, threats are documented that could exploit the vulnerabilities. In general, a threat requires a source and an existing vulnerability to become effective. The threat source can either intentionally or accidentally exploit a vulnerability. The aim of the threat identification is to determine potential threats and their corresponding sources such as human attackers and the attacks they conduct, e.g., network attacks, theft, unintentional data alteration.

Risk Determination   The risk determination determines useful likelihood and impact scales to conduct risk management for assets. The risk determination considers the output of all previous activities and evaluates these results with regard to risk, considering the likelihood and impact scales (see Chap. 7 for an example).

Finally, we explain the *Security Analysis Product* part, which contains *Security Assessment*, *Security Measures*, *Risk Acceptance*, and *Documentation* activities. *Risk Acceptance* is an essential activity of finishing the security analysis product, and if risks are accepted to soon, the entire security analysis product might not be effective. Hence, we aim to document in the template how the standards address this issue. In addition, the certification process of a security standard is usually based on the documentation of the security analysis product. That is why we want to add a description of the demanded documentation in our model.

Security Assessment   The security assessment evaluates if the existing security controls satisfy the security properties of the assets considering the results of the *Vulnerability and Threat Analysis*, as well as the *Risk Determination*. This step also describes how further security controls have to be selected. For example, the ISO 27001 standard (ISO/IEC 2005) has a mandatory Annex A from which controls have to be selected.

Security Measures   The security measures activity specifies a list of new, refined, or existing security controls that are required to improve the protection of the assets. This final result of the selection of controls are the *Security Measures*. For example, the ISO 27001 demands a so-called *Statement of Applicability* that reasons about the necessity of the controls in Annex A.

Risk Acceptance   The risk acceptance evaluates if the *Security Measures* reduce the risk of attacks on assets to acceptable levels. Often a clear-cut criteria has to be defined that is fulfilled or not. For example, the controls prevent threats from attackers with a mediocre skills level and a limited amount of time.

Documentation   The security system description finishes with the security and risk documentation of the security analysis product. The documentation usually has to follow certain guidelines of a standard.

3.4 Application of Our PEERESS Framework

In this section, we illustrate how to apply the PEERESS framework in a UML activity diagram (UML Revision Task Force 2010) in Fig. 3.3. We explained in the previous sections that the establishment of security standard is a problem, due to ambiguous descriptions in standards. We propose to address this problem by creating or using a structured method specifically tailored to refine the ambiguities of the security standard. Hence, the overall approach of the PEERESS framework can be categorized into three groups (see Fig. 3.3): *Use SRE method* if an extension of a well-known[4] SRE method is desired, *Use Knowledge-Area specific RE-Method* if the standard does not concern the security knowledge area,[5] *Use Context-Pattern* if the method shall be tailored to a specific domain, e.g., the cloud computing domain, and focus on reuse which the patterns support. We created at least one proof of concept method for each category. In the following, we explain these separate categories:

   Use SRE Method—We propose to consider the conceptual framework for security requirements engineering by Fabian et al. (2010) (Chaps. 2 and 5) to find a method. Fabian et al.'s framework considers goal-based methods that focus on actors' goals in the system, problem-oriented methods that focus on refining prob-

---

[4] We define a well-known SRE method as follows: An SRE method is well-known if it has been published at least twice in an international venue. This definition is in alignment with Fabian et al. (2010).

[5] We assume that it is also possible to base such a method for other knowledge areas on context-patterns, but a proof of concept is part of our future work and not considered at this point in time.

lem descriptions that a system shall address, and risk-based methods that focus on
risk management. The framework contains mappings from SRE methods to the ISO
27001 and the Common Criteria (ISO/IEC 2012) standards (Chap. 2 and 5). These
mappings help to extend a method to support the establishment of these standards. We
have extended one method of each category to show the applicability of our approach.
Thus, users of our approach can use the goal-based method Si*-ISMS (Chap. 6) to
establish the ISO 27001 standard, the problem-based method UML4PF-CC (Chap. 7)
to establish a security analysis compliant with the Common Criteria standard, and
the risk-based method ISMS-CORAS (Chap. 8) to establish the ISO 27001 standard.
In addition, users can use the conceptual framework and its mappings to the standard
to create their own goal-, problem-, or risk-based methods. The decision of what
method to use depends on the preferences for and familiarities with a method of
the users and the central focus of the resulting standard documentation, and on the
security standard that shall be established. For example, if users are already familiar
with Si*, the resulting analysis shall show the goals of the stakeholders explicitly,
and if the concerned standard is the ISO 27001, the Si*-ISMS might be the right
method.

Use Knowledge-Area specific RE-Method—Our approach can be applied to
other knowledge areas as well. One has to find a requirements engineering method
of that knowledge area and extend it to support the establishment of that standard.
We have extended UML4PF (Chap. 2) for the ISO 26262 standard (Chap. 2) that
concerns automotive safety as a proof of concept. Our method UML4PF-ISO 26262
(Chap. 9) can be used to establish an ISO 26262 compliant hazard analysis. This can
be done in a similar manner for any other knowledge area as well.

Use Context-Pattern—We illustrate how a method works that is based on a
context-pattern specific for the cloud computing domain. Thus, our PACTS method
(Chap. 12) can be used for establishing the ISO 27001 standard specific for the
cloud computing domain. In addition, our catalog of textual security requirements
patterns for clouds can support the establishment even better (Chap. 13). The method
shows how parts of the security analysis can be conducted on only the context-
pattern and, thus, the results of the analysis can be applied to all instantiations of the
patterns. For example, an asset identification can be conducted on the pattern and the
identified assets apply to different kinds of clouds, meaning different instantiations
of the cloud pattern. Moreover, the context-pattern's primary concern is to elicit
domain knowledge in a structured way. Hence, a method has to accompany the
pattern that concerns security analysis. This method can in turn use SRE methods. For
example, our PACTS method relies on misuse cases (Chap. 12) for eliciting security
requirements. In the case that the desired domain is not cloud computing, we have
a catalog of context-patterns (Chap. 10) that can be used as a basis for a method
considering Service-Oriented Architectures (SoA) or Peer-to-Peer (P2P) systems. In
addition, we have initiated a pattern language for context-patterns (Chap. 11) that

can be used to describe context-patterns for further domains such as smart grids.[6] All of these context-patterns require a method to support the security analysis and the specific demands for establishing a standard. These methods can use PACTS as a reference. Furthermore, our pattern language contains a meta-model for context-patterns, and the analysis steps of PACTS that concern a certain element of the meta-model might be used as a reference for a threat analysis concerning the same element, but in another context-pattern. For example, the threat analysis in PACTS concerns threats caused by stakeholders that are part of the cloud pattern. PACTS considers the relations these stakeholders have to the cloud and what assets they could harm. These ideas could be reused for a threat analysis concerning stakeholders in another context-pattern.

In summary, we propose to support the establishment of security standards with either extended SRE methods or context-patterns specifically tailored to support security standards such as ISO 27001 and Common Criteria. SRE methods have the advantage that software engineers are familiar with them and extending them to support security standard establishment enables to teach software engineers to become familiar with security and even to establish security standards. Shostack (2012) from Microsoft argues that teaching software engineers about security is more favorable than using security engineers to conduct the threat analysis, because security engineers have to invest a lot of time to understand the work of the software engineers in order to discover vulnerabilities. In contrast, software engineers are more familiar with the possible vulnerability of their system if they are taught about threat analysis (Shostack 2012). Moreover, we explored the idea of describing domain knowledge in context-patterns and using these context-patterns to create methods to support security standard establishment, which focus on reuse, meaning analyzing as much as possible on the pattern level and reusing the results for all the instantiations of the pattern. Hence, security engineers can focus on the specifics of an instantiation and automate the recurring parts of the security analysis such as identifying assets. At last, we applied our research idea to support standard establishment with requirements engineering methods to the safety knowledge area. The results of supporting the ISO 26262 Hazard Analysis using a requirements engineering method shows that the idea can be adapted for the safety knowledge area. Furthermore, we are planning to use context-patterns to support the establishment of safety standards as well.

3.5 Summary

We presented the PEERESS framework in this chapter, which provides pattern and security requirements engineering-based support for removing the ambiguities of security standards and provides structured methods for standard establishment, including support for their documentation demands. In addition, the framework has

---

[6] A smart grid is an intelligent energy distribution system consisting of multiple information and communication technologies (ICT).

been proven to be adaptable for standards of further knowledge areas, e.g., safety. The core of the framework relies upon the context-pattern approach, which enables the creation of reusable artifacts and analysis results of security standard establishment. We described in this chapter the overall framework and a process of how to apply the framework to gain support for standard establishment.

Our PEERESS framework provides the following main benefits:

- Integration of further security standards with little effort due to the underlying conceptual model for security standards. PEERESS already contains relations to the ISO 27001 and the Common Criteria standards, and the conceptual model allows to provide relations to further standards with little effort.
- Selection of preferred security requirements engineering methods (SRE), because our methods provide the means to extend any SRE method for a standard by relying on a conceptual framework for security requirements engineering methods. Moreover, PEERESS already contains several extensions to well-known SRE methods compliant to security standards.
- Context-specific methods for supporting security standard establishment due to the context-pattern approach. Our context-pattern catalog contains patterns for specific domains such as clouds or peer-to-peer systems and our pattern language allows to describe further context-patterns. PEERESS also contains the PACTS method for cloud-specific support for ISO 27001 establishment.
- Extensibility toward further knowledge areas is possible by transferring the knowledge and methods from one area to another. PEERESS contains so far one transfer of a methodology from a problem-based SRE method that supports a security standard to a method for the safety domain. It is likely that this can be repeated with further SRE methods to different areas.

The PEERESS framework provides a holistic support for extending security requirements engineering methods to support the establishment of a specific security standard and provides the context-pattern approach to strengthen the reusability of domain-specific security standard establishment. The framework relies on generic concepts such as meta-models to provide the means for creating and exploiting relations between security standards, SRE methods, and context-patterns. This enables a broad applicability of our research in order to provide support for security standard establishment in a fashion preferred by the individual security engineer.

References

Ballejos, L., & Montagna, J. (2008). Method for stakeholder identification in interorganizational environments. *Requirements Engineering, 13*(4), 281–297.
Beckers, K., Côté, I., Fenz, S., Hatebur, D., & Heisel, M. (2014). A structured comparison of security standards. In *Advances in engineering secure future internet services and systems* (pp. 1–34). Springer.
BSI. (2008). Standard 100–1 Information Security Management Systems (ISMS). (Version 1.5). Bonn, Germany: Federal Office for Information Security (BSI).

Côté, I. (2012). *A systematic approach to software evolution*. Germany: Deutscher Wissenschafts-Verlag.

Côté, I., Hatebur, D., Heisel, M., & Schmidt, H. (2011). UML4PF—A tool for problem-oriented requirements analysis. In *Proceedings of the International Conference on Requirements Engineering (RE)* (pp. 349–350). IEEE Computer Society.

Fabian, B., Gürses, S., Heisel, M., Santen, T., & Schmidt, H. (2010). A comparison of security requirements engineering methods. *Requirements Engineering—Special Issue on Security Requirements Engineering, 15*(1), 7–40.

Hatebur, D. (2012). *Pattern and component-based development of dependable systems*. Germany: Deutscher Wissenschafts-Verlag.

ISO. (2011). ISO 26262—Road Vehicles—Functional Safety Geneva. International Organization for Standardization (ISO): Switzerland.

ISO/IEC. (2005). Information technology—Security techniques—Information security management systems—Requirements (ISO/IEC 27001). Geneva, Switzerland: International Organization for Standardization (ISO) and International Electrotechnical Commission (IEC).

ISO/IEC. (2012). Common Criteria for Information Technology Security Evaluation (ISO/IEC 15408). Geneva, Switzerland: International Organization for Standardization (ISO) and International Electrotechnical Commission (IEC).

Jackson, M. (2001). *Problem frames: Analyzing and structuring software development problems*. Boston: Addison-Wesley.

Lund, M. S., Solhaug, B., & Søtlen, K. (2010). *Model-driven risk analysis: The CORAS approach* (1st ed.). New York: Springer.

Massacci, F., Mylopoulos, J., & Zannone, N. (2010). Security requirements engineering: The si* modeling language and the secure tropos methodology. *Advances in Intelligent Information Systems, 265*, 147–174.

Pohl, K. (2010). *Requirements engineering fundamentals, principles, and techniques*. New York: Springer.

Pouloudi, A. (1999). Aspects of the Stakeholder Concept and their Implications for Information Systems Development. In *Proceedings of the Hawaii International Conference on System Sciences (HICSS)* (pp. 5–8). IEEE Computer Society.

Sharp, H., Finkelstein, A., & Galal, G. (1999). Stakeholder identification in the requirements engineering process. In *Proceedings of the Dexa Workshop* (pp. 387–391). IEEE Computer Society.

Shostack, A. (2012). Elevation of privilege: Drawing developers into threat modeling Technical Report. Redmond, U.S.: Microsoft. (http://download.microsoft.com/download/F/A/E/FAE1434F-6D22-4581-9804-8B60C04354E4/EoP_Whitepaper.pdf).

Stoneburner, G., Goguen, A., & Feringa, A. (2002). *Risk management guide for information technology systems* (NIST Special Publication No. 800–30). Gaithersburg, U.S.: National Institute of Standards and Technology (NIST).

Sunyaev, A. (2011). *Health-care telematics in Germany—Design and application of a security analysis method*. Wiesbaden: Gabler.

UML Revision Task Force. (2010). OMG unified modeling language: Superstructure.

Chapter 4
The CAST Method for Comparing Security Standards

Abstract Working with security standards is difficult, because these are long and ambiguous texts. The time spent to understand what activities and documents are necessary to establish the standard is significant. Furthermore, comparing standards is even more time consuming, because this process has to be done multiple times. We propose a structured methodology called CAST that helps to understand and compare security standards by using a template derived from existing standards. Our template contains specific sections for each standard activity. Moreover, we defined a common terminology for security standards that serves as a baseline for comparing the terminology of the standards. We show instantiations of the template for the standards ISO 27001:2005, ISO 27001:2013, Common Criteria, and IT Grundschutz. Our results contain an analysis of these instantiations that shows the different approaches of these standards and their differences in terminology. The CAST method can be applied to further standards with little effort.

4.1 Introduction

We illustrate in this chapter how to operationalize the conceptual model for security standard introduced in Chap. 3 in such a way that it can be used in a structured method to compare security standards. A number of different security standards exist and it is difficult to choose the right one for a particular project or to evaluate if the right standards was chosen for a certification. These standards are often long and complex texts, reading and understanding of which takes up a lot of time. We provide a conceptual model for security standards that relies upon existing research and contains concepts and phases of security standards. In addition, we developed a template based upon this model, which can be instantiated for given security standard. These instantiated templates can be compared to understand the differences of security standards. This chapter is based on the works published in Beckers et al. (2014).

© Springer International Publishing Switzerland 2015   51
K. Beckers, *Pattern and Security Requirements*,
DOI 10.1007/978-3-319-16664-3_4

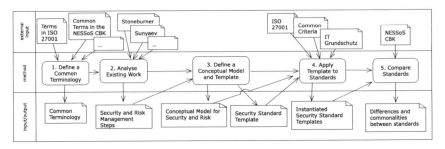

Fig. 4.1 A method for CompAring SecuriTy standards (CAST)

4.2 A Method for Comparing Security Standards

In the following, we present the steps of our method for CompAring SecuriTy standards (CAST) (see Fig. 4.1).

1. **Define a Common Terminology** The Jason institute evaluated the research field of security (JASON 2010) and concluded that the field is missing a common terminology and a basic set of well-defined concepts. We address this concern by defining a common terminology against which the terms of the standards are evaluated. We use the terminology of the ISO 27001 standard and the terms defined in the common body of knowledge (CBK)[1] of the EU project *Network of Excellence on Engineering Secure Future Internet Software Services and Systems (NESSoS)*[2] as a basis.

2. **Analyze Existing Work** We aim to base our work on existing research and analyze approaches that provide meta-models for security and risk standards. In particular, we focus on the works of Sunyaev (2011), who created a security analysis method by identifying common activities in several security standards and the work of Stoneburner et al. (2002), who created a model for risk management as part of the NIST SP 800-30 standard. This analysis results in a set of activities, which are often prescribed in security standards.

3. **Define a Conceptual Model and Template** We use the information from the existing work to create a novel conceptual model, which considers the steps identified by Sunyaev and Stoneburner et al. We propose a novel model based on these related works. Hence, our conceptual model considers the phases of security standards and also considers risk management activities explicitly. In order to apply the conceptual model to security standards, we transform it into a template that can be instantiated. The template contains all phases of security standards considered in the conceptual model, as well as a description on *how* these phases have to be instantiated for a particular standard.

---

[1] http://www.nessos-cbk.org.

[2] http://www.nessos-project.eu/.

4. **Apply Template to Standards** In this phase, we instantiate the template for well-known security standards such as Common Criteria (ISO/IEC 2012), ISO 27001 (ISO/IEC 2005), and the IT Grundschutz standards (BSI 2008c).
5. **Compare Standards** We compare the standards via comparing the different instantiations of our templates. In addition, we consider which of our common terms are considered by the standards and which are not. These insights shall provide a basis for the evaluation of a particular standard.

4.3 CAST Step 1: Define a Common Terminology

We propose a common terminology for security standards and define terms based on different sources. The purpose of the common terminology is to provide fixed definitions of important terms with regard to security standards as a baseline to which the terms in the individual standards can be compared. Using this comparison, it can be analyzed, which terms are used in the standards for the terms with the meaning defined below. We selected relevant terms for security standards in the terminology based on the experience of the authors and their industry contacts. In addition, we used definitions of these terms from well-known sources. In the following, we list the terms related to security defined in the ISO 27001 standard (ISO/IEC 2005).

Asset anything that has value to the organization
Availability the property of being accessible and usable upon demand by an authorized entity
Confidentiality the property that information is not made available or disclosed to unauthorized individuals, entities, or processes
Security Control a control shall reduce the risk of an information security incident occurring. Note that we refer to controls also as security control for the remainder of the paper. Note that the ISO 27001 uses just control, but we use security control instead to make it explicit that the control addresses a security concern.
Information Security Incident a single or a series of unwanted or unexpected information security events that have a significant probability of compromising business operations and threatening information security
Integrity the property of safeguarding the accuracy and completeness of assets

We also include the following terms from the NESSoS Common Body of Knowledge (CBK)'s common terminology (Beckers et al. 2012). These definitions are based on the work of Fabian et al. (2010).

Stakeholder A stakeholder is an individual, a group, or an organization that has an interest in the system under construction. A stakeholder view describes the requirements of a particular stakeholder. The stakeholders may express different types of requirements.

Vulnerability   Stakeholders require a security property to hold for a resource, whose violation implies a potential loss to the stakeholder. This violation can be caused by a vulnerability.

Threat   A vulnerability could potentially be exploited by a threat. A realized threat is an attack that actually exploits a vulnerability and is initiated by an attacker.

Attacker   An attack actually exploits a vulnerability, and the person initiating the attack is an attacker.

Security Goal   A stakeholder's security goal expresses his or her security concerns towards an asset. Security goals are traditionally classified into integrity, confidentiality, and availability goals.

Security requirements   Security requirements capture security goals in more detail. A security requirement refines one or more security goals. It refers to a particular piece of information or service that explicates the meaning of the asset it concretizes in the context of the system under construction.

We also include the following terms to determine the focus of security standards.

Machine   Jackson (2001) defines that the machine is the system or software to be developed. In our context, the machine is the thing in the focus of the security analysis process described in security standards.

Environment   The environment includes a description of all relevant entities in the environment of the machine and, in particular, the interfaces to these entities to the machine.

Policy   Security requirements influence formulating security policies, which contain more information than security requirements. "Security policies state what should be protected, but may also indicate how this should be done." (Gollmann 2005, p. 5). "A security policy is a statement of what is, and what is not, allowed" (Bishop 2003, p. 9) "for us, security boils down to enforcing a policy that describes rules for accessing resources" (Viega and McGraw 2001, p. 14), and "security policy is a [...] policy that mandates system-specific [...] criteria for security" (Firesmith 2003, p. 34).

Security Functions   The machine has descriptions of actual implementable functions that concern the fulfillment of security requirements. The descriptions of these functions are security functions.

4.4 CAST Step 2: Analyze Existing Work

We base our conceptual model for security standards on the HatSec Method (see Sect. 4.4.1) and the NIST SP 800-30 standard (see Sect. 4.4.2).

4.4.1 The HatSec Method

We base our conceptual model for comparing security standards on the HatSec method, because the author analyzed existing security standards and based his method on the resulting common building blocks of the analyzed standards. Only a few standards in the analysis are specific to the health care domain, but most of them are generic security standards such as ISO 27001 (ISO/IEC 2005). Moreover, the HatSec method does not create specific building blocks for the medical domain. Hence, the mining of security-standard-specific building blocks can be reused for our conceptual model. We rely on the HatSec method as a foundation for our conceptual model, but the difference to our work is that the HatSec method provides a means to conduct a security analysis, while we provide a method to compare the processes, documentation demands, and methodologies in security standards.

The Healthcare Telematics Security (HatSec) method by Sunyaev (2011) is a security analysis method developed for the healthcare domain. Sunyaev focuses on security analysis in the investigated standards, even though several of the standards the author investigates concern risk management, as well. However, in these cases the author did not consider the parts in the standards that concern risk in detail. The method consists of seven building blocks, which are derived from the following security and risk management standards: ISO27799 (ISO/FDIS 2007) ISO 27001 (ISO/IEC 2005), IT Grundschutz (BSI 2008c), NIST SP 800-30 (Stoneburner et al. 2002), CRISAM (Stallinger 2004), CRAMM (Farquhar 1991), ISRAM (Karabacak and Sogukpinar 2005), ISMS JIPDEC for Medical Organizations (Japan Information Processing Development Corporation and The Medical Information System Development Center 2004), HB 174-2003 (Standards Australia International; Standards New Zealand 2001), US Department of Health and Human Services—Guideline for Industry, Q9 Quality Risk Management (Food and Administration 2006). Note that only the last four standards are specific to the health care domain.

Schumacher et al. (2006) present a sequence for their security patterns, which are mined from security standards and other sources. The sequence has several steps: (1) security needs identification for enterprise assets, (2) asset valuation, (3) threat assessment, (4) vulnerability assessment, (5) risk determination, etc. This sequence is similar in structure to the HatSec method, but not exclusively built on standards. This sequence considers other sources, as well. We aim to describe a conceptual model focused exclusively on security standards. Hence, we do not consider the sequence of Schumacher et al. as the main foundation for our conceptual model.

The building blocks of the HatSec method are related to the standard as follows. Each building block of the HatSec method occurs also in these standards. However, not all of the steps in the standards occur in the HatSec method. Figure 4.2 shows the seven building blocks of the method. These are further divided into three phases. The *Security Analysis Context and Preparation* phase establishes the context of the security problem. The *Scope Identification* describes the limits of the environment and the system-to-be followed by the *Asset Identification*. The *Security Analysis Process* covers the actual analysis activities of the method. The *Basic Security Check*

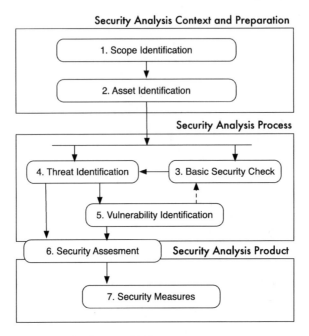

Fig. 4.2 The HatSec method by Sunyaev (2011)

reveals countermeasures already in place and the *Threat Identification* shows dangers resulting from possible attacks on the system-to-be. The *Vulnerability Identification* reveals vulnerabilities to security properties that are potentially exploited by threats. The original HatSec method demands an iteration between the Basic Security Check and the Threat Identification. However, we propose to rather iterate between the Vulnerability Identification and the Basic Security Check, because countermeasures are in place to mitigate vulnerabilities and only subsequent threats. These two building blocks shall be executed in iterations, e.g., if a threat is detected, it shall be checked if a countermeasure for the vulnerability is already in place. The *Security Assessment* concludes the Security Analysis Process by determining the level of security required and the risks remaining. In addition, the Security Assessment also initiates the *Security Analysis Product* phase, because the *Security Measures* activity evaluates the results of the Security Assessment in order to determine if the chosen level of security is adequate or if changes have to be made, e.g., adding additional security controls.

4.4.2 NIST SP 800-30 Standard

The entire information security risk management methodology by Stoneburner et al. (2002) is subdivided into three main phases: (1) risk assessment, (2) risk mitigation, and (3) evaluation. Risk assessment identifies and evaluates potential risks and their impacts, to recommend preventive and risk-reducing controls. In the risk mitigation phase, the identified risks are prioritized and adequate preventive controls are implemented and maintained. After the control implementation, a continual evaluation phase determines whether the implemented risk-reducing controls decrease the risk to an acceptable level or if further controls are required.

We briefly describe the NIST SP 800-30 risk management methodology, which we use as a basis for adding further building blocks to the HatSec method in order to create a conceptual model to compare security standards and also their approaches towards risk management in more detail. The reasons for having chosen the information security risk management methodology by Stoneburner et al. (2002) are: (1) it gives very detailed identification and guidance of what should be considered in the phases of risk assessment, mitigation, and evaluation, (2) the methodology is well accepted and well established, (3) it is freely available, and (4) it supports organizations of all sizes. The comparison of the methodology against others shows that the proposed concepts could be easily applied to similar information security risk management methodologies such as ISO 27005 (ISO/IEC 2008) or EBIOS (DCSSI 2004) due to the similar structures of these methodologies.

4.5 CAST Step 3: Define a Conceptual Model

We extended the HatSec Method with several concepts from the NIST SP 800-30 and refined several concepts to ensure a more detailed comparison of security standards. Moreover, we integrated the conceptual model into a sequence of **Standard Activities**, which are the activities that have to be conducted to establish a security standard. Our conceptual model is shown in Fig. 4.3, we show example instantiations in Sect. 4.6. We structure our conceptual model using the three phases *Security Analysis Context and Preparation*, *Security Analysis Process*, and *Security Analysis Product* (see Sect. 4.4).

We explain the building blocks of the *Security Analysis Context and Preparation* in the following. We split the *scope identification* of the HatSec method into an *environment description* and a *stakeholder description*. The reason is that security is about protection of assets and harm to assets results in a loss to stakeholders. We have to understand the significance of the loss by describing the stakeholder. Moreover, stakeholders can cause threats to assets, and the identification of stakeholders in a scope is a research problem (Pouloudi 1999; Sharp et al. 1999). Moreover, we included the building block *Risk Level Description* to include a mechanism to categorize assets already in the beginning of the security analysis. This is done to focus security analysis on assets with a high risk level, as is suggested by NIST SP 800-30 (Stoneburner et al. 2002) and IT Grundschutz (Bundesamt für Sicherheit in der Informationstechnik (BSI) 2008a).

We describe our building blocks for the *Security Analysis Context and Preparation* phase in the following.

Fig. 4.3 A conceptual
framework for security
standards

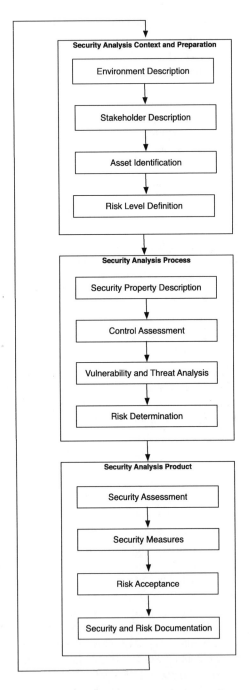

Environment Description   The environment description states the scope of the standard. Hence, the environment in which the security system shall be integrated into should be, e.g., an organization or an Information and Communication Technology (ICT)-based System or combinations of both.

Stakeholder Description   The stakeholder description describes all relevant persons, organizations, and government bodies that have a relation to the environment.

Asset Identification   The asset identification for the stakeholders collects all information or resources that have a value to the stakeholders. The assets shall be protected from harm caused by the environment.

Risk Level Description   For each asset, a risk level description states the impact the loss of an asset has on a stakeholder. Hence, the risk level description classifies the assets into categories according to their significance for the environment. In this building block, the risk level determination is based on the opinion of stakeholders and described on a high level of abstraction.

We explain the building blocks of the *Security Analysis Process* in the following. We divided the building block *Basic Security Check* into a security property definition for assets and an assessment of existing controls. The security properties provide an overview of high-level security goals, which should be separated from the *Control Assessment*, since it considers existing security solutions. Moreover, we combined the threat analysis and vulnerability identification, because threats are exploited vulnerabilities (Fabian et al. 2010) and should be considered together in our view. We add also a *Risk Determination* building block to the *Security Analysis Process* that describes how likelihoods and consequences for the resulting threats are assessed.

Security Property Description   We initiate the *Security Analysis Process* with a high-level security property description, which determines security goals for assets. For example, the ISO 27001 standard uses high-level security objectives to "establish an overall sense of direction and principles for action with regard to information security" (ISO/IEC 2005, p. 4) as part of their ISMS policy, the superset of all security policies that the standard establishment creates.

Control Assessment   The control assessment determines which controls (either technical ones such as encryption mechanisms or nontechnical controls such as security policies) are already in place and their ability to ensure a security property of an assets.

Vulnerability and Threat Analysis   The threat analysis assumes vulnerabilities of an asset. Moreover, threats have to be validated by showing that the potentially exploited vulnerability exists. In general, a threat requires a source and an existing vulnerability to become effective. The threat source can either intentionally or accidentally exploit a potential vulnerability. The aim of the threat identification step is to determine potential threats and their corresponding sources such as human threats (e.g., active network attacks, theft, unintentional data alternation, etc.), or environmental threats (e.g., power failure, water leakage, etc.). On the basis of the threat analysis, the vulnerability analysis shows potential vulnerabilities present in the scope, including the consideration of vulnerabilities in the field of (1) management security (e.g., no assignment of responsibilities, no risk assessment, etc.), (2) operational security (e.g., no external data distribution and labeling, no humidity control, etc.), and (3) technical security (e.g., no cryptography solutions in use, no intrusion detection in place, etc.).

Risk Determination   The risk determination determines useful likelihood and impact scales to conduct risk management for assets. The risk determination considers the output of all previous steps and evaluates these results with regard to risk, considering the likelihood and impact scales. We explain this step based on the NIST 800-30 standard further in the following. Initially, the probability of a threat exploiting a certain vulnerability in the system-to-be-analyzed is determined. For this purpose the organization has to assess:

- motivations and capabilities of possible attackers
- type of the vulnerability, and
- effectiveness and efficiency of existing controls.

Stoneburner et al. (2002) propose a qualitative probability rating as stated in Table 4.1.

Afterwards, the organization's ability to achieve its goals, when a vulnerability is successfully exploited by a threat, is determined in an impact analysis. The NIST SP 800-30 information security risk management methodology recommends measuring the impact in terms of the loss of integrity, availability, and/or confidentiality. NIST recommends the measurement of impacts on a qualitative level (e.g., high, medium, and low) if a quantitative measurement of the impacts in terms of revenue loss is not possible. Quantitative measurement methods have the significant problem that it is very hard to determine if the impact of a certain threat exactly corresponds to a certain amount of money. For example, how can an organization determine that a fire would cause a loss of exactly EUR 443.277 and not EUR 526.473? Usually, organizations tend to use quantitative methods in a qualitative way, for example assigning monetary ranges (e.g., EUR 0–EUR 300.000, EUR 300.000–EUR 500.000, etc.) to the different impact levels.

Next, the organization has the knowledge necessary to determine the actual risk:

- the probability that a possible threat exploits a vulnerability
- the impact caused if the threat exploits the vulnerability, and
- adequacy of the existing controls for reducing or eliminating the risk.

By multiplying the threat probability with the magnitude of the impact, the organization is able to determine the risk level. Thus, the organization can plan the necessary actions as stated in Table 4.2.

Finally, we explain the building blocks of the *Security Analysis Product* phase. We use the *Security Assessment* and *Security Measures* building blocks as described in the HatSec method and we add explicit building blocks for *Risk Acceptance* and *Security and Risk Documentation*. *Risk Acceptance* is an essential step of finishing the security analysis product, and if risks are accepted too soon, the entire security

Table 4.1   NIST 800-30 probabilities (Stoneburner et al. 2002)

Probability level	Probability definition
High	The threat source is highly motivated and sufficiently capable, and controls to prevent the vulnerability from being exercised are ineffective
Medium	The threat source is motivated and capable, but controls are in place that may impede successful exercise of the vulnerability
Low	The threat source lacks motivation or capability, or controls are in place to prevent, or at least significantly impede, the vulnerability from being exercised

Table 4.2   NIST 800-30 risk scale and necessary actions (Stoneburner et al. 2002)

Risk level	Risk description and necessary actions
High	If an observation or finding is evaluated as a high risk, there is a strong need for corrective measures. An existing system may continue to operate, but a corrective action plan must be put in place as soon as possible
Medium	If an observation is rated as medium risk, corrective actions are needed, and a plan must be developed to incorporate these actions within a reasonable period of time
Low	If an observation is described as low risk, the system's administrator must determine whether corrective actions are still required or decide to accept the risk

analysis product might not be effective. Hence, we aim to document in the template how the standards address this issue. In addition, the certification process of a security standard is usually based on the documentation of the security analysis product. That is why we want to add a description of the demanded documentation in our conceptual model and template.

Security Assessment   The security assessment evaluates if the existing security controls satisfy the security properties of the assets considering the results of the *Vulnerability and Threat Analysis*, as well as the *Risk Determination*. This step also describes how further security controls have to be selected. For example, the ISO 27001 standard (ISO/IEC 2005) has a mandatory Annex A from which controls have to be selected.

Security Measures   The security measures activity specifies a list of new, refined, or existing security controls that are required to improve the protection of the assets. This final result of the selection of controls are the *Security Measures*. For example, the ISO 27001 demands a so-called *Statement of Applicability* that reasons about the necessity of the controls in Annex A.

Risk Acceptance   The risk acceptance evaluates if the *Security Measures* reduce the risk of attacks on assets to acceptable levels. Often a clear-cut criteria has to be defined that is fulfilled or not. For example, the controls prevent threats from attackers with a mediocre skills level and a limited amount of time.

Security and Risk Documentation   The security system description finishes with the security and risk documentation of the security analysis product. The documentation has to usually follow certain guidelines of a standard.

We mapped our conceptual model to a template presented in the Appendix D in Tables D.1–D.3. We have elicited a series of questions for each building block, which shall help to fill in the required information. In addition, we stated which common terms are relevant for each part of the template.

4.6 CAST Step 4: Instantiate Template with Standards

We instantiate our template with the ISO 27001 standard (Sect. 4.6.1), IT Grundschutz (Sect. 4.6.3), and Common Criteria (Sect. 4.6.4). Note that there is some overlap with the text in Chap. 2, which we repeat for the convenience of the reader.

4.6.1 ISO 27001

The ISO 27001 defines the requirements for establishing and maintaining an Information Security Management System (ISMS) (ISO/IEC 2005). In particular, the standard describes the process of creating a model of the entire business risks of a given organization and to specify specific requirements for the implementation of security controls. The resulting ISMS provides a customized security level for an organization.

The ISO 27001 standard contains a description of the so-called *ISO 27001 process* (ISO/IEC 2005). The process contains phases for establishing an ISMS, implementing and operating an ISMS, and also monitoring, reviewing, maintaining, and improving it.

In the initial phase, the *scope and boundaries* of the ISMS, its *interested parties*, *environment*, *assets*, and all the *technology* involved are defined. In this phase, also the ISMS *policies*, *risk assessments*, *evaluations*, and *controls* are defined. Controls in the ISO 27001 are measures to *modify risk*.

The ISO 27001 standard demands a set of documents that describe the requirements for the ISMS. Furthermore, the standard demands periodic audits towards the effectiveness of an ISMS. These audits are also conducted using documented ISMS requirements. In addition, the ISO 27001 standard demands that management decisions, providing support for establishing and maintaining an ISMS, are also documented. This support has to be documented via management decisions. This has to be proven as part of a detailed documentation of how each decision was reached and how many resources (e.g., personal, budget, time, etc.) are committed to implement this decision. Moreover, certification of an ISMS according to the ISO 27001 standard is possible, based upon the documentation of the ISMS (Tables 4.3, 4.4, and 4.5).

Table 4.3 Instantiation for ISO 27001 of the security analysis context and preparation part of the template for security standard description

Security analysis context and preparation

Environment description

The machine in this standard is the ISMS and the environment is anything outside the scope of the ISMS. "The standard demands an ISMS scope definition and its boundaries in terms of the characteristics of the business, the organization, its location, assets and technology, and including details of and justification for any exclusions from the scope" (ISO/IEC 2005, p. 4, Section 4.2.1a). The standard mentions the scope explicitly in the following sections. Section 4.2.1 d concerns risk identification and the section recommends to consider the scope definition for identifying assets. Section 4.2.3 demands management reviews of the ISMS that also includes to check for possible changes in the scope of the ISMS. Section 4.3 lists the documentation demands of the standard and Section 4.3.1d requires a documentation of the scope of the ISMS. Moreover, the standard demands an explicit to creating an ISMS. In particular, Section 5.1 Management commitment concerns proof the management shall provide for establishing an ISMS objectives, plans, responsibilities, and accepting risks. Section 5.2 Resource management concerns the provision of resources for establishing the ISMS and the training of the members of the organization for security awareness and competence

Stakeholder description

The stakeholder definition is part of the scope definition. The standard uses the term *Interested Parties* (ISO/IEC 2005, p. vi) instead of stakeholders, who have security "expectations" that are input for the ISMS implementation as well as "security requirements"

Asset identification

The design goal of the ISO 27001 ISMS is to protect assets with adequate security controls and this is stated already on page 1 of the standard. This is relevant in particular in Section 4 that describes the ISMS and in particular in Section 4.2—Establishing and managing the ISMS states the scope definition. Section 4.2.1a demands the definition of assets. Section 4.2.1b concerns the definition of ISMS security policies demands that the policy shall consider assets. Section 4.2.1d that concerns risk identification uses the scope definition to identify assets, to analyze threats to assets, and to analyze the impacts of losses to these assets. Section 4.2.1e concerns risk analysis, which also clearly define to analyze assets and to conduct a vulnerability analysis regarding assets in light of the controls currently implemented

Risk level definition

The standard requires a risk level definition in the steps following the scope definition. Section 4.2.1b states that the ISMS policy has to align with the risk management. Section 4.2.1c demands a risk assessment that includes the criteria for accepting risks and identifying the acceptable risk levels

4.6.2 ISO 27001:2013

The new edition of the ISO 27001 standard from 2013, called ISO 27001:2013, introduces significant changes in comparison to the previous version. For instance, the asset identification, threat, and vulnerability analyzes are not explicitly part of the establishment of an ISMS anymore. We show these differences in the following instantiation of our CAST templates for ISO 27001:2013 (Tables 4.6, 4.7 and 4.8).

Table 4.4 Instantiation for ISO 27001 of the security analysis process part of the template for security standard description

Security analysis process
Security property description
The standard demands the elicitation of high-level security goals in the section after the scope definition; Section 4.2.1b concerns the definition of ISMS policies of which high-level security goals are a part. "The ISMS policy is considered as a superset of the information security policy." (ISO/IEC 2005, p. 4)
Control assessment
The assessment concerns likelihoods of security failures with regard to threats and vulnerabilities. In addition, impacts to assets should be considered of the controls currently implemented according to ISO 27001 Section 4.2.1e2
Vulnerability and threat analysis
The ISO 27001 standard concerns threat analysis in several sections for determining the risks to assets. Section 4.2.1d demands a threat analysis for assets for the purpose of identifying risks and the vulnerabilities that might be exploited by those threats. Section 4.2.1e concerns risk analysis and evaluation and demands to determine likelihoods and consequences for threats. Section 4.2.4d concerns the review process of the ISMS and also demands a threat identification. Section 7.2 that concerns the management review of the ISMS also demands a threat analysis
Risk determination
The standard demands a description of a methodology for risk management and it mentions several related activities explicitly. Section 4.2.1d concerns risk identification and Section 4.2.1e demands risk analysis and evaluation

4.6.3   IT Grundschutz

The German Bundesamt für Sicherheit in der Informationstechnik (BSI) issued the so-called *BSI series of standards for information security* (Bundesamt für Sicherheit in der Informationstechnik (BSI) 2008a) (see left hand side of Fig. 4.4). These are based on the ISO 27001 and ISO 27002 standards and refine them with a new methodology. The series of standards consists of BSI Standard 100-1 that concerns the management issues of the standard such as planning IT processes. The BSI Standard 100-2 (BSI 2008b) describes the methodology of how to build an ISMS, BSI Standard 100-3 (BSI 2008c) concerns of risk management, and BSI 100-4 (BSI 2009) consideration of Business Continuity Management, e.g., data recovery plans. In the following, we focus on BSI 100-2, because it contains the methodology. The BSI standard 100-2 describes how an ISMS can be established and managed. It is compatible to the ISO 27001 standard, meaning that an implementation of the BSI standard 100-2 can be used for an ISO 27001 certification with the German BSI (Bundesamt für Sicherheit in der Informationstechnik (BSI) 2008a, p. 12). In addition, the standard aims towards reducing the required time for an ISMS implementation. This is achieved by provisioning the IT Grundschutz Catalogs (see right-hand side of Fig. 4.4). This catalog contains a significant collection of IT security threats and controls, and a mapping between them. Note that controls are called *safeguards* in the BSI terminology. The standard offers a method depicted in Fig. 4.5 that starts

Table 4.5 Instantiation for ISO 27001 of the security analysis product part of the template for security standard description

Security analysis product

Security assessment

Threats to assets have to be analyzed and existing security controls documented. The risk has to be evaluated of these threats according to the criteria set previously, considering the existing security controls

For all unacceptable risks security controls have to be selected to reduce the risk to acceptable level. The control selection is based on security requirements, which are refinements of the high-level security goals. This is explained in the following

Security measures

The ISO 27001 standard concerns high-level ISMS policies during the establishment of the ISMS to guide the focus of security and security policies as controls that define in detail what a specific security control should achieve. In particular, the Annex A of the ISO 27001 standard describes the normative controls of the standard. This is stated in Section 4.2.1f concerning risk treatment and Section 4.2.1g discussing controls for risk treatment

Risk acceptance

Criteria for acceptable risk have to be defined in the beginning of the risk analysis (Section 4.2.1c) and after the control selection it has to be shown that the criteria for acceptable risk are fulfilled. The standard also demands management approval for acceptable levels of risk (see Section 4.2.1h)

Security and risk documentation

The ISO 27001 standard demands the following documents:

• ISMS policies and objectives

• Scope and boundaries of the ISMS

• Procedures and controls

• The risk assessment methodology

• Risk assessment report

• Risk treatment plan

• Information security procedures

• Control and protection of records that can provide evidence of compliance to the requirements of the ISMS

• Statement of applicability describing the control objectives and controls that are relevant and applicable to the organization's ISMS

In addition, the ISO 27001 standard demands the documentation of management decisions that provide support for establishing and maintaining an ISMS

with a structural analysis of the organization and the environment. The standard suggests a focus on at least the areas organization, infrastructure, IT systems, applications, and employees. The next step is to determine the required security level, followed by modeling the security measures, and a basic security check. This security check classifies the assets and executes a risk analysis for the 20 % of assets with the highest security level. The remaining 80 % are not considered in a risk analysis and simply suggested safeguards in the IT Grundschutz Catalogues for these assets are implemented. After the security check, the measures are consolidated and another basic security check is executed. The last step is realizing the measures (Tables 4.9, 4.10 and 4.11).

Table 4.6 Instantiation for ISO 27001:2013 of the security analysis context and preparation part of the template for security standard description

Security analysis context and preparation

Environment description

In ISO 27001:2013 the machine in this standard is still the *ISMS* and the environment is the *context of the organization*

"The organization shall determine external and internal issues that are relevant to its purpose and that affect its ability to achieve the intended outcome(s) of its information security management system." (ISO/IEC 2013, p. 1)

Moreover, the new version of the standard ISO27001:2013 demands a more detailed description of the context of the organization and the environment in which the organization operates. In addition, the standard demands to include the elicitation of external and internal issues relevant to the purpose of the ISMS. It includes the demands for context description of ISO 31000 (ISO 2009), which demands an elicitation of the external context that shall contain descriptions of numerous forces such as social, cultural, and technological ones

Stakeholder description

The ISO 31000 context description demands "relationships with, perceptions and values of external stakeholders." (ISO 2009, p. 15). The ISO 31000 standard also demands similar analysis for the organization, its stakeholders and "objectives and criteria of a particular project, process or activity should be considered in the light of objectives of the organization as a whole" (ISO 2009, p. 15). Note that ISO27001:2005 does not contain any reference to the ISO 31000 Standard and that both ISO27001:2005 and ISO27001:2013 use the term *interested party* instead of stakeholder. Overall the ISO27001:2013 demands an analysis of interested parties relevant for the information security management system and the elicitation of "the requirements of these interested parties relevant to information security" (ISO/IEC 2013, p. 1). Afterwards the scope, boundaries, and applicability of the information security management system (ISMS) shall be defined based on the previously elicited requirements

Asset identification

ISO 27001:2013 does not use the word asset in the sections of the standard, but the word asset appears in several controls of the Appendix A such as control *A.6.1.2—Segregation of duties* and *A.8—Asset management*. In addition, the terminology ISO 27001:2013 refers to in ISO 27000:2014 also does not contain the word *asset*, but the word asset appears in the definition of other terms such as *vulnerability* or *information system*. Hence, asset identification seems still to be relevant for ISO 27001:2013, but it is not stated explicitly

The ISMS security policy in Section 5.2 refers to security objects as the initial steps for defining the policy. Furthermore, the risk assessment is initiated by defining risk acceptance criteria and criteria for performing information security risk assessments. None of these refer explicitly to assets

Risk level definition

The risk level definition is done during the *Information security risk assessment* demanded in Section 6.1.2 (ISO/IEC 2013, p. 4). The risk level definition is a subsequent step to risk identification, and the consequences and likelihoods scales

Table 4.7 Instantiation for ISO 27001:2013 of the security analysis process part of the template for security standard description

Security analysis process

Security property description

The standard demands to elicit security requirements as part of the context description in Section 4.2 (ISO/IEC 2013, p. 1). The leadership has to align the information security objectives with the strategic direction of the organization as part of the ISMS policy in Section 5 (ISO/IEC 2013, p. 2)

Control assessment

The risk assessment shall lead to appropriate selection of controls to implement risk treatment options "Organizations can design controls as required, or identify them from any source" (ISO/IEC 2013, p. 1). These controls have to be compared with the controls defined in Appendix A in the standard. A reasoning has to be done that shows that no relevant controls are missing (ISO/IEC 2013, p. 4)

Vulnerability and threat analysis

The standard does not prescribe a vulnerability or threat analysis in the standard. Section 6.1.2 contains a information security risk identification step, which could possibly contain vulnerability and threat analysis activities, but is not explicitly stated in the standard

Risk determination

ISO 27001:2013 also demands a description of a methodology for risk management and it mentions several related activities explicitly. Section 6 describes a process for information security risk assessment that contains activities such as defining risk criteria, identify information security risks, analyze risks, evaluate risks, and treat risks

Table 4.8 Instantiation for ISO 27001:2013 of the security analysis product part of the template for security standard description

Security analysis product

Security assessment

The standard demands an *information security risk treatment process* that selects risk treatment options based on the outcome of the risk assessment results. The standard demands to select controls from every possible source and map these controls to the controls defined in the Annex A of the standard. The mapping is the basis for a reasoning that all relevant controls are considered (ISO/IEC 2013, p. 4)

Security measures

ISO 27001:2013 starts from high-level security objectives elicited during the context description and stated in the ISMS Policy. These security objectives are the basis for selecting controls. In particular, the Annex A of the ISO 27001 standard describes the normative controls of the standard stated in Section 6.1.3. of the information security risk treatment section of the standard

Risk acceptance

Section 6.1.2 demands an explicit definition of risk acceptance criteria and criteria for performing information security risk assessments. The acceptance of residual risk has been explicitly documented and the approval of the risk owners has to be granted

(continued)

Table 4.8 (continued)

Security analysis product

Security and risk documentation

ISO 27001:2013 does not define document types that have to be written, but defines so-called *documentation information* in its sections. The following sections contain documentation information:

- Section 4.3 demands documentation information about the *Scope of the ISMS*
- Section 5.2 demands documentation information about the *Information Security Policy*
- Section 6.1 demands documentation information about the *Risk Assessment* and the *Risk Treatment* including the *Statement of Applicability*
- Section 6.2 demands documentation information about the *Information Security Objectives*
- Section 7.2 demands documentation information about the *Competence Records*
- Section 8.2 demands documentation information about the *Risk Assessment Results*
- Section 8.3 demands documentation information about the *Risk Treatment Results*
- Section 9.1 demands documentation information about the *Monitoring and Measuring Results*
- Section 9.2 demands documentation information about the *Audit Programme and Results*
- Section 9.3 demands documentation information about the *Management Review Results*
- Section 10.1 demands documentation information about the *Evidence of Corrective Actions*

The documentation can contain further documents if some topics have to be elaborated further, as suggested by, e.g., Section 7.5 and 8.1

BSI-Standards of Information Security Information Security and IT-Grundschutz	**IT-Grundschutz Catalogues** Loose-leaf-collection and Internet
BSI-Standard 100-1 Information Security Management System (ISMS)	**Chapter 1** Epilogue **Chapter 2** Layers and Modelling

BSI-Standard 100-2
IT-Grundschutz Methodology
	Modules Catalogues
	            Generic Aspects
	            Infrastructure
	            IT Systems

BSI-Standard 100-3 Risk Analysis on the Basis of IT-Grundschutz	Networks Applications
BSI-Standard 100-4 Business Continuity Management	**Threats Catalogues** **Safeguards Catalogues**

Fig. 4.4 BSI IT Grundschutz overview inspired by Bundesamt für Sicherheit in der Informationstechnik (BSI) (2008a)

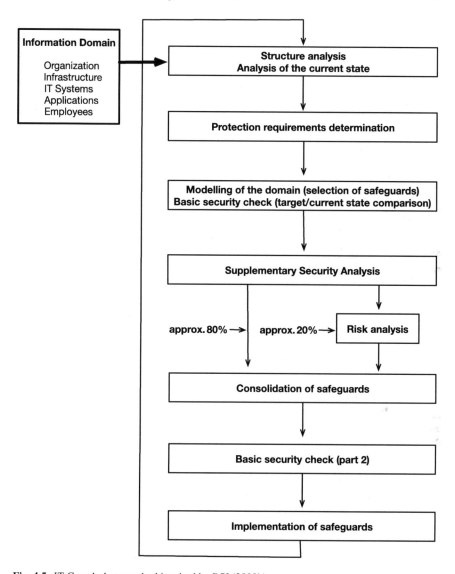

Fig. 4.5 IT Grundschutz method inspired by BSI (2008b)

4.6.4 The Common Criteria

The ISO/IEC 15408—Common Criteria for Information Technology Security Evaluation is a security standard that can achieve comparability between the results of independent security evaluations of IT products. These are so-called *targets of evaluation (TOEs)*.

Table 4.9 Instantiation for BSI 100.2 of the security analysis context and preparation part of the template for security standard description

Security analysis context and preparation
Environment description
The standard demands a description of the scope and in particular (BSI 2008b, p. 37):
• "Specify which critical business processes, specialized tasks, or parts of an organization will be included in the scope
• Clearly define the limits of the scope
• Describe interfaces to external partners"
The machine in this standard is an ISMS and the environment is described via interfaces to external partners
The scope definition is accompanied by a structure analysis, which demands a separate documentation of the following parts of the scope: information, application, IT systems, rooms, communication networks
Stakeholder description
The employees of the organization that take part in the business processes have to be documented. Moreover, the users of the scope elements such as applications are documented, as well. These are both part of the scope definition. The standard refers to users or employees of the organization instead of stakeholders
Asset identification
For each business process in the scope, a level of protection has to be determined. The entire processes and, in particular, the information technology used and processing of the information it contained are considered as assets
Risk level definition
The standard uses the protection requirements as an indicator for high-level risks

The Common Criteria (CC) is based upon a general security model (see Fig. 4.6). The model considers TOE owners that value their assets and wish to minimize risk to these assets via imposing countermeasures. These reduce the risk to assets. Threat agents wish to abuse assets and give rise to threats for assets. The threats increase the risk to assets. The concepts of the Common Criteria consider that potential TOE owners infer their security needs for specific types of TOEs, e.g., a specific firewall. The resulting documents are called Security Targets (ST). Protection profiles (PP) state security needs for an entire class of TOEs, e.g., client VPN application. The evaluators check if a TOE meets its ST. Protection profiles (PP) state the security requirements of TOE owners. TOE developers or vendors publish their security claims in security targets (ST). A CC evaluation determines if the ST is compliant to a specific PP. The standard relies upon documents for certification, which state information about security analysis and taken measures (Tables 4.12, 4.13, 4.14, and 4.15).

Table 4.10 Instantiation for BSI 100.2 of the security analysis process part of the template for security standard description

Security analysis process
Security property description
All general security concerns are specified in an information security policy, which describes the general direction of information security in the organization. In addition, for each asset security goals have to be determined in terms of confidentiality, integrity, and availability. The standard calls them protection requirement, which have to be categorized in the levels: normal, high, and very high (BSI 2008b, p. 48). These categories have the meaning (BSI 2008b, p. 48):
Normal "The impact of any loss or damage is limited and calculable."
High "The impact of any loss or damage may be considerable."
Very High "The impact of any loss or damage may be of catastrophic proportions which could threaten the very survival of the organization."
Note that the standard also allows to define a different scale, but this is the scale recommended
The protection requirements are refined with damage scenarios (BSI 2008b, p. 48):
"Violations of laws, regulations, or contracts
Impairment of the right to informational self-determination Physical injury
Impaired ability to perform the tasks at hand
Negative internal or external effects
Financial consequences"
These damage scenarios have to be put in relation to the protection requirement for each organization that establishes the standard. This means it has to be defined for each category what the damage scenario means, e.g., what means normal financial consequences
Control assessment
The standard relies on the security controls listed in the IT Grundschutz catalog. These are categorized into (BSI 2008b, p. 48):
S 1 Infrastructure,
S 2 Organization,
S 3 Personnel,
S 4 Hardware and software,
S 5 Communication,
S 6 Contingency planning
Several of the threats listed in the IT Grundschutz Catalogues have existing mappings to possibly relevant safeguards. These have to be considered as relevant if a threat is selected. The safeguards have to be refined for the scope. The standard refers to safeguards instead of security controls
Vulnerability and threat analysis
The standard demands a model of the scope. The IT Grundschutz catalog provides modules that support this modeling. These modules are categorized in the following domains (BSI 2008b, p. 48): *General aspects Infrastructure IT systems Networks Application*. The modules contain a mapping to the following threat categories:
T 1 Force majeure,
T 2 Organizational shortcomings,
T 3 Human error,

(continued)

Table 4.10 (continued)

Security analysis process

Vulnerability and threat analysis

T 4 Technical failure,

T 5 Deliberate acts

All the threats in each category of the IT Grundschutz catalog have to be analyzed with regard to the scope and the relevant threats have to be documented. The threats have to be refined for the scope of analysis

Risk determination

A risk analysis can be conducted either after the basic security check or the supplementary security check. The management has to make a choice, for which assets a risk analysis has to be conducted. The standard does not prescribe a strict methodology for risk management, but provides rather advice for how to consider threats and safeguards and in use which step of the method to apply the threat analysis. It is not providing a method for, e.g., eliciting likelihood or consequences scales

Table 4.11 Instantiation for BSI 100.2 of the security analysis product part of the template for security standard description

Security analysis product

Security assessment

A security assessment is done using a so-called *basic security check*. The model of the scope and the protection requirements are used to develop a security test plan, which determines the effectiveness of existing security controls. Each test has to describe a target state and after conducting the test it is determined if a control is effective by analyzing the state of the tested scope elements. In a sense, the security testing plans are based on security requirements, which refine the protection requirements

This basic security check consists of three different steps. "The first step consists of making the organizational preparations and, in particular, selecting the relevant contact people for the target/actual state comparison. In Step 2, the target state is compared to the actual state by conducting interviews and performing random checks. In the final step, the results of the target/actual state comparison are documented together with the reasoning behind the results." (BSI 2008b, p. 66)

Security measures

After considering the threats and safeguards in the IT Grundschutz catalog, a supplementary security analysis is conducted.

"The supplementary security analysis is to be performed on all target objects in the information domain to which one or more of the following applies:

- The target objects have high or very high protection requirements in at least one of the three basic values—confidentiality, integrity, or availability
- The target objects could not be adequately depicted (modeled) with the existing modules in the IT Grundschutz Catalogues
- The target objects are used in operating scenarios (e.g., in environments or with applications) that were not foreseen in the scope of IT Grundschutz" (BSI 2008b, p. 66)

Risk acceptance

Accepted risks have to be documented with a reasoning

Security and risk documentation

Each step of the methodology presented in the standard has to be documented

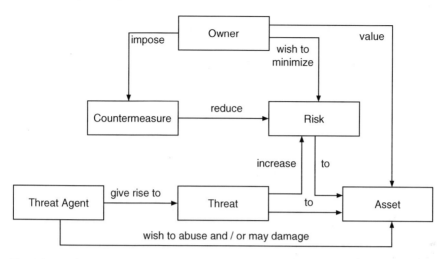

Fig. 4.6 The common criteria basic security model taken from ISO/IEC (2012) (Reproduced by permission of DIN Deutsches Institut für Normung e.V. The definitive version for the implementation of this standard is the edition bearing the most recent date of issue, obtainable from Beuth Verlag GmbH, Burggrafenstraße 6, 10787 Berlin, Germany)

4.7 CAST Step 5: Compare Standards

We analyze the instantiated templates (see Sect. 4.6) of the ISO 27001 standard (Sect. 4.6.1), IT Grundschutz (Sect. 4.6.3), and Common Criteria (Sect. 4.6.4) in Sect. 4.7. In addition, we describe the tool support for our method in Sect. 4.7.

Comparison

We compared the terminology of the security standards ISO 27001, IT Grundschutz, and Common Criteria in Table 4.16 with the terminology introduced in Sect. 4.3. The symbol "∼" means that the term is equal to the definition in our terminology (Sect. 4.3). A "−" states that the standard does not consider that term explicitly.

Furthermore, we show the results of our comparison in the following by illustrating relevant statements for each of our building blocks of our security standard template instances.

Security Analysis Context and Preparation
Environment description—The ISO 27001 demands a scope definition including assets and justifications. The standard explicitly indicates to use the scope in subsequent steps such as risk identification. Moreover, the scope is also referred to in the documented management commitment.

ISO27001:2013 demands a context description of the organization and the scope of the ISMS within this context. Assets are not explicitly mentioned.

Table 4.12 Instantiation for common criteria of the security analysis context and preparation part of the template for security standard description

Security analysis context and preparation
Environment description
The common criteria demands a description of the TOE in its environment. Hence, the TOE is the machine. The environment contains stakeholders, other software components the TOE requires, e.g., a specific operating system. The standard discusses the environment simply as outside the TOE
"An ST introduction containing three narrative descriptions of the TOE" (ISO/IEC 2012, p. 64, Part 1: Introduction and general model). The TOE reference provides a description of unique identifications for an ST that describes the TOE such as a version number for the revision of the ST. The TOE overview describes the intended functionality of the TOE and security features on a high level of abstraction. The standard describes the TOE and its environment, which is simply referred to as *outside* or *operational environment* of the TOE. Hence, the system consists of the TOE and its *operational environment*
Stakeholder description
The Common Criteria focuses on describing a software product and it describes stakeholders just as much as they are required to understand the TOE's functionality or security features. For example, a TOE shall display certain information to a user
The standard uses the term external entity for all stakeholders that interact with the TOE from the outside. It explicitly states that a user is an external entity. Note that the term external entities also includes IT entities (ISO/IEC 2012, pp. 16 and 20, Part1:Introduction and general model)
Asset identification
"Security is concerned with the protection of assets." (ISO/IEC 2012, p. 38, Part1: Introduction and general model). Stakeholders consider assets valuable (see below), which is highly subjective. Thus, the identification of assets depends upon information from stakeholders, because "almost anything can be an asset" (ISO/IEC 2012, p. 38, Part1: Introduction and general model). Hence, assets should have a description and also some information regarding the need for protection. This is aligned with descriptions of existing PPs such as (BSI 2011c). Furthermore, in PPs the concept of a SecondaryAssets is used (BSI 2011c), whose loss do not cause harm to the ToE Owner directly, but the harm can cause harm to an Asset. This in turn can cause a loss to a ToE Owner
The standard defines "assets entities that the owner of the TOE presumably places value upon." (ISO/IEC 2012, pp. 16 and 20, Part1:Introduction and general model)
Risk level definition
The Common Criteria concerns risks arising from attacks and the standard does not define basic risk levels, but attack potentials. The scale is basic, enhanced-basic, moderate, high

The IT Grundschutz demands also explicitly to document external partners and to document certain parts of the scope separately, such as applications.

The Common Criteria focuses on functionalities of the TOE and its environment in the scope description.

Stakeholder description—The ISO 27001 demands stakeholder description as part of the scope description including their security expectations. ISO 27001 demands stakeholder description as part of the context description. The IT Grundschutz considers all employees and external staff involved in relevant business processes as stakeholders. The Common Criteria concerns all users of the TOE as stakeholders.

Table 4.13 Instantiation for common criteria of the security analysis process part of the template for security standard description

Security analysis process
Security property description
Security needs of assets are expressed in terms of confidentiality, integrity, and availability or other not specified security goals. "Security-specific impairment commonly includes, but is not limited to: loss of asset confidentiality, loss of asset integrity and loss of asset availability." (ISO/IEC 2012, p. 39)
These terms are not defined in the general term definition section of Part 1, but refined terms are defined in Part 2: security functional components. For example, *FDP_UCT* describes the meaning of user data confidentiality
Control assessment
"Subsequently countermeasures are imposed to reduce the risks to assets. These countermeasures may consist of IT countermeasures (such as firewalls and smart cards) and non-IT countermeasures (such as guards and procedures)." (ISO/IEC 2012, p. 39, Part1:Introduction and general model). The standard uses the term countermeasure for security control
Vulnerability and threat analysis
The common criteria considers threats from malicious attackers and also from attackers that present unintentional threats such as accidental disconnecting a server from a power supply. "The Common Criteria is applicable to risks arising from human activities (malicious or otherwise) and to risks arising from nonhuman activities." (ISO/IEC 2012, pp. 16 and 20, Part1: Introduction and general model).
The common criteria suggests further to describe the *attack potential* that "measure of the effort to be expended in attacking a TOE, expressed in terms of an attacker's expertise, resources and motivation." (ISO/IEC 2012, p. 14, Part1: Introduction and general model). The description of attackers leads to threats the attacker present by exploiting vulnerabilities
Risk determination
The Common Criteria focuses on identifying vulnerabilities and attackers that might exploit these vulnerabilities. "These threats therefore give rise to risks to the assets, based on the likelihood of a threat being realized and the impact on the assets when that threat is realized." (ISO/IEC 2012, p. 39, Part1:Introduction and general model). However, the standard does not follow a risk management approach like ISO 31000, but focuses on documenting vulnerabilities and countermeasures of a TOE. An ST shall help to decide if a stakeholder is willing to accept the risk of using a TOE. "Once an ST and a TOE have been evaluated, asset owners can have the assurance (as defined in the ST) that the TOE, together with the operational environment, counters the threats. The evaluation results may be used by the asset owner in deciding whether to accept the risk of exposing the assets to the threats." (ISO/IEC 2012, p. 58, Part1: Introduction and general model)

Asset identification—ISO 27001 demands the definition of assets, but does not provide methodological support for it. ISO 27001:2013 does not demand asset identification explicitly. The IT Grundschutz considers all information technology and information in the business processes as assets. The Common Criteria considers also the concept of a secondary asset. But the standard does not provide a method for identifying them, either.

Table 4.14 Instantiation for common criteria of the security analysis product part of the template for security standard description (1/2)

Security analysis product

Security assessment

Each of the threats previously identified leads to the formulation of a *security objective*, which is equal to a security requirement in the common terminology. The Common Criteria distinguishes between security objectives, which concern the TOE, and the ones concerning the environment. The latter ones are so-called *security objectives for the environment*. Moreover, the Common Criteria considers organization security policies, which are equal to the policy term

The Common Criteria uses cross-tables that present a mapping of all identified threats to security objectives: security objectives for the environment, assumptions, or organization security policies. Each threat has to mapped to at lease one security objective: security objectives for the environment or assumptions

Security measures

Security objectives are refined by security functional requirements, which are gap texts that concern specific security functions such as access control functions. Security objectives are on a high abstraction level, while security functional requirements concern concrete implementable security functionalities

All security objectives have to be refined using security functional requirements. A cross-table has to show that all security objectives are refined by at least one security functional requirement

Risk acceptance

"Owners of assets may be (held) responsible for those assets and therefore should be able to defend the decision to accept the risks of exposing the assets to the threats." (ISO/IEC 2012, p. 39, Part1: Introduction and general model)

Risk level determination—The ISO 27001 demands a high-level risk definition in alignment with the risk management of the organization, a similar step is part of ISO 27001:2013. The IT Grundschutz standards use protection requirements as high-level risk indicators. The Common Criteria standard does not consider high-level risks, but it does define attack potentials.

Security Analysis Process

Risk Determination—The ISO 27001 demands a description of the risk management methodology, and ISO 27001:2013, as well. The IT Grundschutz proposes a categorization of assets and to conduct a risk analysis only for the assets with significant security concerns. The standard does not demand a specific method for risk management, but it provides advice for considering risk, threats, and security controls. The Common Criteria focuses on documenting vulnerabilities and security controls of the TOE. It does not consider risk management per se, but rather provides the information about threats and countermeasures to stakeholders. Afterwards the stakeholders can use this information to conduct a risk analysis.

Security Property Description—ISO 27001 demands high-level security goals as part of the ISMS policy, which defines the focus of security of the ISMS and is described right after the scope. ISO 27001:2013 allows also to describe a framework

Table 4.15 Instantiation for common criteria of the security analysis product part of the template for security standard description (2/2)

Security analysis product
Security and risk documentation

The concepts of the Common Criteria consider that potential ToE owners infer their security needs for specific types of ToE, e.g., a specific database. The resulting documents are called Security Targets (ST). Protection profiles (PP) state security needs for an entire class of ToEs, e.g., client VPN application. The evaluators check if a ToE meets its ST. PPs state the security requirements of ToE owners. ToE developers or vendors publish their security claims in an ST. A CC evaluation determines if the ST is compliant to a specific PP. The standard relies upon documents for certification, which state information about security analysis and taken measures

The structure of a CC security target starts with an ST *Introduction* that contains the description of the ToE and its environment. The *Conformance Claims* describe to which PPs the ST is compliant. The *Security Problem Definition* refines the external entities, e.g., stakeholders in the environment and lists all assets, assumptions about the environment and the ToE, and threats to assets and organizational security policies. The *Security Objectives* have to be described for the ToE and for the operational environment of the ToE. The *Extended Component Definitions* describe extensions to security components described in the CCs part 2. The *Security Requirements* contain two kinds of requirements. The security functional requirements (SFR) are descriptions of security functions specific to the ToE. The security assurance requirements (SAR) describe the measures taken in development of the ToE. These are evaluated against the security functionality specified in the SFR. The Evaluation Assurance Level (EAL) is a numerical rating ranging from 1 to 7, which states the depth of the evaluation. Each EAL corresponds to an SAR package. EAL 1 is the most basic level and EAL 7 the most stringent

The Common Criteria defines a set of *Security Assurance Components* that have to be considered for a chosen *Evaluation Assurance Level (EAL)*. For these components, developer activities, content of corresponding components, and actions for an evaluator are defined. The Common Criteria defines security assurance components for the following *Assurance classes*:

- Protection Profile Evaluation (APE)
- Security Target Evaluation (ASE)
- Development (ADV)
- Life-Cycle support (ALC)
- Tests (ATE)
- Vulnerability Assessment (AVA)

In the Security Target, Security Objectives are defined for the TOE on for the TOE's environment. The Security Objectives are related to Security Functional Requirements. Part of the assurance classes for the development documentation (ADV) is the functional specification (ADV_FSP). In this document, the security functions (SFs) are defined. According to the security architecture (as required in ADV_ARC), the TOE design with details about the subsystems and modules are documented in the TOE design (ADV_TDS). This design document brakes down the security functions (SFs) and relates all subsystems and modules to the security functional requirements (SFRs) they implement. Vulnerabilities are assessed in the corresponding document according to the claimed attack potential (high, medium, low)(AVA_VAN)

for eliciting high-level security goals instead of stating high-level security goals. The IT Grundschutz demands to describe protection requirements using confidentiality, integrity, and availability. In addition, the standard demands a categorization into the levels: normal, high, very high. The Common Criteria demands that security

Table 4.16 Term comparison between security standards

Terms\standards	ISO 27001	ISO 27001:2013	IT grundschutz	Common criteria
Machine	ISMS	ISMS	ISMS	TOE
Environment	Outside the boundaries of the ISMS	Context of the organization	Interfaces to external partners	Operational environment
Stakeholder	Interested party	Interested party	Employee or users	TOE owner or user
Asset	~	No explicit definition	~	~
Security control	Control	Control	Safeguard	Countermeasure
Attacker	–	–	–	threat agent
Vulnerability	~	~^{c}	~	~
Threat	~^{a}	~^{c}	~^{a}	~
Policy	ISMS policy, security policy	Information security policy	Information security policy	Organizational security policy
Security goals	Security objectives	Security objectives	Protection requirements	Security needs
Security requirements	~	~	(Security test plans)^b	Security objective
Security functions	–	–^{d}	–	Security functional requirements

[a]Note that attackers can be seen as threats.
[b]Note that the security test plans are not requirements, but are based on refined protection requirements.
[c]Note these terms are not mentioned explicitly in ISO 27001:2013, but are defined in ISO 27000:2014.
[d]Note that the ISO27001:2013 demands to consider security objectives for relevant functions. However, the standard does not discuss security functions or uses the term functions later on in the standard

concerns are described in terms of confidentiality, integrity, and availability. The standard contains a catalog of refinements of these terms, which have to be used in TOE descriptions.

Control Assessment—The ISO 27001 focuses on likelihoods of threats exploiting existing vulnerabilities and the effect already implemented controls have on these likelihoods. ISO270001:2013 has a similar demand. The IT Grundschutz has mappings from threats to security controls and it has to be checked if the recommended security controls are implemented for all identified threats. The Common Criteria documents existing security controls by describing existing security functionalities of the TOE. The gap texts in the security functional requirements of the standard have to be used for these descriptions.

Vulnerability and Threat Analysis—The ISO 27001 concerns threat analysis in order to determine risks for assets. ISO 27001:2013 has no explicit demands for threat and vulnerability analysis. The threat analysis is based on a vulnerability identification. The IT Grundschutz standard relies on a list of threats for the identified scope parts, e.g., applications from the IT Grundschutz Catalogues. The Common Criteria demands to describe threats from malicious and from unintentional attackers. The capabilities of these attackers have to be described in terms of expertise, resources, and motivation.

Security Analysis Product

Security Assessment—The ISO 27001 demands to evaluate the risks to assets considering threats and existing security controls. For all assets with unacceptable risks, additional security controls have to be selected from the normative ANNEX A of the standard. ISO27001:2013 also refers to selecting controls and mapping them to the normative ANNEX A of the standard, but does not refer to assets. The IT Grundschutz standards begin with a basic security check, which is based on security tests derived from the protection requirements. The tests are used for an effectiveness evaluation of the existing security controls. The Common Criteria relies on cross-tables that map threats to security objectives. All threats have to be addressed by at least one security objective or assumption.

Security Measures—The ISO 27001 demands first high-level security policies, which are refined into a set of relevant security controls considering the controls listed in the mandatory ANNEX A of the standard. ISO 27001:2013 allows to use any controls, but demands a mapping to the ANNEX A controls. The IT Grundschutz demands using the mapping from scope elements to threats, and subsequently to security controls in the IT Grundschutz Catalogues. Only assets that are not considered adequately in the IT Grundschutz Catalogues demand a separate security analysis. The Common Criteria refines security objectives using a catalog of security functional requirements. A further cross-table has to prove that each security objective is addressed by at least one security functional requirement.

Risk Acceptance—The ISO 27001 demands to define criteria for risk acceptance in the management approval document. The standard demands a reasoning why the selected security controls reduce the risk to acceptable limits for each asset. ISO 27001:2013 states a similar demand. The IT Grundschutz simply demands a documentation of accepted risks including a reason why these risks are accepted. The Common Criteria demands risk acceptance decisions from asset owners. They have to make an informed decision to accept the risks of the identified threats.

Security and Risk Documentation—The ISO 27001 demands documentation about the scope and security policies, and extensive documentation of the risk management. The demands are similar in ISO27001:2013. The IT Grundschutz standards simply demand to document all the steps of the method. The Common Criteria demands an extensive documentation of the security reasoning and the resulting software product, and in particular the security functions of the product.

To sum up, the ISO 27001 and ISO 27001:2013 concern a high-level process with regard to security. The IT Grundschutz refines the ISO 27001 process and provides further guidances for identifying threats and security controls based on the IT Grundschutz Catalogues. In contrast, the Common Criteria focuses on documenting a software or hardware product including details of its implementation. The reasoning about which security standard is applicable should be based on the concerned application domain. A vendor of a hardware router might want to select the Common Criteria, due to the detailed security analysis of its implementation. A cloud computing provider who offers scalable IT resources and particular business processes concerning these resources might favor ISO 27001. A reason could be that documenting a high-level security process allows changes within the cloud implementation, because the process does not consider the implementation in detail. Using the Common Criteria would demand a documentation of its implementation and a recertification each time the implementation changes.

CAST Tool Support

We base our tool support on the NESSoS CBK (Sect. 12.4) that aims to collect knowledge on engineering secure systems. The structure of the *CBK* relates Knowledge Objects (KOs) for specific fields (referred to as Knowledge Areas—KAs). We define the following four types of KOs. *Methods* define a set of activities used to tackle problems in engineering secure software and services in a systematic way. *Techniques* describe an approach that contains only a single activity. *Tools* support a software engineer in achieving a development goal in an (at least partially) automated way. A *Notation* defines symbols, a syntax, and semantics to express relevant artifacts (Schwittek et al. 2011). We included security standards as a fifth type of KO, meaning we implemented the security standard template in its underlying ontology. In addition, the CBK offers the functionality to compare KOs by displaying their attributes next to each other. Hence, we can display two instantiated security standard templates next to each other. This way the comparison of them is supported. Furthermore, a search functionality allows to search the instantiated templates for specific search terms. In the future, we are planning to implement an automatic search for supporting KOs for security standards and a comparison of security standard support methodologies.

Our method provides the means to describe three building blocks of security standards. The first block states how context description and preparation of a security analysis has to be done in a standard. This provides an overview of the level of detail demanded for a security standard compliant system documentation. For example, the IT Grundschutz standards demand to treat every item in the scope as an asset and conduct a security analysis for it, while the ISO 27001 demands a reasoning about which are the assets in the scope. Hence, the ISO 27001 allows more flexibility in the security analysis.

The security analysis process shows how existing controls, risk, threats, and vulnerabilities have to be analyzed. For example, the IT Grundschutz demands a characterization of the existing controls according to certain categories, while the ISO 27001 simply refers to a statement of how the existing controls reduce the likelihoods of security failures. This is another indication that the ISO 27001 demands a less-structured documentation than the IT Grundschutz standards. In contrast, the Common Criteria controls are clearly separated into IT and non-IT countermeasures. For this reason, the standard can be applied especially for product development.

Finally, the security analysis product shows the overall security assessment and in particular how security measures have to be described, risk acceptance to be determined, and what documentation is required for a certification. As an example, the ISO 27001 demands a specific set of a few documents, while the IT Grundschutz simply demands to document the entire process.

4.8 Discussion

Our method creates the following main benefits:

- A simple overview of the core activities of security standards.
- Enabling a structured comparison of standard activities by storing the knowledge about standards in defined template fields.
- Providing indication of the focus, level of detail, and effort for providing or even reading a system documentation according to a specific standard.

We could identify the following points for improvement of our work:

- The approach could also be extended to compare support tools for standard compliant system documentation and analysis.
- Our templates can be analyzed for reusing artifacts and results of the certification of one standard for another. This could lead to a possible optimal sequence of certifications of different standards with regard to resources spent in terms of time and money.
- The overview is provided on an abstract level and the engineers still have to read the standards to compare these on a more granular level. Our method could be extended to support a more detailed analysis of the standard documents.

4.9 Summary

We contributed a conceptual model of security standards based on existing research such as the HatSec method and the NIST SP 800-30 standard. Furthermore, we derived a template from the conceptual model that can be instantiated for different security standards. Our approach offers the following main benefits:

- A common terminology and a conceptual model of security standards, and a template that supports the structured collection of knowledge by using common security standard activities, e.g., asset identification
- A set of instantiated security standard templates for the standards ISO 27001, IT Grundschutz, and Common Criteria. The templates provide an overview of the most relevant standard activities.

- Improving the understanding of commonalities and differences of security standards by analyzing the difference in the common standard activities, e.g., how do ISO 27001 and Common Criteria identify assets?
- Supporting security engineers in the decision which certification scheme to pursue and what kind of information to expect from a security standard documentation.

We applied this idea to several security standards and compared the resulting template instances.

References

Beckers, K., Eicker, S., Faßbender, S., Heisel, M., Schmidt, H., & Schwittek, W. (2012). Ontology-based identification of research gaps and immature research areas. In *Proceedings of the International Cross Domain Conference and Workshop (CD-ARES 2012)* (pp. 1–16). Berlin: Springer.

Beckers, K., Côté, I., Fenz, S., Hatebur, D., & Heisel, M. (2014). A structured comparison of security standards. In M. Heisel, W. Joosen, J. Lopez, & F. Martinelli (Eds.), *Advances in engineering secure future internet services and systems* (pp. 1–34). Berlin: Springer.

Bishop, M. (2003). *Computer security: Art and science* (1st ed.). Upper Saddle River: Pearson.

BSI. (2008a). Standard 100–1 Information Security Management Systems (ISMS), Version 1.5. Bonn Germany: Bundesamt für Sicherheit in der Informationstechnik (BSI).

BSI. (2008b). IT-Grundschutz-Vorgehensweise (BSI Standard 100-2). Bonn Germany: Bundesamt für Sicherheit in der Informationstechnik (BSI).

BSI. (2008c). Standard 100-3 Risk Analysis Based on IT-Grundschutz, Version 2.5 Technical Report. Bonn Germany: Bundesamt für Sicherheit in der Informationstechnik (BSI).

BSI. (2009). BSI Standard 100-4 Business Continuity Management, Version 1.0 (BSI Standard 100-4). Bonn Germany: Bundesamt für Sicherheit in der Informationstechnik (BSI).

BSI. (2010). IT-Grundschutzkataloge. Bonn Germany: Bundesamt für Sicherheit in der Informationstechnik (BSI). http://www.bsi.bund.de

BSI. (2011a). BSI Grundschutz Homepage. Bonn Germany: Bundesamt für Sicherheit in der Informationstechnik (BSI). https://www.bsi.bund.de/DE/Themen/ITGrundschutz/itgrundschutznode.html

BSI. (2011b). BSI Grundschutz Standards Homepage. Bonn Germany: Bundesamt für Sicherheit in der Informationstechnik (BSI). https://www.bsi.bund.de/EN/Publications/BSIStandards/BSIStandardsnode.html

BSI. (2011c). Protection Profile for the Gateway of a Smart Metering System (Gateway PP) (Version 01.01.01(final draft)). Bonn, Germany: Bundesamt für Sicherheit in der Informationstechnik (BSI)—Federal Office for Information Security Germany. https://www.bsi.bund.de/SharedDocs/Downloads/DE/BSI/SmartMeter/PP-SmartMeter.pdf?blob=publicationFile

DCSSI. (2004). Expression des Besoins et Identification des Objectifs de Sécurité (EBIOS)—Section 2—Approach. General Secretariat of National Defence Central Information Systems Security Division (DCSSI).

Fabian, B., Gürses, S., Heisel, M., Santen, T., & Schmidt, H. (2010). A comparison of security requirements engineering methods. *Requirements Engineering—Special Issue on Security Requirements Engineering*, 15(1), 7–40.

Farquhar, B. (1991). One approach to risk assessment. *Computers and Security*, 10(10), 21–23.

Firesmith, D. (2003). *Common concepts underlying safety, security, and survivability engineering* Technical report SEI-2003-TN-033). Pittsburgh, United States: Carnegie Melon University.

Food, & Administration, D. (2006). Guideline for Industry, Q9 quality Risk Management. (In US Department of Health and Human Services).

Gollmann, D. (2005). *Computer security* (2nd ed.). Hoboken: Wiley.

ISO. (2009). *ISO 31000 risk management—Principles and guidelines*.

ISO/FDIS. (2007, November). ISO/IEC 27799:2007(E), Health informatics—Information security management in health using ISO/IEC 27002. Geneva, Switzerland: International Organization for Standardization (ISO) and International Electrotechnical Commission (IEC).

ISO/IEC. (2005). Information technology—Security techniques—Information security management systems—Requirements (ISO/IEC 27001). Geneva, Switzerland: International Organization for Standardization (ISO) and International Electrotechnical Commission (IEC).

ISO/IEC. (2008). *Information technology—security techniques—information security risk management* (ISO/IEC 27005). Geneva, Switzerland: International Organization for Standardization (ISO) and International Electrotechnical Commission (IEC).

ISO/IEC. (2012). Common criteria for information technology security evaluation (ISO/IEC 15408). Geneva, Switzerland: International Organization for Standardization (ISO) and International Electrotechnical Commission (IEC).

ISO/IEC. (2013). Information technology—Security techniques—Information security management systems—Requirements (ISO/IEC 27001). Geneva, Switzerland: International Organization for Standardization (ISO) and International Electrotechnical Commission (IEC).

Jackson, M. (2001). Problem Frames. *Analyzing and structuring software development problems*. Addison-Wesley.

Japan Information Processing Development Corporation and The Medical Information System Development Center. (2004). ISMS User's Guide for Medical Organizations.

JASON. (2010). Science of Cyber-Security. Technical Report. Bedford Massachusetts and McLean Virginia United States: The MITRE Corporation. Retrieved from http://www.fas.org/irp/agency/dod/jason/cyber.pdf (JSR-10-102)

Karabacak, B., & Sogukpinar, I. (2005). ISRAM: information security risk analysis method. *Computers and Security, 24*(2), 147–159.

Pouloudi, A. (1999). Aspects of the stakeholder concept and their implications for information systems development. In *Proceedings of the Hawaii International Conference on System Sciences (HICSS)* (pp. 5–8). IEEE Computer Society.

Schumacher, M., Fernandez-Buglioni, E., Hybertson, D., Buschmann, F., & Sommerlad, P. (2006). *Security patterns: Integrating security and systems engineering*. Wiley.

Schwittek, W., Schmidt, H., Eicker, S., & Heisel, M. (2011). Towards a common body of knowledge for engineering secure software and services. In *Proceedings of the International Conference on Knowledge Management and Information Sharing (KMIS)* (pp. 369–374). SciTePress—Science and Technology Publications.

Schwittek, W., Schmidt, H., Beckers, K., Eicker, S., Faßbender, S., & Heisel, M. (2012). A common body of knowledge for engineering secure software and services. In *Proceedings of the International Conference on Availability, Reliability and Security (ARES)—1st International Workshop on Security Ontologies and Taxonomies (SecOnT 2012)* (pp. 499–506). IEEE Computer Society.

Sharp, H., Finkelstein, A., & Galal, G. (1999). Stakeholder identification in the requirements engineering process. In *Proceedings of the Dexa Workshop* (pp. 387–391). IEEE Computer Society.

Siemens. (2003). CRAMM—The total information security toolkit. http://www.cramm.com/

Stallinger, M. (2004). CRISAM—Corporate risk application method—Summary V2.0.

Standards Australia International; Standards New Zealand. (2001). Guidelines for managing risk in healthcare sector: Australian/ New Zealand handbook. (Standards Australian International).

Stoneburner, G., Goguen, A., & Feringa, A. (2002). *Risk management guide for information technology systems* (NIST Special Publication No. 800-30). Gaithersburg, U.S.: National Institute of Standards and Technology (NIST).

Sunyaev, A. (2011). *Health-care telematics in Germany—design and application of a security analysis method*. Gabler.

Viega, J., & McGraw, G. (2001). *Building secure software: How to avoid security problems the right way* (1st ed.). Boston: Addison-Wesley.

Chapter 5
Relating ISO 27001 to the Conceptual Framework for Security Requirements Engineering Methods

Abstract The establishment of the ISO 27001 security standards is a difficult endeavor, due to ambiguity in its natural language text and sparse descriptions of the needed system analysis procedures. Security engineering methods provide guidance for a detailed and step-by-step security analysis within a given software engineering project. We propose not to limit security requirements engineering methods to software engineering, but to extend their usefulness to the establishment of security standards. Moreover, we do not aim to create just one extension of a security engineering method for ISO 27001. We want to provide the means to extend any possible security requirements engineering method to be compliant to ISO 27001. For this purpose we rely on an existing conceptual framework for security requirements engineering introduced by Fabian et al. and identify relations from this framework to the ISO 27001 standard. These relations are fundamental to understand which parts of a given security requirements engineering method already support ISO 27001 and what parts have to be extended. Furthermore, we discuss the required documentation for ISO 27001 compliance and how to extend security requirements engineering methods to be able to produce the needed documentation.

5.1 Introduction

Security standards are ambiguous on purpose, because they should serve a multitude of different domains and stakeholders. In order to clarify their ambiguity, we propose to use security requirements engineering methods to support system development and document demands of security standards. We make our work applicable to different security requirements engineering methods by relying on the conceptual framework for security requirements engineering methods introduced in Sect. 2.4. In this book, we consider two security standards: The Common Criteria (Sect. 2.2.4) and the ISO 27001 standard (Sect. 2.2.2). We contribute relations from the framework to the conceptual model, which we published in (Beckers et al. 2012a, b). We are the main author of both publications. Stephan Faßbender and Maritta Heisel helped to refine

© Springer International Publishing Switzerland 2015
K. Beckers, *Pattern and Security Requirements*,
DOI 10.1007/978-3-319-16664-3_5

Fig. 5.1 Wordcloud ISO 27001

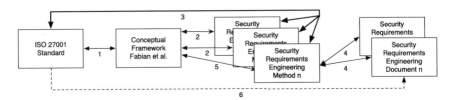

Fig. 5.2 Relating security requirements engineering methods with the ISO 27001 standard

the description of the concept for relating security requirements engineering methods
and the ISO 27001 standard (see Fig. 5.2). Stephan Faßbender also supported the
description of the practical applications of our research (see Fig. 5.10). Jan-Christoph
Küster and Holger Schmidt provided valuable feedback on our work. This chapter
is based upon the aforementioned publications. The relations from the Common
Criteria to the conceptual framework were published in the original publication of
the framework (Sect. 2.4) (Fig. 5.1).

5.2 Relating ISO 27001 to Security Requirements Engineering Methods

Fulfilling organizations' security requirements is a challenging task. Security standards, e.g., the ISO 27000 series of standards, offer a way to attain this goal. The normative standard of the aforementioned series, the ISO 27001, contains the requirements for an *Information Security Management System (ISMS)* (ISO/IEC 2009). The standard prescribes a process, which tailors security to the needs of any kind of organization. The remaining standards of the ISO 27000 series describe parts, or usage scenarios, of the ISMS in detail (ISO/IEC 2009; Kersten et al. 2011; Calder 2009). For example, the ISO 27005 (ISO/IEC, 2008) describes information security risk management (see also Sect. 2.2.2). The ISO 27005 has a certain significance as the ISO 27001 is risk-centered in many sections, and the ISO 27005 describes the risk assessment process and the risk documentation and management in detail. However, the ISO 27005 is not normative.

The ISMS consists of processes, procedures, and resources that can be software. Ambiguous descriptions in the standard are a problem during the establishment of an ISMS. For example, the required input for the *scope and boundaries* description is to consider "characteristics of the business, the organization, its location, assets, and technology" (ISO/IEC 2005, p. 4). This is a problem because the security analyst has to decide how to describe the business, organization, etc., and to what level of detail. In addition, the security analyst has to find a method that allows her/him to achieve completeness of assets, etc.

Moreover, the standard does not provide a method for assembling the necessary information or a pattern on how to structure that information. The importance of these steps becomes apparent when one realizes that essential further steps of the ISO 27001 depend upon them, e.g., the identification of *threats*, *controls*, and *vulnerabilities*.

Security requirements engineering (SRE) methods, on the other hand, provide structured elicitation and analysis of security requirements.[1] SRE methods can be part of the early phases of a given software development process. However, we propose not to limit SRE methods to software development. The structured elicitation and analysis of security requirements of SRE methods are also useful for different security engineering contexts. Therefore, we propose to use SRE methods to support security engineers in the development and documentation of an ISMS compliant with ISO 27001. In addition, the ISMS is a process for security that may also rely on secure software. Thus, SRE methods can also support software engineers in building secure software for an ISMS.

A word cloud is a picture of words in different sizes.[2] The algorithm to create word clouds is counting words and displaying them in a size according to the number of appearances of the word in the text. The more often the word appears, the bigger it is displayed. The word "requirements" appears in a significant size, which is because

---

[1] In this book we classify also security extensions of more general requirements engineering methods, e.g., Tropos and i* as SRE methods, because only SRE is the focus of this work.

[2] The figure was created using the wordle tool: http://www.wordle.net.

the standard is defining the requirements for building an ISMS (Fig. 5.1). Hence, the consideration of requirements is of utmost importance to the standard and the word appears quite often in the standard. This was a first hint for us to analyze the relations between security requirements engineering and the ISO 27001 security standard.

The outcome of this analysis answers the research question, if and to what extent SRE methods can support the development of an ISO 27001 compliant ISMS. Moreover, it answers the question in what way SRE methods provide the required documentation for an ISMS and how existing SRE documentation can be reused for an ISMS.

Our work starts with a *top-down* method. We systematically analyze the ISO 27001 standard in order to determine *where* and *how* SRE methods can support the development and documentation of an ISMS according to ISO 27001. We depicted the results of the analysis in Fig. 5.2. First, we create a relation between the ISO 27001 standard and the conceptual framework (CF) of Fabian et al. (2010) (see the arrow marked 1). Second, we use the relation of terminologies and notions from the CF to numerous SRE methods already provided by Fabian et al. (2010) (2). Third, combining the relations of Steps 1 and 2 we can relate the ISO 27001 to different SRE methods (3).

The second part of our work is a *bottom-up* method. We support reusing documents created by an SRE method, so-called *SRE documents* (4). We propose to use these documents as part of an ISMS specification. We can reuse the relation between the used SRE method and the CF to figure out *what* ISO 27001 section the SRE documents support. If this relation does not exist, we have to create it (5). It is sufficient to create a relation between the CF and the SRE method, because of the existing relation between the CF and ISO 27001. Any new relation between the CF and an SRE method, not considered by Fabian et al., results in a further relation between the ISO 27001 and this method, because of the existing relation between the CF and the ISO 27001. Thus, transitive relations from the ISO 27001 to existing SRE documents are possible (6).

Table 5.1 relates relevant terms for security from the CF by Fabian et al. (2010) to the ISO 27001 standard. The matching benefits from the fact that both documents rely on ISO 13335 (ISO/IEC 2004) definitions for several terms.

ISO 27001 Section 4 describes the ISMS. Hence, we focus on this section in particular. Table 5.2 lists relations between subsections of ISO 27001 Section 4 and the CF's building blocks. We present all subsections of ISO 27001 Section 4.2, because these describe the establishment of the ISMS. In addition, we show risk management as a separate column, even though it is part of the CF's building block *threat analysis*. The reason is that some subsections of ISO 27001 Section 4 and some SRE methods specifically focus on risk management. Moreover, the importance of risk in the ISO 27000 series of standards resulted in the standard ISO 27005 (ISO/IEC, 2008) for information security risk management that specifies the risk management of the ISO 27001. A "+" in Table 5.2 marks a part of the section that can be supported by a building block of the CF. However, the free cells of the table do not imply that a method could not support that section of the ISO 27001. An *gray* row indicates that there are no explicit matches between the ISO 27001 section and the CF.

Table 5.1 Correspondence between ISO 27001 terms and terms of the CF

CF Fabian et al.	ISO 27001
System	The *organisation* is the "scope" of the standard (ISO/IEC 2005, p. 1)
Machine	The *Information Security Management System (ISMS)* is the machine to be built (ISO/IEC 2005, p. v)
Environment	The *scope and boundaries* of the "organization" (ISO/IEC 2005, p. 4, Sect. 4.2.1a) relevant for the ISMS
Security goal	The standard uses *security objectives* (ISO/IEC 2005, p. 4, Sect. 4.2.1b) instead of security goals
Security requirement	*Security requirement* is also used in ISO 27001 as a description of the "organization" after the "ISMS" is introduced (ISO/IEC 2005, pp. v and vi)
Specification	The ISMS's policy, controls, processes and procedures (ISO/IEC 2005, p. vi) are the specification of the machine
Stakeholder	The *Interested Parties* (ISO/IEC 2005, p. vi) have security "expectations" that are input for the ISMS implementation as well as "security requirements"
Domain knowledge	The *characteristics of the business, the organization, its location, assets, and technology* (ISO/IEC 2005, p. 4)
Availability	The definition in ISO/IEC 13335 (ISO/IEC 2004) is also used (ISO/IEC 2005, p. 2)
Confidentiality	The definition in ISO/IEC 13335 (ISO/IEC 2004) is also used (ISO/IEC 2005, p. 2)
Integrity	The definition in ISO/IEC 13335 (ISO/IEC 2004) is also used (ISO/IEC 2005, p. 2)
Asset	The definition in ISO/IEC 13335 (ISO/IEC 2004) is also used (ISO/IEC 2005, p. 2)
Threat	The definitions match. Threats are defined toward assets and threats exploit vulnerabilities (ISO/IEC 2005, p. 4)
Vulnerability	The definitions match (ISO/IEC 2005, p. 4)
Risk	The CF defines risk as "the potential loss of a stakeholder" (Fabian et al. 2010, p. 13), while in ISO 27001 risk is not defined explicitly. However, the risk identification evolves around identifying asset, threat, vulnerability, and the impact a loss of availability, confidentiality, and availability has on an asset (ISO/IEC 2005, p. 4). Hence, we can conclude that the meaning is similar

Relating ISO 27001 Section 4.2.1a to SRE Methods

ISO 27001 Sect. 4.2.1a demands to "Define the scope and boundaries of the ISMS in terms of the characteristics of the business, the organization, its location, assets and technology, and including details of and justification for any exclusions from the scope." (ISO/IEC 2005, p. 4).

Table 5.2 Relating ISO 27001 Sect. 4 to CF building blocks

Section	Description	SV	SR	SDK	TA	RM
Section 4.1	General requirements	+	+	+	+	+
Section 4.2	Establish and manage the ISMS	+	+	+	+	+
Section 4.2.1	Establish the ISMS	+	+	+	+	+
Section 4.2.1a	Define scope and boundaries	+		+		
Section 4.2.1b	Define ISMS policy	+	+		+	+
Section 4.2.1c	Define risk assessment					+
Section 4.2.1d	Identify the risk	+			+	+
Section 4.2.1e	Analyze and evaluate risk			+	+	+
Section 4.2.1f	Identify risk treatment				+	+
Section 4.2.1g	Select controls				+	+
Section 4.2.1 h, i	*Obtain management approval*					
Section 4.2.1j	Prepare a statement of applicability				+	+
Section 4.2.2	Implement and operate the ISMS				+	+
Section 4.2.3	Monitor and review the ISMS	+	+	+	+	+
Section 4.2.4	*Maintain and improve the ISMS*					
Section 4.3	Documentation requirements	+	+	+	+	+

SV Stakeholder Views, *SR* System Requirements, *SDK* Specification and Domain Knowledge, *TA* Threat Analysis, *RM* Risk Management

The characteristics of the business include *interested parties*, which are stakeholders in the CF terminology (see Table 5.1). The security expectations of the stakeholders are input for the ISMS (ISO/IEC 2005, p. vi). Moreover, ISO 27001 Sect. 4.2.1a states that assets shall be defined. The *Stakeholder Views* provide a description of stakeholders, their assets and security goals. In addition, the *characteristics of the business* are the functional and nonfunctional goals of the stakeholders (Fig. 2.3). Functional, nonfunctional, and security goals are refined into functional, nonfunctional, and security requirements by security experts. These experts add details to the goals to arrive at requirements.

ISO 27001 Sect. 4.2.1a requires information about the location and technology of the organization. Furthermore, it requires information about *exclusions from the scope* of the ISMS. The *Specification and Domain Knowledge* contains information about the environment in the *Domain Knowledge* (see Fig. 2.3). This information includes location and technology of the organization. The information about the environment enables also decisions for *exclusions from the scope* of the ISMS.

This relation between ISO 27001 Sect. 4.2.1a and the CF's building blocks *Stakeholder Views* and *Specification and Domain Knowledge* provides in consequence a relation to SRE methods, because the work of Fabian et al. (2010) already contains a relation between the CF and SRE methods. We illustrate our matching between the conceptual framework and the ISO 27001 standard (see Fig. 5.3). We underlined specific terms and parts of ISO 27001 sections and showed arrows that mean "maps to". The arrows point at a part of the conceptual framework the part of the ISO 27001 section has a relation to. Table 5.3 shows the relation of several different SRE methods to the CF taken from Fabian et al. (2010). The "x" in a field of the table means

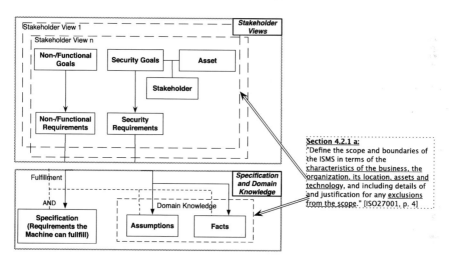

Fig. 5.3 Matching the CF and ISO 27001 Sect. 4.2.1a

Table 5.3 SRE methods supporting ISO 27001 Sect. 4.2.1a

Methods	Stakeholder views	Domain knowledge
Goal-based methods		
KAOS	x	x
Secure tropos	x	
Si*	x	
Problem-based methods		
SEPP		x
Abuse frames		
Risk-based methods		
CORAS		x
ISSRM		x

that there is a relation to a CF building block. However, the free cells of the table do not imply that a method could not support a section of the ISO 27001. Table 5.3 presents the goal-oriented SRE methods KAOS (van Lamsweerde 2007; Bertrand et al. 1998), Secure Tropos (Susi et al. 2005; Mouratidis and Giorgini 2007), and Si* (Massacci et al. 2010), the problem-oriented methods SEPP (Schmidt et al. 2011), and Abuse Frames (Lin et al. 2004), and the risk analysis-based methods CORAS (Lund et al. 2010), and ISSRM (Mayer et al. 2005).

KAOS' realization of the *Stakeholder Views* considers multiple stakeholders in and multiple views toward a system-to-be (Fabian et al. 2010). These views are different models of the system, e.g., *goal, object, agent,* and *security threat* models. Goals of stakeholders are modeled in a tree that refines the goals until a goal can be assigned to an agent. At this stage the goal becomes a requirement (Fabian et al. 2010). To sum up, KAOS describes the *organization* in the ISO 27001 standard via the views on it. It also realizes *Specification and Domain Knowledge*. The information about *assets, location,* and *technology* relevant for the ISO 27001 is included in different views.

Secure Tropos realizes *Stakeholder Views*. The notation models actors and their goals. An actor in this notation is also equivalent to a stakeholder (Fabian et al. 2010). Secure Tropos uses goals of stakeholders to model the *organization* in the ISO 27001 standard. Secure Tropos does not realize *Specification and Domain Knowledge*. The notation focuses on analyzing the trade-off relations between different security goals from stakeholders and their perspectives. The same statement is also true for the Si* method (Fabian et al. 2010).

SEPP is a problem-based method and does not realize *Stakeholder Views* explicitly. It centers around a description of the problem that the *machine* to be built shall solve. The problem is described in terms of the *environment* around it. SEPP realizes *Specification and Domain Knowledge* (Fabian et al. 2010). The method captures the domain knowledge in problem diagrams and natural language. This method models the *environment* of the ISMS in a context diagram.

Abuse frames is also a problem-based method, which models the machine in its environment. The method focuses on describing Anti-requirements, which describe the threats to the assets. In contrast to SEPP, this method does not capture domain knowledge.

CORAS bases its scope model upon stakeholder interviews and investigations from risk experts. Risk analysis experts generate a model from the resulting information. However, the interviews do not focus on the perspective of stakeholders, but solely on the target of the risk evaluation. CORAS realizes *Specification and Domain Knowledge*. The method develops a model of the risk target, including assumptions about the target and its environment. The ISO 27001 description of the location, assets, and technologies are modeled as assumptions. The ISSRM method considers the steps context analysis, asset identification, security requirements description, and the reasoning about countermeasures. The security requirements and countermeasure selection is based upon risk analysis. ISSRM also realizes *Specification and Domain Knowledge*. The method describes the scope of the analysis in a context model.

The ISO 27001 standard dictates a documentation that proves the relationship between chosen controls to the ISMS policies and objectives. This documentation has to contain a description of the *scope and boundaries* of the ISMS (ISO/IEC 2005, p. 13, Sect. 4.3.1). Therefore, the output of *Stakeholder Views* and *Specification and Domain Knowledge* of the different SRE methods in Table 5.3 supports the ISO 27001 documentation of the *scope and boundaries* of the ISMS.

Relating Section 4.2.1b to SRE Methods

ISO 27001 Sect. 4.2.1b stipulates an ISMS policy. The policy shall take into account the *characteristics of the business*. Moreover, the stakeholder's attributes *location*, *assets*, and *technologies* have to be described. In addition, ISO 27001 Sect. 4.2.1b prescribes the establishment of a framework for security objectives and an "overall sense of direction" of security (ISO/IEC 2005, p. 4).

We determined that security objectives are security goals in the CF (see Table 5.1). Moreover, in order to establish "an overall sense of direction and principles for action with regard to information security" (ISO/IEC 2005, p. 4), nonfunctional and functional requirements have to be elicited. These are part of the *Stakeholder Views* of the CF (see Fig. 5.4). Furthermore, we interpret the "overall sense of direction" for security as a set of coherent security requirements. The CF part *System Requirements* includes a reconciliation step of security requirements in order to get a consistent set of these and resolves possible conflicts among security requirements (see Fig. 5.4). Hence, we link the Sect. 4.2.1b to the *Stakeholder Views* and the *System Requirements* parts of the CF.

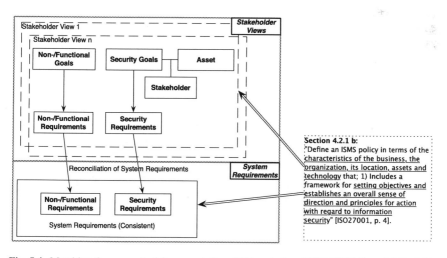

Fig. 5.4 Matching the conceptual framework from Fabian et al. and ISO 27001 Sect. 4.2.1b (1/2)

Nevertheless, ISO 27001 Sect. 4.2.1b also stipulates an alignment with the organization's risk management context and to establish risk evaluation criteria. We link this part to the *Threat Analysis* part of the CF, because the entire *Threat Analysis* part is necessary to describe the context of risk (Sect. 2.4). Risk evaluation criteria have to define potential loss(es) of a security property of assets. In addition, they at least have to consider vulnerabilities of security properties and countermeasure. All of these are part of the *Threat Analysis* part of the CF (Sect. 2.4). Hence, we conclude that the *Threat Analysis* part has also a relation to the ISO 27001 Sect. 4.2.1b (see Fig. 5.5).

KAOS covers *System Requirements* and *Threat Analysis* (Fabian et al. 2010). The method provides an obstacle and threat analysis for goals. This is accompanied by a conflict analysis between goals. KAOS explores solutions to problems and evaluates the results. KAOS also implements *Stakeholder Views* as noted in Sect. 5.2.

Secure Tropos implements *Stakeholder Views* and *System Requirements* (Fabian et al. 2010). The notation contains security constraints and dependencies of assets based on the goal-based Tropos notation. Security experts can model the effect of these security constraints, dependencies, and entities to the goals of actors, which are equivalent to stakeholders (Fabian et al. 2010). It also considers *Stakeholder Views*, as established previously in this section (Table 5.4).

Si* also implements *Stakeholder Views* and *System Requirements*, as well as *Threat Analysis* (Fabian et al. 2010). The method uses a meta-model of security concepts, which considers actors who want to achieve goals and in particular security goals. These goals require assets (or resources) for their fulfillment and these assets are targeted by threats (or attacks).

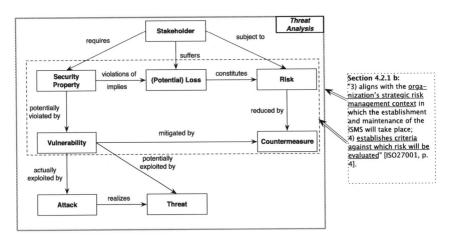

Fig. 5.5 Matching the conceptual framework from Fabian et al. and ISO 27001 Sect. 4.2.1b (2/2)

Table 5.4   SRE methods supporting ISO 27001 Sect. 4.2.1b

	Stakeholder views	System requirements	Threat analysis	Risk management
Goal-based methods				
KAOS	x	x	x	
Secure tropos	x	x		
Si*	x	x	x	
Problem-based methods				
SEPP		x		
Abuse frames		x	x	
Risk-based methods				
CORAS		x	x	x
ISSRM		x	x	x

SEPP implements *System Requirements* (Fabian et al. 2010). The method uses security problem frames to model security requirements in terms of the *environment*. The security problem frames have pre- and postconditions that can be evaluated after their instantiation. SEPP evaluates if a solution fulfills a security requirement by checking if the pre- and postconditions hold after the instantiation. Hence, SEPP makes it possible to check if a potential solution achieves a security requirement.

Abuse frames concern *System Requirements* and threats (Fabian et al. 2010). The method uses problem frames, which consider functional requirements in terms of the environment. Security needs are defined as constraints on functional requirements. Threats for these security needs are identified and documented in abuse frames. The method identifies security vulnerabilities that the threats can exploit and derives security requirements that address these vulnerabilities.

The CORAS implements *System Requirements*, *Threat Analysis*, and Risk Management (Fabian et al. 2010). The CORAS method uses workshops to collect assumptions about the scope of a risk analysis. The next steps are workshops for vulnerability and threat identification. The collected data from the workshops is used for a detailed risk analysis and a treatment plan.

The ISSRM method implements also *System Requirements*, *Threat Analysis*, and Risk Management (Fabian et al. 2010). The method uses a combined technique for security requirements elicitation and risk management. Based on the security requirements and associated risks for assets, the method supports the countermeasure selection.

The documentation of the standard ISO 27001 also requires a description of "the risk assessment methodology," "risk assessment report," "risk treatment plan," and "procedures and controls" for risk management (ISO/IEC 2005, p. 13). The resulting models and documents of the CORAS methods support this part of the ISMS documentation.

Relating Section 4.2.1d to SRE Methods

ISO 27001 Sect. 4.2.1d considers risk identification, which includes several typical elements of security requirements engineering methods, e.g., identification of stakeholders, assets, vulnerabilities, and threats (ISO/IEC 2005, p. 4).

The first part of the ISO 27001 Sect. 4.2.1d demands an "identification of assets within the scope of the ISMS, and the owners of these assets." (ISO/IEC 2005, p. 4). The CF part *Stakeholder Views* describes stakeholders and their assets. Figure 5.6 shows this relation. In addition, the relation of the scope and boundaries of the ISMS and the CF property *Stakeholder Views* is already elicited previously in this section.

Figure 5.7 presents the remaining relations between the ISO 27001 Sect. 4.2.1d and the *Threat Analysis* part of CF. ISO 27001 Sect. 4.2.1d prescribes the identification of threats in relation to assets (ISO/IEC 2005, p. 4). Assets are not explicitly part of the *Threat Analysis* part of the CF. However, they are considered in the *Stakeholder Views* part and linked to a stakeholder. The stakeholder in turn is an element of the *Threat Analysis* part. In addition, stakeholders require security properties that can be violated by attacks. Threats are potential exploits of vulnerabilities, and attacks are realized threats. Thus, we have a relation to the *Threat Analysis* part, and ISO 27001 Sect. 4.2.1d demands an identification of vulnerabilities and threats (ISO/IEC 2005, p. 4) as well.

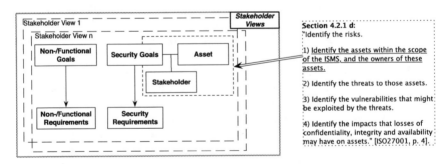

Fig. 5.6 Matching the conceptual framework from Fabian et al. and ISO 27001 Sect. 4.2.1d (1/2)

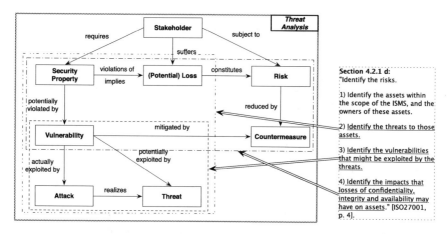

Fig. 5.7 Matching the conceptual framework from Fabian et al. and ISO 27001 Sect. 4.2.1d (2/2)

Table 5.5 SRE methods supporting ISO 27001 Sect. 4.2.1d

	Stakeholder views	Threat analysis	Risk management
Goal-based methods			
KAOS	x	x	
Secure Tropos	x		
Si*	x	x	
Problem-based methods			
SEPP			
Abuse frames		x	
Risk-based methods			
CORAS		x	x
ISSRM		x	x

We established a relation of ISO 27001 Sect. 4.2.1d and the CF properties *Stakeholder Views* and *Threat Analysis* in Table 5.2. Table 5.5 presents which of our considered SRE methods support the CF building blocks relevant for ISO 27001 Sect. 4.2.1d.

Furthermore, ISO 27001 Sect. 4.2.1d requires an identification of impacts and losses (ISO/IEC 2005, p. 4). The loss to a stakeholder of a security property constitutes a risk in the *Threat Analysis* property of the CF. This is also an obvious connection to risk management, which is a major property of ISO 27001 Sect. 4.2.1d. In addition, ISO 27001 Sect. 4.2.1d considers using security goals, the CIA tirade (confidentiality, integrity, availability), to describe the loss of a stakeholder that can be caused by a risk. This is not part of the graphical representation of the CF. However, the authors state that security goals in the CF have to be written down in terms of the CIA triad (Fabian et al. 2010, p. 12). Moreover, a security property contains security goals among others (Fabian et al. 2010, pp. 11 and 14).

We discussed the realization of *Stakeholder Views* using the methods KAOS, Secure Tropos, and Si* in this section. We also described the realization of *Threat Analysis*. Therefore, we now focus on *Risk Management* and the CORAS and ISSRM methods that implement it. The CORAS method identifies assets, vulnerabilities and subsequent threats, followed by a risk analysis and treatment. Hence, the method is an almost perfect match for implementing ISO 27001 Sect. 4.2.1d. The ISSRM method in contrast focuses on aligning security requirements and risk, and does not consider risk management as a separate step from the security requirements elicitation and analysis. Both methods do not consider *Stakeholder Views*. However, *Stakeholder Views* are included in the goal-oriented methods KAOS, Secure Tropos, and Si*. These methods, on the other hand, do not explicitly consider *Risk Management*. The problem-oriented methods SEPP and Abuse Frames also do not consider *Risk Management*. Nevertheless, both inspect a problem in detail under consideration of the environment.

ISO 27001:2013

The ISO 27001:2013 standard does not have an own terminology section, but refers to the terminology defined in ISO 27000:2014 (ISO/IEC 2014) (see Table 5.6).

We show a mapping from Sects. 4–8 of ISO 27001:2013 to the conceptual framework for security requirements engineering methods in Table 5.7. The standard states that the requirements for an ISMS are written in Sects. 4–10 of ISO 27001:2013 (ISO/IEC 2013, p. 1). Our conceptual framework does not concern leadership, management reviews, and monitoring. Hence, we cannot provide mappings to Sections 5.1, 9.1 and 9.3. In addition, our conceptual methods do not focus on the evaluation process during runtime, so the subsections of Sect. 10 cannot be mapped, as well.

5.3   Insights

We presented a relation between SRE methods and the ISO 27001 standard. The relations were obtained via the CF of Fabian et al. (2010). This CF presents four distinct building blocks of SRE methods. Table 5.2 relates the ISO 27001 standard to these building blocks. The *Stakeholder Views* building block has multiple relations to ISO 27001 sections. The reason is that the counterparts in the standard focus on the view of the organization including its stakeholders. The *Stakeholder Views* are part of numerous goal-oriented approaches, e.g., Secure Tropos (Mouratidis and Giorgini 2007) and KAOS (van Lamsweerde 2007). This is no surprise because these methods often derive goals considering the views of stakeholders (Figs. 5.8 and 5.9).

Table 5.6 Correspondence between ISO 27001:2013 terms and terms of the CF

CF Fabian et al.	ISO 27001:2013
System	The standard considers the organisation as context. "This International Standard specifies the requirements for establishing, implementing, maintaining, and continually improving an information security management system within the context of the organization." (ISO/IEC 2013, p. 1)
Machine	The *Information Security Management System (ISMS)* is the machine to be built (ISO/IEC 2013, p. 1)
environment	ISO27001:2013 demands a detailed description of the *context of the organisation*. In particular, the standard demands a description of the environment in which the organisation operates (ISO/IEC 2013, p. 1)
Security goal	The standard uses *security objectives* (ISO/IEC 2013, p. 2) instead of security goals. The standard ISO27001:2013 demands security objectives for specific parts of the organization. "The organization shall establish information security objectives at relevant functions and levels." (ISO/IEC 2013, p. 5)
Security requirement	ISO27001:2013 simply states that security requirements are explicit statements about the needs regarding security of the organisation. These are the basis for an assessment if an acceptable security level is reached. "This International Standard can be used by internal and external parties to assess the organization's ability to meet the organization's own information *security requirements*." (ISO/IEC 2013, p. v)
Specification	The ISMS specification in ISO 27001:2013 is based on a "a risk management process and gives confidence to interested parties that risks are adequately managed." (ISO/IEC 2013, p. v). The standard acknowledges the importance to integrate this process with the "organization's processes and overall management structure and that information security is considered in the design of processes, information systems, and controls." (ISO/IEC 2013, p. v). In addition, the standard states that the descriptions shall be scalable according to the organization's needs
Stakeholder	The standard "gives confidence to *interested parties* that risks are adequately managed." (ISO/IEC 2013, p. v). In addition, the standard concerns *internal and external parties* (ISO/IEC 2013, p. v) Furthermore, ISO27001:2013 references ISO27000 for terminology definitions. In ISO27000 an *interested party* is defined as "person or organization [...] that can affect, can be affected by, or perceive themselves to be affected by a decision or activity." (ISO/IEC 2014, p. 5)
Domain knowledge	The standard demands to understand the organisation and its *context*, as well as the needs and expectations of interested parties. This domain knowledge should be the basis for reasoning about the scope of the ISMS, considering all external and internal issues relevant for the ISMS (ISO/IEC 2013, p. 1)
Availability	The definition in ISO 27000 is "property of being accessible and usable upon demand by an authorized entity." (ISO/IEC 2014, p. 2)

(continued)

Table 5.6 (continued)

CF Fabian et al.	ISO 27001:2013
Confidentiality	The definition in ISO 27000 is "property that information is not made available or disclosed to unauthorized individuals, entities, or processes" (ISO/IEC 2014, p. 2)
Integrity	The definition in ISO 27000 is "property of accuracy and completness" (ISO/IEC 2014, p. 5)
Asset	Neither the ISO 27001 nor the ISO 27000 standards define the term asset. However, the term asset appears in several definitions of other terms in the ISO 27000 (ISO/IEC 2014) such as attack and vulnerability. An attack is defined as "attempt to destroy, expose, alter, disable, steal, or gain unauthorized access to or make unauthorized use of an asset" (ISO/IEC 2014, p. 1). The use of asset in this definition implies that it is an item of value. Hence, there is no explicit definition, but the usage of the term in other definitions implies that it is used in a similar meaning as in the conceptual framework
Threat	A threat is defined as a "potential cause of an unwanted incident, which may result in harm to a system or organization" (ISO/IEC 2014, p. 11). The definition of vulnerability states that the exploit of a vulnerability causes a threat and that a vulnerability is a weakness of an asset. Hence, the definition is similar to the conceptual framework
Vulnerability	A vulnerability is a "weakness of an asset or control [...] that can be exploited by one or more threats [...]" (ISO/IEC 2014, p. 12). This definition matches the one of the conceptual framework
Risk	The CF defines risk as "the potential loss of a stakeholder"(Fabian et al. 2010, p. 13), while the ISO 27000:2014 (ISO/IEC 2014) defines it as an "effect of uncertainty on objectives" (ISO/IEC 2014, p. 8). Several notes state further interpretations of risk. Note 6 states "Information security risk is associated with the potential that threats [...] will exploit vulnerabilities [...] of an information asset or group of information assets and thereby cause harm to an organization" (ISO/IEC 2014, p. 9). This note is aligned with the definition of the conceptual framework

Also the *Threat Analysis* building block has multiple counterparts in the ISO 27001. The cause is the strong emphasis of the standard on risk, which is part of that building block. Hence, risk management-oriented approaches, such as CORAS (Lund et al. 2010), play a crucial role in an ISO 27001 assembly. The problem-oriented approaches, e.g., SEPP (Schmidt et al. 2011), are useful for the structured collection of knowledge about the environment that must be considered.

Table 5.8 presents the mandatory documents for an ISO 27001 documentation according to (ISO/IEC 2005, p. 13). In addition, Table 12.30 shows the kinds of SRE methods that support the assembly of these documents. The table is based upon our analysis in Sect. 5.2.

Table 5.7 Relating ISO 27001:2013 sections to CF building blocks

Section	Description	SV	SR	SDK	TA	RM
Section 4: Context of the organization						
Section 4.1	Understanding the organization and its context	+	+	+		
Section 4.2	Understanding the needs and expectations of interested parties	+		+		
Section 4.3	Determining the scope of the information security management system		+	+		
Section 4.4	Information security management system	+	+	+		
Section 5 Leadership						
Section 5.1	**Leadership and commitment**					
Section 5.2	Policy	+	+	+	+[a]	
Section 5.3	Organizational roles, responsibilities and authorities	+	+	+	+[a]	+
Section 6 Planning						
Section 6.1 Actions to address risks and opportunities						
Section 6.1.1	General				+[a]	+
Section 6.1.2	Information Security Risk Assessment	+		+	+[a]	+
Section 6.1.3	Information Security Risk Treatment				+[a]	+
Section 6.2	Information security objectives and planning to achieve them	+	+	+	+[a]	+
Section 7 Support						
Section 7.1	**Resources**					
Section 7.2	Competence	+				
Section 7.3	Awareness	+				
Section 7.4	Communication	+				
Section 7.5	Documented Information	+	+	+	+[a]	+

(continued)

Table 5.7 (continued)

Section	Description	SV	SR	SDK	TA	RM
Section 8 Operation						
Section 8.1	Operational planning and control	+	+	+	+[a]	+
Section 8.2	Information security risk assessment					+
Section 8.3	Information security risk treatment					+
Section 9 Performance evaluation						
Section 9.1	**Monitoring, measurement, analysis and evaluation**					
Section 9.2	Internal audit	+	+	+	+	+
Section 9.3	**Management review**					
Section 10 Improvement						
Section 10.1	**Nonconformity and corrective action**					
Section 10.2	**Continual improvement**					

SV Stakeholder Views, *SR* System Requirements, *SDK* Specification and Domain Knowledge, *TA* Threat Analysis, *RM* Risk Management

[a]The standard does not refer to threat analysis in these sections, but refers to security attributes such as confidentiality, integrity, and availability, or security objectives in general, or even internal and external issues. All of these are the outcome of a threat analysis

We introduced the idea of combining several requirements engineering methods to support a standard. In discussions with security experts on the scientific conferences this work was presented, and according to our own experience, e.g., in (Beckers et al. 2013), we recognize that the combination of several methods leads to a more complex method. This complexity is difficult even for experts. Hence, we pursued a direction in our research where we rather tried to extend one existing method than combining two.

We illustrate the mapping of the required documented information of ISO 27001:2013 to SRE methods in Table 5.9. The insights for ISO 27001 still hold for the new version of the standard. In addition, ISO 27001:2013 has several documentation demands that are not supported by SRE methods such as the *Evidence of Corrective Actions*.

Relating Section 4 to SRE methods

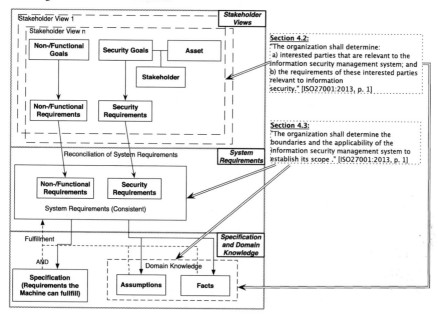

Fig. 5.8 Matching the CF and ISO 27001:2013 Sect. 4

Relating Section 6 to SRE methods

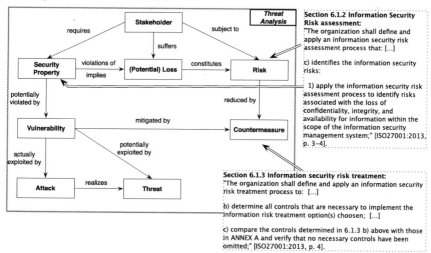

Fig. 5.9 Matching the conceptual framework from Fabian et al. and ISO 27001:2013 Sect. 6

Table 5.8 Support of SRE Methods for ISO 27001 Documentation

Documentation requirements ISO 27001	Support from SRE methods
ISMS policies and objectives	Goal-/problem-/risk-oriented methods
Scope and boundaries of the ISMS	Goal-/problem-/risk-oriented methods
Procedures and controls	Risk-oriented methods
The risk assessment methodology	Risk-oriented methods
Risk assessment report	Risk-oriented methods
Risk treatment plan	Risk-oriented methods
Information security procedures	Goal-/problem-oriented methods
Control and protection of records	No support from SRE methods
Statement of applicability	Goal-/problem-/risk-oriented methods

Table 5.9 Support of SRE Methods for ISO 27001:2013 Documentation Information

Documentation information ISO 27001:2013	Support from SRE methods
Scope of the ISMS	Goal-/problem-/risk-oriented methods
Information security policy	Goal-/problem-/risk-oriented methods
Risk assessment and the risk treatment including the statement of applicability	Risk-oriented methods
Risk assessment results	Risk-oriented methods
Risk treatment results	Risk-oriented methods
Information security objectives	Goal-/problem-oriented methods
Competence records	No support from SRE methods
Monitoring and measuring results	No support from SRE methods
Audit programme and results	Goal-/problem-/risk-oriented methods
Management review results	No support from SRE methods
Evidence of corrective actions	No support from SRE methods

5.4 Practical Application of Our Results

Developing an ISMS for a scenario, given a setting in the real world, is difficult due to the sparse descriptions in the ISO 27001 standard. The security experts have to find methods for asset identification, do a threat analysis on their own, and reason why the application of these methods result in a sufficient security analysis. We support the establishment of an ISMS for a given scenario with two different use cases, derived from our results. The use cases, as discussed in Sect. 5.2, are a top-down and a bottom-up procedure. The first use case is to apply SRE methods to systematically support an ISO 27001 establishment and implementation, hence a *top-down* approach. The second use case is to reuse SRE documentation generated in former activities which applied SRE methods, hence a *bottom-up* approach. Combinations of these use cases are possible.

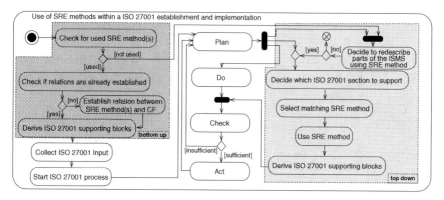

Fig. 5.10 Practical application of this work within an ISO 27001 implementation

The results of our work can be applied within an ISO 27001 compliant ISMS establishment and implementation, which is depicted in Fig. 5.10. The white area of the UML activity diagram contains the basic steps of an ISO 27001 process with the additional pre-phase from the ISO 27005, as described in Sect. 2.2.2. The dark gray area depicts the bottom-up use case, and the light gray area the top-down use case.

The **top-down** use case supports most parts of the ISO 27001 process, using SRE methods. Therefore, we select the sections of the ISO 27001 we want to refine via SRE methods. Next, we select matching SRE methods by using Table 5.2 as we presented previously in this section.

As a next step, we use the selected SRE methods. We use the resulting documents from applying the SRE method as part of an ISO 27001 compliant ISMS documentation. In addition, we can also use the selected SRE methods to describe already existing software. The resulting documentation can also be part of the ISMS documentation.

For the **bottom-up** use case, we have to check if there were any SRE methods used in previous activities, e.g., when developing software, related to the ISMS to be built. For existing SRE documents, we have to check if the applied SRE method is already considered within this work. If this is not the case, we have to establish the relation between the SRE method and the CF using the work of Fabian et al. (2010). After the relation is established, we are able to derive ISO 27001 supporting parts from the SRE documentation according to Table 5.2. Furthermore, we can map the relevant ISO 27001 sections to the derived parts of the documentation. By doing so, we get the information where a derived part is of use within the ISO 27001 sections.

Note that for ISO 27001:2013 the Plan-Do-Check-Act process is not mandatory anymore and other structures are allowed as well. However, it is still in compliance to use the Plan-Do-Check-Act process. Hence, the practical applications presented in this section are valid also for ISO 27001:2013. Furthermore, the ideas in this section can be mapped to other process structures with little effort.

5.5 Related Work

Calder (2009) and Kersten et al. (2011) provide advice for an ISO 27001 realization. In addition, Klipper (2010) focuses on risk management according to ISO 27005. The author also includes an overview of the ISO 27000 series of standards. However, none of these works consider to use security requirements engineering methods.

Cheremushkin and Lyubimov (2010), Lyubimov et al. (2011) present a UML-based meta-model for several terms of the ISO 27000, e.g., assets. These meta-models can be instantiated and, thus, support the mapping from abstract concepts to parts of the process. However, the authors do not present a holistic approach to information security. The work mostly constructs models around specific terms in isolation. The CF of Fabian et al. (2010), on the other hand, presents a holistic framework for information security.

Mondetino et al. investigate possible automation of controls that are listed in the ISO 27001 and ISO 27002 (Montesino and Fenz 2011). Their work can complement our own.

5.6 Summary

We have established a relation between the ISO 27001 standard and SRE methods. Thereby, we build on the CF of Fabian et al. (2010), which already established relations between the CF's terms and notions of several SRE methods. We contribute further relations from the ISO 27001 standard to the CF. The two sets of relations, from the ISO 27001 standard to the CF and from the CF to SRE methods, can be combined into transitive relations from the ISO 27001 standard to SRE methods. These transitive relations can support the identification of suitable SRE methods for establishing an ISMS compliant with ISO 27001.

Our approach offers the following main benefits:

- Reusing SRE methods to support the development and documentation of security standards (here: ISO 27001) compliant systems.
- Systematic identification of relevant SRE methods that can support the establishment of a specific part of the ISO 27001 standard.
- The resulting documentation of an ISO 27001 compliant ISMS benefits from the use of proven SRE methods, because these result in documentation of a security analysis that has been proven before in software engineering. For several methods, the quality of the proven documentation has been evaluated in empirical experiments or even in industry, e.g., CORAS (Lund et al. 2010).
- Reusing structured SRE methods for analyzing and eliciting security requirements to support the refinement of sparsely described sections of the ISO 27001 standard. These methods define in steps each task to do to analyze the security concerns of a system, which supports engineers in conducting a security analysis.

The results of this initial work is the basis for the research in the following chapters. We present examples of how security requirements engineering methods can be extended to support the establishment of security standards.

References

Beckers, K., Faßbender, S., Heisel, M., Küuster, J.-C., & Schmidt, H. (2012a). Supporting the development and documentation of ISO 27001 information security management systems through security requirements engineering approaches. In *Proceedings of the International Symposium on Engineering Secure Software and Systems (ESSoS)* (pp. 14–21). Springer.

Beckers, K., Heisel, M., Faßbender, S., & Schmidt, H. (2012b). Using security requirements engineering approaches to support ISO 27001 information security management systems development and documentation. In *Proceedings of the International Conference on Availability, Reliability and Security (ARES)* (pp. 243–248). IEEE Computer Society.

Beckers, K., Faßbender, S., Heisel, M., & Paci, F. (2013). Combining goal-oriented and problem-oriented requirements engineering methods. In *Proceedings of the International Cross Domain Conference and Workshop (CD-ARES 2013)* (pp. 178–194). Springer.

Bertrand, P., Darimont, R., Delor, E., Massonet, P., & van Lamsweerde, A. (1998). GRAIL/KAOS: an environment for goal driven requirements engineering. In *Proceedings 20th International Conference on Software Engineering (ICSE)*, IEEE Computer Society.

Calder, A. (2009). Implementing information security based on ISO 27001/ISO 27002: A management guide. Van Haren Publishing.

Cheremushkin, D. V., & Lyubimov, A. V. (2010). An application of integral engineering technique to information security standards analysis and refinement. In *Proceedings of the international conference on security of information and networks* (pp. 12–18). ACM.

Fabian, B., Gürses, S., Heisel, M., Santen, T., & Schmidt, H. (2010). A comparison of security requirements engineering methods. *Requirements Engineering—Special Issue on Security Requirements Engineering, 15*(1), 7–40.

ISO/IEC. (2004). Information technology—Security techniques—Management of information and communications technology security—Part 1: Concepts and models for information and communications technology security (ISO/IEC 13335-1). Geneva, Switzerland: International Organization for Standardization (ISO) and International Electrotechnical Commission (IEC).

ISO/IEC. (2005). Information technology—Security techniques—Information security management systems—Requirements (ISO/IEC 27001). Geneva, Switzerland: International Organization for Standardization (ISO) and International Electrotechnical Commission (IEC).

ISO/IEC. (2008). *Information technology—security techniques—information security risk management* (ISO/IEC 27005). Geneva, Switzerland: International Organization for Standardization (ISO) and International Electrotechnical Commission (IEC).

ISO/IEC. (2009). Information technology—Security techniques—Information security management systems—Overview and Vocabulary (ISO/IEC 27000). Geneva, Switzerland: International Organization for Standardization (ISO) and International Electrotechnical Commission (IEC).

ISO/IEC. (2013). Information technology—Security techniques—Information security management systems—Requirements (ISO/IEC 27001). Geneva, Switzerland: International Organization for Standardization (ISO) and International Electrotechnical Commission (IEC).

ISO/IEC. (2014). Information technology—Security techniques—Information security management systems—Overview and Vocabulary (ISO/IEC 27000). Geneva, Switzerland: International Organization for Standardization (ISO) and International Electrotechnical Commission (IEC).

Kersten, H., Reuter, J., & Schrüoder, K.-W. (2011). IT-sicherheitsmanagement nach ISO 27001 und Grundschutz. Vieweg+Teubner.

Klipper, S. (2010). Information Security Risk Management mit ISO/IEC 27005: Risikomanagement mit ISO/IEC 27001, 27005 und 31010. Vieweg+Teubner.

Lin, L., Nuseibeh, B., Ince, D. C., & Jackson, M. (2004). Using abuse frames to bound the scope of security problems. In *Re* (pp. 354–355).

Lund, M. S., Solhaug, B., & Stølen, K. (2010). *Model-driven risk analysis: The coras approach* (1st ed.). Springer.

Lyubimov, A., Cheremushkin, D., Andreeva, N., & Shustikov, S. (2011). Information security integral engineering technique and its application in isms design. In *Proceedings of the international conference on availability, reliability and security (ARES)* (pp. 585–590). IEEE Computer Society.

Massacci, F., Mylopoulos, J., & Zannone, N. (2010). Security requirements engineering: The SI* modeling language and the secure tropos methodology. *Advances in Intelligent Information Systems, 265,* 147–174.

Mayer, N., Rifaut, A., & Dubois, E. (2005). Towards a risk-based security requirements engineering framework. In *Proceedings of the 5th International Working Conference on Requirements Engineering: Foundation for Software Quality (REFSQ)*.

Montesino, R., & Fenz, S. (2011). Information security automation: How far can we go? In *Proceedings of the international conference on availability, reliability and security (ARES)* (pp. 280–285). IEEE Computer Society.

Mouratidis, H., & Giorgini, P. (2007). Secure tropos: A security-oriented extension of the tropos methodology. *International Journal of Software Engineering and Knowledge Engineering, 17*(2), 285–309.

Schmidt, H., Hatebur, D., & Heisel, M. (2011). A pattern- and component-based method to develop secure software. *Software engineering for secure systems: Academic and industrial perspectives* (pp. 32–74). IGI Global.

Susi, A., Perini, A., Mylopoulos, J., & Giorgini, P. (2005). The tropos metamodel and its use. *Informatica, 29,* 401–408.

Van Lamsweerde, A. (2004). Elaborating security requirements by construction of intentional anti-models. In *Proceedings 26th International Conference on Software Engineering,* (pp. 148–157). IEEE Computer Society.

van Lamsweerde, A. (2007). Engineering requirements for system reliability and security. *Software System Reliability and Security, NATO Security Through Science Series-D: Information and Communication Security, 9,* 196–238.

Chapter 6
Supporting ISO 27001 Compliant
ISMS Establishment with Si*

Abstract The establishment of an ISO 27001 security standard demands a description of the environment including its stakeholders and their security goals. Hence, goal-based security requirements engineering methods are an ideal fit for creating a methodology for supporting ISO 27001 compliance. In particular one type of stakeholder is of interest for a security analysis, the attacker. We show how a threat and risk analysis can be conducted by focusing on the attacker's goals and subsequent actions. In particular, we rely on the Si* method as an example for a goal-based security requirements engineering method. Our structured method explains for each step how a model of the system-to-be has to be created, analyzed, and refined to include all required actions for creating an ISO 27001 compliant information security management system. We describe validation conditions to find missing information and fix mistakes in the model and explain how the created artifacts can fulfill the ISO 27001 documentation demands during the application of our method. We illustrate the application of our method by a smart grid example.

6.1 Introduction

The challenge of every threat analysis is to identify attackers and provide the basis for structured security reasoning. We propose a goal-based threat analysis and risk management method that considers attacker motivation in combination with attacker types. This modeling language allows us to model an attacker's motivation by using goals. This is one of the reasons why we consider the goal-based Si* modeling language in this chapter. Further reasons are the ability of Si* to express trust relationships (Sect. 2.5.1 for details). We model an attacker's motivations as Si* soft goals, which we refine into goals. These attacker goals initiate threats to the assets of a stakeholder. We model threats as events meaning uncertain circumstances. Hence, we provide a security reasoning from attacker motivation, to attacker goals, to threats for assets. For example, an attacker can have the motivation of financial gain and a goal to steal cars. An event refines the goal for a specific scenario, e.g., if our scenario is to analyze the security of a hospital, the event can be *steal car from hospital parking lot*. The explicit consideration of combinations of attacker types and motivations allows a structured identification of attackers and also an explicit

© Springer International Publishing Switzerland 2015
K. Beckers, *Pattern and Security Requirements*,
DOI 10.1007/978-3-319-16664-3_6

elimination of not relevant attacker motivations and goals. By eliminating attackers, we also reduce the number of threats to consider for security experts. Hence, security experts can focus their efforts in protecting the system-to-be on particular threats. The parts of the system the threats target must be carefully checked for vulnerabilities, which might be exploited by the considered attackers. Our method provides a structured refinement of the IT system's and stakeholders' information to assess the threats for a particular system. We also provide a goal-based risk management process as part of our method to fulfill the ISO 27001 demands in this regard. Our risk management process is based on existing research. Our method uses this information for risk assessment and security control selection according to the ISO 27001 standard. We also provide the required documentation of an ISMS for certification. We illustrate our method by an example of a smart grid, which provides scalable energy infrastructure to consumers.

In summary, we show how the security requirements engineering (SRE) method Si* (Sect. 2.5.1) can support the establishment of an ISO 27001 compliant information security management system (ISMS) (Sect. 2.2.2). We call our Si* extension *Si*-ISMS*. We consider the goal-based Si* modeling language in this chapter, because we show how to use goals to model an attacker's motivation. In addition, the Si* language also contains security specific modeling elements such as trust relations. This chapter is based on our work presented in Beckers (2014).

6.2 A Method for Goal-Based ISMS Establishment

We propose a goal-based method (see Fig. 6.1) for creating an ISMS compliant to the ISO 27001 standard. The description of the method is based on the agenda principle (Sect. 2.6) and contains the following steps:

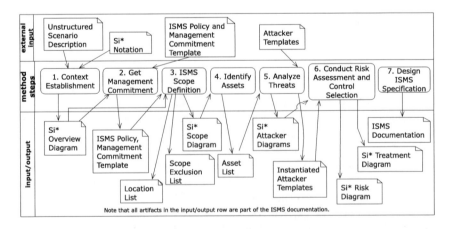

Fig. 6.1 The Si*-ISMS method

Step 1: Context Establishment

Input	Unstructured Scenario Description, Si* Notation
Activities	Create an Si* diagram that includes all relevant elements of the scenario and its environment. First, consider all the stakeholders in the scenario and model the roles first and the agents that play the roles afterwards. Then, elicit the goals of the roles and agents and model them, as well. Refine the goals into subgoals if needed using AND and OR refinements. Model the resources and tasks needed to fulfill the goals. Use the Si* Eco model to express which agent/role owns, provides, or requests a resource. Express also all trust relationships that exist in the scenario using the Si* trust notions. The resulting diagram that contains all the information above is the so-called *Si* Overview Diagram*
Output	Si* Overview Diagram
Validation Conditions	– Check that all stakeholders appear in the model – Check that all stakeholders have at least one goal – Check that all relevant resources and tasks appear in the model – Check that all resources have an explicit owner relation modeled

Changes for ISO 27001:2013 Compliance

An external and internal context of the ISMS has to be elicited including the relevant interested parties to the ISMS and their security requirements. This information is the basis for determining the *boundaries and applicability of the ISMS* and finally to establish the scope of the ISMS based on this information (ISO/IEC 2013, pp. 1–2).

Step 2: Get Management Commitment

Input	Si* Overview Diagram, ISMS Policy, and Management Commitment Template
Activities	The precondition for building an ISMS is that the management commits to it. Thus, we dedicated this step of our method to elicit the management commitment of the project and the provision of adequate resources to do so. Use the previously created Si* Overview Diagram to document the concerned scenario in which an ISMS shall be established. The ISMS policy has to contain high-level security concerns for the scenario referring to the elements of the Si* Overview Diagram. The management commitment should be given in written form. Without this commitment insufficient resources will be provided for the ISMS, and with insufficient resources the results will be an insufficient ISMS. Instantiate our *ISMS Policy and Management Commitment Template* to describe the security concerns and document the management support for the ISMS
Output	Instantiated ISMS Policy and Management Commitment Template
Validation Conditions	– The security objectives have to refer to Si* model elements – Is the ISMS commitment done in writing and signed by management? – Does the ISMS policy refer only to elements that are in the Si* model?

> ### Changes for ISO 27001:2013 Compliance
> ISO27001:2013 demands an explicit reasoning of why the policy supports the purpose of the organization. In addition, ISO27001:2013 allows during its establishment to substitute the step of stating security objectives with the definition of a framework for how security objectives can be elicited and documented. The policy also requires statements of how the satisfaction of the security requirements will be ensured and how the policy is communicated within the organization and to interested parties outside the organization (ISO/IEC 2013, pp. 1–2.)

Step 3: ISMS Scope Definition

Input	Si* Overview Diagram, Instantiated ISMS Policy, and Management Commitment Template
Activities	The ISMS Policy and Management Commitment Template defines security concerns the ISMS shall address. Define a scope of the ISMS. A scope is a clearly defined area in the Si* Overview Diagram that contains all elements relevant for the security concerns defined in the ISMS policy. Mark the scope in the Si* Overview Diagram with a clear boundary to all elements not in the scope. The resulting diagram is the so-called *Si* Scope Diagram*. Document all elements that are not part of the scope in the *Scope Exclusion List* including a reason why they are not part of the scope of the ISMS. Define the location of all elements in the scope in the *Location List*
Output	Si* Scope Diagram, Scope Exclusion List, Location List
Validation Conditions	– Is a scope entered in the Si* Scope Diagram? – Is the scope in alignment with the statements in the ISMS policy? – Are all Si* elements referenced in the ISMS policy part of the scope? – Are all elements outside of the scope listed in the Scope Exclusion List? – Do all entries in the Scope Exclusion List have a reason why these are excluded from the scope?

> ### Changes for ISO 27001:2013 Compliance
> The scope description for ISO 27001:2013 demands a description of the external and internal context. This makes the scope exclusion list obsolete, because all items in the external context are outside the scope. A list of all relevant interfaces between the external and internal context has to be explicitly documented. The standard demands the elicitation of all stakeholders and their security requirements, which can be done as it is explained in the Si*-ISMS method. Furthermore, ISO 27001:2013 demands a reasoning about the scope borders based on the previously elicited information. This can be done as descried in the Si*-ISMS method, as well (ISO/IEC 2013, pp. 1–3).

Step 4: Identify Assets

Input	Si* Scope Diagram
Activities	The entire Si* Scope Diagram is the input for the asset identification. Identify all items of value to stakeholders by analyzing all relevant relations in the Si* model. These range from resource, goal, and stakeholder relations to the trust relations in the Si* Scope Diagram, and the relations also help to clearly define the need for protection of the identified assets. In addition, a high-level risk assessment of the assets is conducted. This step results in an *Asset List*, the stakeholders that own them, and initial risk levels for assets as an output
Output	Asset List
Validation Conditions	– Are all resources in the Si* Scope Diagram referenced in the Asset List? – Do all assets in the Asset List reference an Si* element in the Si* Scope Diagram? – Do all assets have a location? – Do all assets have a need for protection? – Do all assets have a risk level?

Changes for ISO 27001:2013 Compliance

The asset identification is not a mandatory step in ISO27001:2013 (ISO/IEC 2013, p. 4). However, the risk identification step in the standard could be supported by identifying assets and their vulnerabilities and threats. Note that several controls in Appendix A explicitly refer to assets such as control *A8—Asset management.*

Step 5: Analyze Threats

Input	Si* Scope Diagram, Asset List, Attacker Template
Activities	Conduct a threat analysis using the Si* Scope Diagram. Model attackers that threaten the assets documented in the Asset List. The diagram with the attackers is a so-called *Si* Attacker Diagram*. Instantiate an attacker template for each attacker added to the Si* Attacker Diagram. These templates have to contain further information about the attackers such as the attacker's motivation and assumptions about the attacker
Output	Si* Attacker Diagram, Instantiated Attacker Templates
Validation Conditions	– Do all assets have at least one threat caused by an attacker? – Do all attackers have a motivation? – Do all attackers cause at least one threat? – Are any attacker types or motivations not considered?

Changes for ISO 27001:2013 Compliance

Threat analysis is not a mandatory step in ISO27001:2013, as well (ISO/IEC 2013, p. 4). However, the risk identification, analysis, evaluation, and treatment steps in the standard could be supported by identifying vulnerabilities and threats. However, controls in Appendix A refer to vulnerabilities and threats such as control *A 12.6—Technical vulnerability management.*

Step 6: Conduct Risk Assessment and Control Selection

Input	Si* Attacker Diagram, Instantiated Attacker Templates
Activities	Analyze the Si* Attacker Diagram and the Instantiated Attacker Templates. Look into the leave goals, meaning the goals that are not further decomposed into subgoals. Consider the assets these goals threaten. Model at least one event per leave goal that describes the realization of an attacker goal in a concrete action. Describe the likelihoods and consequences for each event in an *Si* Risk Diagram*
	Consider the Si* Risk Diagram and decide for each event if the risk the event represents should be treated. The decision shall be based on the likelihoods and consequences of each event. Document the risk treatment in an *Si* Treatment Diagram*. The risk treatment considers controls to reduce the likelihoods or consequences of an event. These controls have to be part of the normative Annex A of the ISO 27001 standard. Only if none of these controls are sufficient to address the risk, other controls can be chosen
Output	Si* Risk Diagram, Si* Treatment Diagram
Validation Conditions	– Do all events have likelihoods and consequences?
	– Are all risks modeled in the Si* Risk Diagram?
	– Are all unacceptable risks addressed with a control?
	– Do all controls in the Si* Treatment Diagram reference an ISO 27001 Annex A control?

Changes for ISO 27001:2013 Compliance

ISO 27001:2013 allows to select any control to treat a risk, but the controls have to be mapped to controls in Annex A of the standard (ISO/IEC 2013, p. 4). We provide a mapping between the controls of ISO 27001 and ISO 27001:2013 in Appendix C.

Step 7: Design ISMS Specification

Input	All artifacts created in the previous steps
Activities	The final step of our method concerns the ISO 27001 specification, an implementable description of the ISMS. We consider the ISO 27001 documentation demands and use the information elicited and documented in the previous steps of our method. This information is mapped to the required document types for certification. The standard demands several documents for each part of the ISMS, but the standard states no demands for the form or medium. Hence, we propose a mapping in Table 6.1 of the generated artifacts from our method to the documentation demands
Output	ISMS Documentation
Validation Conditions	– Are all controls listed in Table 6.1 documented?
	– Are all the documents complete?

Changes for ISO 27001:2013 Compliance

The documentation demands for ISO 27001:2013 can be found in Table 6.2.

Table 6.1   Support of our method for ISO 27001 documentation demands

ISO 27001 documentation requirements	Artifacts created while using our method
ISMS policies and objectives	Instantiated ISMS policy and management commitment template
Scope and boundaries of the ISMS	Si* scope diagram
Procedures and controls	Documentation of selected security controls and their implementation
The risk assessment methodology	Description of our risk management process method (p. 138)
Risk assessment report	Si* risk diagram
Risk treatment plan	Si* treatment diagram
Information security procedures	Documentation of security processes for controls, which are refinements of the controls in the Si* treatment diagram
Control and protection of records	Documentation of selected measures to control documents
Statement of applicability	Si* treatment diagram and further texts that explain the reasoning for selecting the chosen controls

Mapping to the Conceptual Framework for Security Standards—We present a mapping in Table 6.3 from this method to the conceptual framework for security standards, which is the foundation for our PEERESS framework (Chap. 3). The table lists the activities of the conceptual framework for security standards on the left column and the steps of the Si*-ISMS method that concern these activities in the right column.

6.3 Application of Our Method to a Smart Grid Scenario

We illustrate our method on a case study of a Smart Grid system. The case study was provided by the industrial partners of the EU project NESSoS.[1]

Step 1: Context Establishment

A smart grid provides energy on demand from distributed generations to consumers. The grid intelligently manages the behavior and actions of its participants using information and communication technologies (ICT). A novelty compared to previous energy networks is the two-way communication between consumers and energy providers. The benefits of the smart grid are envisioned to be a more economic, sustainable, and reliable supply of energy. However, significant security concerns have to be addressed for this scenario, due to the possible dangers of missing availability of energy for customers, as well as threats to the integrity and confidentiality of

---

[1] http://www.nessos-project.eu/.

Table 6.2 Support of our method for ISO 27001:2013 documented information

ISO 27001:2013 documented information requirements	Artifacts created while using our method
Scope of the ISMS	Si* scope diagram
Information security policy	Instantiated ISMS policy and management commitment template
Information security objectives	Instantiated ISMS policy and management commitment template
Risk assessment and the risk treatment methodology	Description of our risk management process method (p. 138)
Risk assessment results	Si* risk diagram
Risk treatment results	Si* treatment diagram
Statement of applicability	Si* treatment diagram and further texts that explain the reasoning for selecting the chosen controls
Competence records	Documentation of the actors in the Si* scope diagrams and their training, education, and experience records
Monitoring and measuring results	Documentation of security processes and their monitoring results
Audit program and results	Documentation of the audits and their results
Management review results	Documentation of the management review activities and their results
Evidence of corrective actions	Documentation of selected measures of corrective actions

consumer's data. These concerns are of particular relevance, because energy grids have a significantly longer lifespan than telecommunication networks (Aloula et al. 2012). In addition, privacy concerns have risen, such as the possibility of creating behavioral profiles of customers if their energy consumption is transmitted over the smart grid in small time intervals (Lin and Fang 2013).

Figure 6.2 presents an Si* diagram of our smart grid scenario. The *Energy Supplier* provides electricity to the smart grid. The actor *Energy Alpha* plays the role of the *Energy Supplier*. The *Energy Alpha* has the goals *Sell Energy* for which the subgoals *Collect Prosumer Data* and *Calculate Bill* need to be fulfilled. The *Prosumer Information* is a means to the end of achieving the goal *Collect Prosumer Data*, similar to the *Aggregated Billing Data* for the *Calculate Bill* goal. The *Provide Grid Services* goal requires the goal *Manage ESS*. The *Energy Supplier Server (ESS)* collects *Aggregated Billing Data* from the *Home Gateway* of the *Prosumer*. The *Energy Alpha* company delegates the goals *Calculate Bill* and *Manage EMS* to a *Billing Operator*. Moreover, the *Energy Alpha* company also aims to *Manage EMS* and *Manage Smart Meters*. Both goals rely on a *Network Gateway* to ensure communicate between Smart Meters, EMS, and the ESS. These goals are delegated to a *Meter Point Operator*.

Table 6.3 A mapping between the standard activities in the conceptual framework for security standards and the steps of the Si*-ISMS method

Activities in the conceptual framework for security standards	Steps in the Si*-ISMS method
Environment description	Step 1: Context establishment
	Step 2: Get management commitment
	Step 3: ISMS scope definition
Stakeholder description	Step 1: Context establishment
	Step 2: Get management commitment
	Step 3: ISMS scope definition
Asset identification	Step 4: Identify assets
Risk level description	Step 6: Conduct risk assessment and control selection
Security property description	Step 6: Conduct risk assessment and control selection
Control assessment	Step 6: Conduct risk assessment and control selection
Vulnerability and threat analysis	Step 5: Analyze threats
Risk determination	Step 6: Conduct risk assessment and control selection
Security assessment	Step 6: Conduct risk assessment and control selection
Security measures	Step 6: Conduct risk assessment and control selection
Risk acceptance	Step 6: Conduct risk assessment and control selection
Documentation	Step 7: Design ISMS specification

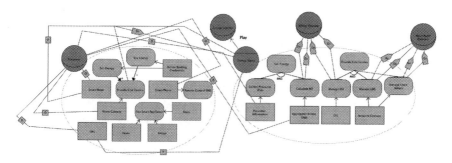

Fig. 6.2 Si* overview diagram

The *Prosumer* wants to *Buy Energy* and also to *Sell Energy* energy to *Energy Supplierss*. Both goals have a positive effect on the *Provide Grid Services*, which the *Prosumer* delegates to the *Energy Alpha*. The *Prosumer* owns the *Home Gateway*

Table 6.4 ISMS policy and management commitment template

High-level security concerns	State the high-level security goals in relation to the Si* model
Criteria for risk acceptance	Define worst-case scenarios
Establish responsibilities	State which person is responsible for the overall ISMS establishment
Conduct ISMS audits	Define audit responsibilities
ISMS management reviews	Define ISMS management audit responsibilities

and also the *Smart Meter*. The *Smart Meter* is an ICT system that measures energy consumption and contribution to the smart grid. The *Smart Meter* sends aggregated *Energy Consumption Data* to the *Energy Provider* using the internet connection of the *Home Gateway*. The *Energy Management System (EMS)* is a computer that allows the *Prosumer* to check which data the *Smart Meter* collected. The *EMS* is owned by the *Prosumer*, as well. It also allows to control *Smart Appliances*. These are electronic components that communicate with the *Home Gateway* and the *Smart Meter* to offer services. An example for a *Smart Appliance* is a controllable air conditioning. *Smart Appliances* are owned by the *Prosumer*. The *Prosumer* owns a *Smart Phone*, as well. The phone enables to *Remote Control the EMS*. The *Prosumer* uses three *Smart Appliances*, a *Heater*, *Stove*, and a *Fridge*.

Step 2: Get Management Commitment

The ISO 27001 standard demands documentation of management commitment for the establishment of an ISMS. The demands are described in Section 5 of the standard. *Section 5.1 Management Commitment* concerns the proof that the management has done a reasonable effort to establish an ISMS. In particular, the management has to document the objectives and responsibilities in a document. *Section 5.2 Resource Management* concerns the provision of resources for establishing the ISMS and the training of the members of the organization for security awareness and competence. The management commitment for implementing an ISMS according to the ISO 27001 standard is of utmost importance, because without the provision of sufficient staff and resources by the management the ISMS establishment is doomed to fail. In addition, the publicly available sources of examples of ISMS implementations, e.g., the ISMS toolkit,[2] define this also as the first step when implementing an ISMS. The management commitment should be based upon a high-level description of the part of an organization for which the management commits resources to build an ISMS. We use our Si* model for this purpose (see Fig. 6.2). Furthermore, we propose a template for ISMS policies and Management Commitment (see Table 6.4) that we instantiate for our example (see Table 6.5).[3]

---

[2]http://www.iso27001security.com/html/iso27k_toolkit.html.

[3]Note that we do not consider resources in the template in this example to keep the example simple.

Table 6.5 Instantiated ISMS policy and management commitment template

High-level security concerns	– The confidentiality and integrity of all information about the prosumer shall be kept including his/her energy consumption
	– The integrity of the configuration data of smart appliances shall be preserved
	– The availability of the grid services shall not be prevented by an attacker
Criteria for risk acceptance	– Few cases of prosumer information occur
	– Only minor parts of grid services are unavailable for a small amount of time
	– Only minor availability issues for a short time and a small amount of users occur
Establish responsibilities	Mr. Jiggleby is responsible for establishing the ISMS
Conduct ISMS audits	Mr. Smith is responsible for audits
ISMS management reviews	Mr. Jackson is responsible for ISMS management reviews

Step 3: ISMS Scope Definition

After acquiring the management commitment, we have to provide a more detailed scope definition. Section 4 of the ISO 27001 standard describes the ISMS and in particular in Section 4.2—Establishing and managing the ISMS—the scope definition is stated. Section 4.2.1a demands to "Define the scope and boundaries of the ISMS in terms of the characteristics of the business, the organization, its location, assets and technology, and including details of and justification for any exclusions from the scope" (ISO/IEC 2005b, p. 4). In Sect. 4.2.1d, which concerns risk identification, the scope definition is used to identify assets. Section 4.3 lists the documentation demands of the standard and Sect. 4.3.1d requires a documentation of the scope of the ISMS. The ISMS scope definition of the ISO 27001 standard is a vital step for its successful implementation, because all subsequent steps use it as an input.

Our ISMS focuses on protecting the prosumer and is established by the Energy Alpha company. Hence, the scope is focused on the prosumer's data. We marked the scope as a kind of rectangle in the Si* model with the word scope in it, depicted in Fig. 6.3.

The standard further demands the definition of "location." We propose to attach templates to the Si* model. The location template, shown in Table 6.6, lists the location of all Si* resources and agents/roles. Goals are not listed here, because these do not have physical locations.

Moreover, the standard demands "details of and justification for any exclusions from the scope." We propose to use a scope exclusion template for that purpose that lists all resources and agents/roles that are excluded from the scope. In Fig. 6.3 we already excluded the *Meter Point Operator* from the scope of the ISMS. The template in Table 6.7 states the reasoning behind this and further scope exclusions.

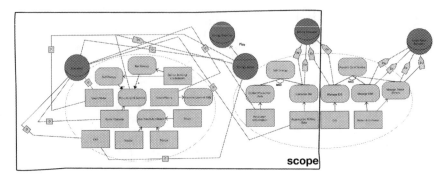

Fig. 6.3 Si* scope diagram

Table 6.6 Location list

Si* element	Location
Prosumer	Hannover, Germany
Energy alpha	Hannover, Germany
Billing operator	Bremen, Germany
EMS	Hannover, Germany
Home gateway	Hannover, Germany
Smart meter	Hannover, Germany
Heater	Hannover, Germany
Stove	Hannover, Germany
Fridge	Hannover, Germany
Aggregated billing data	Bremen and Hannover, Germany
Prosumer information	Bremen and Hannover, Germany
ESS	Bremen, Germany
Network gateway	Bremen, Germany

Table 6.7 Scope exclusion list

Si* element	Reason for scope exclusion
Meter point operator	Energy alpha has an ISMS in place that concerns the meter point operator
ESS	The ESS is also considered in the ISMS that concerns the operations of the meter point operator
Network gateway	The network gateway is considered in the ISMS that concerns the operations of the meter point operator, as well
Meter point operator	The meter point operator is also considered in the other ISMS

Step 4: Identify Assets

The main goal of the ISO 27001 ISMS is to protect assets with proper security controls, which is stated already on p. 1 of the standard. Section 4.2.1a of the standard demands the definition of assets. Section 4.2.1d that concerns risk identification uses the scope definition to identify assets, to analyze threats to the identified assets, and to analyze the impacts of losses to these assets. Section 4.2.1e concerns risk analysis, which concerns assets, as well. The section demands further to conduct a vulnerability analysis regarding assets and to consider controls currently implemented. Thus, identification and analysis of assets is a vital part of establishing an ISO 27001 compliant ISMS. An asset is defined in the standard as "anything that has value to the organisation" (ISO/IEC 2005b, p. 2.) We propose the following steps for identifying assets, which concern resources in our Si* model. Thus, the following process aims to find resources and if the resources have a value for a stakeholder in the ISMS scope, the resources are assets.

a. Investigate the Eco Model Relations   The relations of the *Eco Model*: *request*, *own*, and *provide* that consider a resource at one end reveal possible assets and in case of the *own* relation, also the asset owner.

b. Investigate Goal Relations   Means-end relations between a goal and a resource have to be investigated. In addition, for each goal we have to check if not a resource is missing that might be an asset.

c. Iterate over all Resources   In order not to miss any assets, an iteration of all resources in the model is performed and a check is conducted if this resource is an asset.

For an accurate description of assets, the following information has to be elicited for each asset.

State the Stakeholder that owns the Asset   Check if the *own* relation of the Eco Model is defined on an asset. If this is the case, the agent or role on that relation is the Stakeholder that owns the Asset.[4] If this relation is not set, it has to be included into the model.

Define the Need of Protection   We want to state the need for protection of an asset. This information can help to assess an initial risk level for an asset and serves as an input to the threat analysis. At this stage, only the trust relations in the Si* Overview Diagram are considered. Any assets (resources) that have an *execution dependency* or *permission delegation* relation have an interaction with another agent or role. These can indicate a need of protection, which has to be described. The trust relations *trust of execution* or *trusting in execution* result in a limited need for protection, while a *distrust relation* requires a significant protection.

Assess Initial Risk   The description of assets and their need for protection entries in the asset list (e.g., Table 6.8) shall be analyzed by domain experts, and initial risk values shall be assigned. These values are meant to categorize assets by risk

---

[4]Note that in ISO 27001 the asset owner means the person that is responsible for the assets' security. But in this case, we mean the property rights holder of the asset.

Table 6.8 Asset list

Si* resource	Owner	Need for protection	Initial risk	Asset owner
Smart meter	Prosumer	The smart meter measures the energy consumption and its manipulation can cause significant financial damages. In addition, the loss of energy consumptions is a violation of the privacy of the prosumer	3	Mr. Smith
Home gateway	Prosumer	All communications of entities in the smart home, e.g., the smart meter has to cross the home gateway. This makes it a central point of failure	2	Mr. Blake
EMS	Prosumer, but provided by energy alpha	The EMS allows to (the energy consumption data and change energy plans. Hence, if the data integrity is not ensured it can cause financial harm to the prosumer	2	Mr. Blake
Heater	Prosumer	A change in the heater configuration can cause health problems in winter	2	Mrs. Wong
Stove	Prosumer	A change in the stove configuration can cause a fire in the smart home	3	Mrs. Wong
Fridge	Prosumer	A change in the fridge configuration can let food go bad and thus cause health problems or financial damage	1	Mrs. Wong
Smart phone	Prosumer	The prosumer uses the smart phone to check the energy consumption and configure smart appliances. Threats can cause changes of the configuration of smart appliances or even energy consumption data	2	Mr. Schuster
Prosumer Information	Prosumer	The Prosumer information such as his/her address are personal information. Their leakage leads to a privacy issue	3	Mrs. Cheng
Aggregated billing data	Prosumer	The billing data is comprised of aggregated energy consumption data. Depending on the aggregation it may be possible to derive behavioral profiles of the prosumer from these data	3	Mrs. Silvester

Table 6.9 Classifying Si* elements into potential attacker targets

Attacker type	Threatened Si* elements
Software attackers target any kind of software	Resources which represent software
Network attackers are reading or manipulating network traffic and network related devices, e.g., routers	Resources that represent network traffic and network related devices
Physical attackers targeting hardware installations	Resources which represent hardware
Social engineering attackers targeting stakeholders	Roles or agents which represent stakeholders

level. We propose to limit the possible labels to low (1), medium (2), and high (3) as proposed by the NIST 800-30 (Stoneburner et al. 2002) standard for risk management. These values are refined later in the process in order to assess if an asset has an acceptable risk level in light of its threats.

We illustrate the resulting asset list in Table 6.8. Note that the *owner* in the table is the person holding the property rights of that asset. The column *asset owner* in the ISO 27001 is responsible for the protection of the asset. These are employees of Energy Alpha that are responsible for addressing the security concerns for these assets.

Step 5: Analyze Threats
Section 4.2.1b concerns the definition of ISMS security policies and it demands that the policy shall consider assets. The ISO 27001 standard concerns threat analysis in several sections for determining the risks to assets. Section 4.2.1d demands a threat analysis for assets for the purpose of identifying risks and the vulnerabilities that might be exploited by those threats. Section 4.2.1e concerns risk analysis and evaluation and demands to determine likelihoods and consequences for threats. In particular, the standard mentions the importance of physical and network threat analysis. We consider four basic kinds of attackers for our threat analysis as proposed in Beckers et al. (2013). We map these attackers to Si* elements they can exploit and types of assets they can harm in Table 6.9. A study of the SANS Institute from 2006[5] revealed four fundamental motivations of social engineering attackers: Financial gain, self-interest, revenge, and external pressure. We believe these motivations are generic enough to serve all types of IT attackers. We also added the motivation curiosity, which we identified in discussions with the industrial partners of the NESSoS project.

Financial gain   The attacker aims for monetary gain for various reasons.
Self-interest   An attacker wants to change or destroy information about a person with whom the attacker has a relation. For example, a friend or family member.

---

[5] http://www.sans.org/reading_room/whitepapers/engineering/social-engineering-means-violate-computer-system_529.

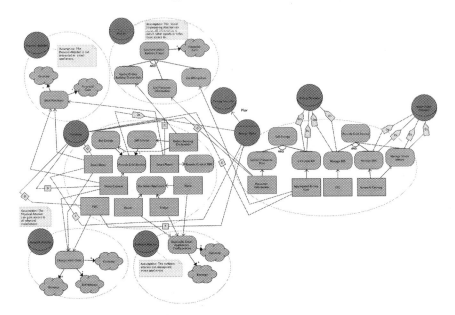

Fig. 6.4 Si* attacker diagram

Revenge   An attacker has the emotional desire for vengeance for reasons only the
attacker is aware of. This results in an attack on, e.g., an employer or a friend.

External pressure   An attacker can receive pressure from, e.g., friends, family,
or an organization to satisfy their motivations for financial gain, self-interest, or
revenge.

Curiosity   We included also the attacker motivation curiosity. An attacker wants
to know the content of a certain data set or simply to find out if his or her skills
are sufficient to penetrate a system.

We model the above-mentioned attacker motivations as soft goals of attackers. The
assumptions about each attacker are annotated using UML notes. The refined goals
of attackers from their soft goals are threats. This refinement is modeled with a *contributes* relationships, because the threats (attacker's goals) contribute to fulfilling
the attacker's soft goals (attacker's motivations). We use the *means-end* relationship to model the relation between threats (attacker's goals) and threatened assets
(resources). We illustrate this in Fig. 6.4. We show how each of our attacker types
threaten the smart home. We describe the attacker in attacker templates, which we
introduce in Table 6.10. These are inspired by the ISMS-CORAS attacker templates
(Chap. 7). The templates provide a structured way to elicit information about the
attackers, which are also depicted in Fig. 6.4. The templates contain further information about attackers such as assets these attackers can possibly reach. The assets are
instantiated based on the experience of the customer and all other information available, e.g., expert security analysis. We instantiate the template for a physical attacker

Table 6.10 Attacker template

Attacker type	State the type of attacker
Exploitable Si* elements	Explain which Si* elements the attacker can exploit in the system. These possible exploits are modeled as *means-end* relations from resources to goals of attackers
Threatened assets	State the assets that the attacker can possibly reach when exploiting the threatened Si* elements. These assets are assumptions by security experts
Motivation	State the possible attacker motivations for this type of attacker. Explain the reasons for choosing these motivations in the relational field
Skills	State the skills required by the attacker
Assumptions	State the assumptions about the attacker
Reasoning	State the arguments for selecting the threatened assets

(see Table 6.11), a social engineering attacker (see Table 6.12), for a software attacker (see Table 6.13), and for a network attacker (see Table 6.14).

Step 6: Conduct Risk Assessment and Control Selection

Risk management is mentioned in numerous sections of the ISO 27001 standard (Chap. 5). In our method, risk is used to assess if an asset requires an additional control or not. We use a risk management technique inspired by Asnar et al. (2007, 2011) for goal-based requirements engineering. The authors propose the tropos goal risk framework, which consists of the *asset layer*, the *event layer*, and the *treatment layer*. The asset layer concerns goals of the stakeholders. Impacts on assets are caused by events modeled in the event layer. Events are depicted by pentagons and represent uncertain circumstances that have an impact on the fulfillment of goals. The impact relation are modeled with dashed line-arrows. The treatment layer concerns tasks that are modeled as hexagons. These tasks are sequences of actions, which can influence an event in a positive or negative way. This is modeled using a contributes relationship, a line with a "+" (positive contribution) or "−" (negative contribution) sign next to it.

In Asnar et al. (2007, 2011) an event contains two properties: The consequences of an event[6] and the likelihood of an event occurring. When an event has a negative impact (or consequence) on a goal it is a risk. Negative impacts are denoted with the sign "−" and for significant negative impact with the sign "−−." Positive impacts from an event to a goal are opportunities and denoted with the sign "+" and for significant positive impact with the sign "++." The work uses a qualitative assessment of the likelihood of an event and is expressed with the following values: likely,

---

[6] Note that Asnar et al. call this severity. However, we will refer to it as consequences for the remainder of this work.

Table 6.11 Instantiated attacker template for physical attacker

Attacker type	Physical attacker
Exploitable Si* elements	Smart meter, home gateway, smart phone
Threatened assets	Smart meter, home gateway, smart phone
Motivation	Financial gain, revenge
Skills	The attacker needs to be able to enter the smart home and remove the smart meter and home gateway from the installations
Assumptions	We assume the physical attacker does not steal or break hardware out of curiosity. External pressure is also unlikely, because we assume that criminal organizations would be more interested of manipulating software or network interfaces, as the pay off is likely to be much higher. Self-interest is also unlikely, because the energy consumption data is forwarded in small time intervals and the theft of the physical metering device or its surroundings. We also think a physical attacker would not focus on kitchen appliances in this scenario, because these are not particular for a smart home
Reasoning	The physical attacker could gain access to the smart home and steal the smart meter, home gateway, and smart phone. These are hardware devices that are not difficult to carry and their theft can cause a serious disruption in the smart grid operations

occasional, rare, and unlikely with the meaning: *likely < occasional < rare < unlikely*. Afterwards, all risks for each goal have to be considered and a decision has to be made, which risks are acceptable and which are not. Asnar et al. (2007, 2011) propose four treatments for unacceptable risks. *Prevention* demands to select a mechanism to reduce the likelihood of a risk. These treatments are modeled using tasks and contribute relationships as explained above. *Attenuation* means to introduce a mechanism to reduce the consequences of a risk. *Retention* demands to accept the risk or transfer it, e.g., to an insurance company. *Removal/avoidance* of risk means to remove the goal the risk concerns. For the remainder of the chapter, we focus on prevention and attenuation of risks in order to illustrate the ISO 27001 control selection.

Our risk management method differs from Asnar et al.'s method. The first change is the way assets are represented. We use resources to model assets, because these represent information or physical things that stakeholder's value. Asnar et al. represent assets as goals. The ISO 27001 standard demands that risks are expressed in relation to assets. We use resources to model assets, because these represent information or physical things that stakeholders' value. Goals are not appropriate to model

Table 6.12 Instantiated attacker template for social engineering attacker

Attacker type	*Social engineering attacker*
Exploitable Si* elements	Prosumer, energy supplier, energy alpha, billing operator
Threatened assets	Prosumer information, aggregated billing data
Motivation	Financial gain
Skills	The social engineering attacker needs to be able to deceive the roles mentioned above and convince them to provide the prosumer information and aggregated billing data
Assumptions	We assume social engineering attackers aim to enrich themselves. Self-interest, revenge, curiosity, and external pressure (m unlikely, because the attack requires a certain soft skill like theatrics and preparation to manipulate people
Reasoning	The attacker can manipulate the stakeholder energy alpha to release billing information and prosumer information. The attacker uses this information to impersonate an employee of energy alpha and explains that there is a problem with the financial transactions to pay for energy. The attacker claims that online banking credentials are needed to fix this, otherwise the prosumer will not receive further energy

Table 6.13 Instantiated attacker template for software attacker

Attacker type	*Software attacker*
Exploitable Si* elements	Heater, fridge, stove
Threatened assets	Heater, fridge, stove
Motivation	Revenge, curiosity
Skills	The software attacker needs the skills to use software exploits on the smart appliances to change their configuration
Assumptions	We assume the software attacker aims to manipulate smart appliance. Moreover, we did not consider financial gain as a motivation, because manipulation of the smart appliances configuration is unlikely to provide financial gain. The same can be said for self-interest and external pressure
Reasoning	The software attacker aims to manipulate smart appliances and use them to cause financial or physical harm to the prosumer

Table 6.14 Instantiated attacker template for network attacker

Attacker type	*Network attacker*
Exploitable Si* elements	Home gateway, EMS, smart meter
Threatened assets	Prosumer information, aggregated billing data
Motivation	Curiosity, revenge, self-interest
Skills	The attacker needs to be able to access and manipulate the network traffic routed between the home gateway, EMS, and smart meter
Assumptions	We assume that the network attacker can change billing data, but the financial gain (ms to be very difficult to achieve for an attacker by changing the billing data, because the financial benefit would go to energy alpha. For the same reason, we exclude external pressure
Reasoning	The network attacker can change the billing data for revenge, because it causes financial harm to the prosumer. Moreover, to change this data is difficult and it supports the attacker's curiosity. In addition, for self-interest an attacker might change the billing data to save the prosumer money

assets in the sense of the ISO 27001 standard. This change demands further modifications of Asnar et al.'s risk management method. Hence, we propose our modified risk management method in the following.

a. Model Events

Input: Si* Attacker diagram and attacker templates

Output: Si* Risk Diagram with events

Describe at least one event for each goal of every attacker. These events are circumstances that can happen when the attacker attempts to fulfill his/her goal. We model a so-called *custom association* between a goal and an event (a black line). The event can cause harm to an asset, which is modeled as an impact relation (a hashed line with white arrowhead) (see Fig. 6.5 for an example). Remove also the means-end relations between assets and the attacker goals that threaten them. These are redundant, because the common associations to events and from events to assets refine these relations. Describe also existing controls that have an impact on events in this step.

b. Assess Consequences and Likelihoods

Input: Si* Risk Diagram with events

Output: Refined Si* Risk Diagram with likelihoods and consequences

The values for likelihoods and consequences for each event have to be determined. This determination should be based on past experience, expert reports, security testing, and all further information available relevant for the risk assessment. The resulting values are annotated in the model as follows (see Fig. 6.5 for an example). Likelihoods are annotated as text on a custom association with one of

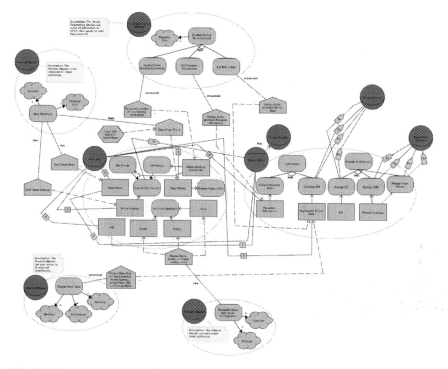

Fig. 6.5 Si* risk diagram

the likelihood values introduced by Asnar et al. The consequences are modeled according to the description of Asnar et al. (see above).

c. Reason about Risk Treatment

Input: Refined Si* Risk Diagram with likelihoods and consequences

Output: Si* Treatment Diagram

For each event, the likelihoods and consequences have to be discussed. The discussion has to determine if the risks are acceptable or not. Each risk that is unacceptable has to be treated with one of the measures introduced by Asnar et al. (see above). If a prevention or attenuation of risks is selected to treat the risk, risk measures (or controls) from Annex A of the ISO 27001 standards have to be chosen. The numbering of the controls in ISO 27001 Annex A starts with A.5 and ends with A.15. The reason for not starting the numbering with A.1 is that the control numbering shall align with the controls listed in the ISO/IEC 17799:2005 standard. This standard provides guidelines on how to implement the controls, but it is not normative. The selection of these controls is followed by selecting concrete measures. For example, we have to conduct security awareness trainings in order to prepare the employees of Energy Alpha not to distribute personal information about the prosumer. These are modeled as task and an impact relation (dashed line with a white arrowhead). The impact relations state the task helps to

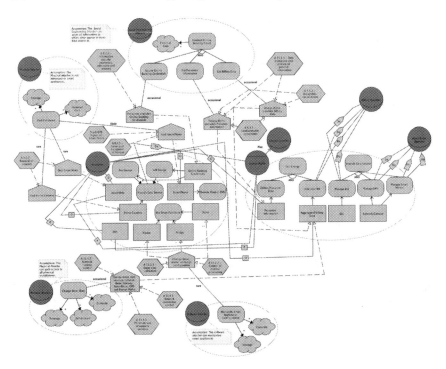

Fig. 6.6 Si* treatment diagram

reduce the risk that the event represents, which is depicted by a "−" sign next to the arrowhead of the impact relation (see Fig. 6.6 for an example).

We show an example for our method in the following.

a. Model Events   Figure 6.5 illustrates the events elicited for our smart grid example. The physical attacker has the goal to *StealHardware*, which results in the events *Steal Smart Meter*, *Steal Home Gateway*, and *Steal Smart Phone*. The social engineering attacker aims to *Conduct Online Banking Fraud*, which is refined into the subgoals *Aquire Online Banking Credentials*, *Get Prosumer Information*, and *Get Billing Data*. The attacker gets *Energy Alpha to provide Billing Data* and *Energy Alpha to provide Prosumer Information*, as well. The attacker uses this information to impersonate an Energy Alpha employee to get the *Prosumer to provide Online Banking Credentials*. The attacker uses the credentials to transfer money from the account of the Prosumer to his/her own, maybe via indirect ways via money laundering or other means for hiding the identity of the attacker. The network attacker has the goal to *Change Meter Data* and causes the event *Change Meter Data on route between Home Gateway, Smart Meter, and EMS*. The software attacker aims to *Manipulate Smart Appliances Configuration*, which causes the event *Change Stove, Heater, or Fridge configuration*.

b. **Assess Consequences and Likelihoods** We describe the likelihoods and consequences (or impacts) of the events described in the previous step, depicted in Fig. 6.5. The experience and educated guesses of the customer and statistics are used to determine the values. The likelihood of the events *Steal Home Gateway* and *Steal Smart Meter* is *rare*. If the events occur, the theft of the smart meter has a significant negative impact for the smart home, because it is a specialized device and its replacement is not instantaneous and might result in an interruption of smart grid services. On the other hand, the home gateway is a more generic device and can be replaced faster. The *Steal Smart Phone* event is likely, but the impact is not significant, because it can be replaced fast and the smart phone has a GPS tracking signal activated. This signal makes its retrieval likely.

The social engineering attacker causes the events *Energy Alpha to provide Billing Data* and *Energy Alpha to provide Prosumer Information*. Both events occur occasional and have a negative impact on their respective asset. The event *Prosumer to provide Online Banking Credentials* is occasional, as well. Nevertheless, the impact on the asset is significant, because the event can cause significant financial harm to the prosumer.

The software attacker event *Change Stove, Heater, or Fridge configuration* has a rare likelihood. The event has a negative impact on the fridge and heater. In addition, the event causes a significant negative impact on the stove, because a misconfiguration has the potential to cause a fire and in turn this fire can cause financial and physical harm to the prosumer.

The network attacker causes the event *Change Meter Data on route between Home Gateway, Smart Meter, and EMS* with an occasional likelihood. The negative impact to the prosumer is significant, because the energy consumption data in combination with the personal information of the prosumer can be used to derive behavioral profiles of the consumer. For example, the usage of a shower has a distinctive energy signature and the network attacker could analyze to check when the Prosumer is showering and how many people are living in the smart home, based on the amount of showers taken.

c. **Reason about Risk Treatment** In this example, we show the controls the customer selected to treat the risks introduced previously, depicted in Fig. 6.6. The selected treatment for the events *Steal Home Gateway* and *Steal Smart Meter* is the ISO 27001 control *A 9.2.7—Removal of property*. The control states that equipment shall not be removed from the site without authorization. Mechanisms such as physical locks or fixed installation on walls require some effort to remove. The *Steal Smart Phone* event is treated with the control *A 9.2.5—Security of Equipment of Premises* and the already existing control *Track GPS Signal of Smart Phone*. *A 9.2.5* defines that special measures have to be taken for carrying equipment off side. A refinement of this mechanism can be a fingerprint lock, so if the smart phone is stolen, the thief cannot use it.

The event *Prosumer to provide Online Banking Credentials* is addressed by the control *A 8.2.2-Information security awareness, education, and training*. The teaching mechanism should help prosumer to identify social engineering attacks. Moreover, the *Energy Alpha to provide Billing Data* and *Energy Alpha to provide*

Prosumer Information events are addressed by the control *A 8.2.2*, as well. The control *A 15.1.4—Data protection and privacy of personal information* concerns the definition of privacy goals. The Energy Alpha stores and processes personal information of the prosumer and it will implement *A 6.1.5—Confidentiality agreements*. These state that appropriate procedures and technologies are used to prevent data loss or leakage of these data. In addition, the control *A 7.1.3—Acceptable use of Assets* demands that the life cycle of these data is documented: When does Energy Alpha ask for the personal information, what exactly are the steps for processing it, and when and how is it deleted?

The *Change Stove, Heater, or Fridge configuration* event is addressed by *A 12.2.1—Control of internal processing* and *A 12.2.2—Input data validation*, which the vendors of the smart appliances stove, heater, or fridge have to apply. This should be checked before purchasing these. The Energy Alpha company should provide lists of products and companies that apply these mechanisms.

The event *Change Meter Data on route between Home Gateway, Smart Meter, and EMS* should be addressed by *A 11.4.6—Network connection control* implemented in the home gateway. Energy Alpha will provide guidelines on how to configure routers accordingly and also sell routers that are configured accordingly. Moreover, the control *A 11.4.1—Policy on use of network services* describes proper use of networks and explains what should not be done, e.g., not letting persons that you do not trust use your network connection. The control *A 11.4.7—Network routing control* demands that it has to be checked that the home gateway only provides authorized entities connections to the smart meters.

Step 7: Design ISMS Specification
The ISMS is documented with the artifacts generated in the previous step using the mapping presented in Table 6.1.

6.4 Discussion

The procedure presented in this chapter was discussed with three different practitioners. The security consultants mentioned that this structured procedure

- supports the identification of assets,
- guides the elicitation of threats for the identified assets,
- helps to describe the attackers' abilities in more detail based on attacker types and motivations,
- uses one model for all analysis steps, which helps to maintain the focus of the analysis on the system levels and not to get lost in details,
- increases the use of models instead of texts in standards, which eases the effort of understanding the system documentation,
- provides the means for abstraction of a complex system and structured reasoning for security based upon this abstraction.

One issue that needs further investigation is that of scalability, both in terms of the effort needed by the requirements engineer in order to enter all information about the organization and the threat analysis proposed. We will use the method for different scenarios to investigate if the method scales for complex goal models. We used the Si* tool for creating the models in our method,[7] which reduces the effort in creating the Si* models. A shortcoming of the tool support are missing consistency checks between different models. For example, a threat analysis model should not contain any roles or agents that are not in the scope definition. We will develop such a support in the future.

Our method does not consider legal obligations. The consideration of legal obligations is already mentioned in the first paragraph on p. 1 of the standard. It states that the compliance to the standard does not confer immunity for legal obligation. Thus, legal obligations have to be known in order to be able to follow this demand. In particular, ISO 27001 Sect. 4.2.1b concerns the definition of an ISMS policy state and Sect. 4.2.1g concerns the consideration of legal and regulatory demands. We propose to look into the integration of goal-based legal analysis frameworks such as Ghanavati et al. (2009) to compensate this shortcoming.

6.5 Related Work

Our method uses goal-based requirements engineering, in particular the Si* notation to establish an ISO 27001 ISMS. We are not aware of a similar effort. Nevertheless, we structure our related work into techniques that support ISO 27001 establishment and goal-based requirements engineering methods that support security analysis.

6.5.1 Techniques that support ISO 27001 compliant ISMS Establishment

Jayawickrama (2006) present a high-level framework for implementing an ISO 27001 ISMS based on practical experience. The framework focuses on implementing, monitoring, managing, and improving the security of process control systems in particular to critical infrastructures such as electricity or water plants. The framework consists of a series of guidelines for these critical infrastructures. It differs from our work in this regard.

Watson and Jones (2013) focus on digital forensic with regards to ISO 27001. That means identifying and documenting proof that a sufficient ISO 27001 compliant ISMS was established. The work provides best practices from practitioners to aquire this proof. Freeman (2007) advocates to address system security with an ISO 27001 ISMS. The author provides numerous best practices toward this effort. Calder (2009)

---

[7] http://sistar.disi.unitn.it/index.php/Si*_Tool.

and Kersten et al. (2011) provide advice for an ISO 27001 realization. In addition, Klipper (2010) focuses on risk management according to ISO 27005. None of these works consider to use security requirements engineering methods.

Cheremushkin et al. (2010) present a UML-based meta-model for several terms of the ISO 27000, e.g., assets. These meta-models can be instantiated and, e.g., support the identification of assets (Lyubimov et al. 2011). The method is based on instantiating isolated fragments such as assets, but does not present a structured method to establish an ISO 27001 compliant ISMS.

Mondetino et al. investigate possible automation of controls that are listed in the ISO 27001 and ISO 27002 (Montesino and Fenz 2011). Their work can complement our own.

6.5.2 Goal-based Requirements Engineering for Security Analysis

Van Lamsweerde and Letier (2000) extend KAOS with the concept of an obstacle in order to analyze the security goals of a system, as well as the concept of an antigoal (Van Lamsweerde 2004). An obstacle is an undesired state of affairs that threaten security goals, and antigoals describe the intentions of an attacker. Van Lamsweerde and Letier (2000), Van Lamsweerde (2004) propose a formal framework to identify the obstacles to a goal in given domain properties and to generate resolutions to those obstacles. Liu et al. (2003) extend the i* framework (Yu 1995) to include attackers, and analyze vulnerabilities these attackers can exploit. In this framework, all actors are considered as potential attackers. Therefore, the skills to exploit a vulnerability of all actors are analyzed. Li et al. (2010) contribute a formal framework to support their attacker analysis. In addition, Elahi et al. (2009) also extend i* to analyze vulnerabilities in relation to systems requirements. None of these approaches support the establishment of an ISO 27001 compliant ISMS.

Buyens et al. (2011) have proposed an analysis technique to identify violations of the least privilege principle in software architectures. The analysis leverages (a) a task execution model, which includes all the elements of an architecture that are relevant for the least privilege analysis, and (b) a security policy specifying the assigned permissions. The analysis identifies the violations of the principle, and the architectural components that are concerned with the violations. This method differs from our own, because our work applies to the analysis part of a software engineering process and the authors apply theirs to the design phase of a software engineering process. Their work has the potential to complement our own. SQUARE (Mead and Stehney 2005) is a requirements engineering method that supports risk assessment as an explicit step to identify security requirements. Matulevičius et al. (2012) extend the goal-based security requirements notation Secure Tropos (Mouratidis and Giorgini, 2007) to support modeling of security concerns including risks and countermeasures. None of the methods do present explicit support for establishing an ISO 27001 ISMS.

Li et al. (2005) propose a security analysis that verifies that a set of security properties such as availability considers also the delegation of access to resources to partially trusted stakeholders. Similarly to us, they consider that delegating permissions on resources to not fully trusted entities can be a source of threats. The consideration of access delegation to partially trusted entities can be used to extend our work in the future.

6.6 Summary

We have presented a structured method to establish an Information Security Management System (ISMS) according to the ISO 27001 standard, which builds upon the security requirements engineering notation Si*.

Our method offers the following main benefits:

- A structured method that relies on the goal-based requirements engineering for analyzing attackers' motivations and goals toward the assets of a system. The motivations and goals of attackers are refined into events the attacker causes. These events are oriented toward assets and represent threats.
- The events an attacker causes are the basis for a risk management process that is part of our method. The process includes risk treatment considering the controls specified in Annex A of the ISO 27001 standard.
- Reusing security requirements engineering methods to support the development and documentation of an ISO 27001 compliant ISMS
- Support for generating consistent ISMS documentation compliant to ISO 27001
- Reusing the structured techniques of SRE methods for analyzing complex systems and eliciting security requirements to support the refinement of sparsely described sections of the ISO 27001 standard

Our method provides the means to elicit the context of an ISMS considering management commitment, threat and risk analysis, as well as security requirements-based control selection.

References

Aloula, F., Al-Alia, A. R., Al-Dalkya, R., Al-Mardinia, M., & El-Hajj, W. (2012). Smart grid security: Threats, vulnerabilities and solutions. *International Journal of Smart Grid and Clean Energy, 1*(1), 1–6.

Asnar, Y., Giorgini, P., Massacci, F., & Zannone, N. (2007). From trust to dependability through risk analysis. In *Proceedings of the Second International Conference on Availability, Reliability and Security, ARES 2007* (pp. 19–26). IEEE Computer Society.

Asnar, Y., Giorgini, P., & Mylopoulos, J. (2011). Goal-driven risk assessment in requirements engineering. *Requirements Engineering, 16*(2), 101–116.

Beckers, K. (2014). Goal-based establishment of an information security management system compliant to ISO 27001. In *Proceedings of SOFSEM 2014: Theory and Practice of Computer Science*. LNCS (Vol. 8327, pp. 102–113). Springer.

Beckers, K., Côté, I., Hatebur, D., Faßbender, S., & Heisel, M. (2013). Common criteria compliAnt software development (CC-CASD). In *Proceedings of the 28th Symposium on Applied Computing* (pp. 937–943). ACM.

Buyens, K., Scandariato, R., & Joosen, W. (2011). Least privilege analysis in software architectures. *Software and Systems Modeling*, 1–18.

Calder, A. (2009). *Implementing information security based on ISO 27001/ISO 27002: A management guide*. Zaltbommel: Van Haren Publishing.

Cheremushkin, D. V., & Lyubimov, A. V. (2010). An application of integral engineering technique to information security standards analysis and refinement. In *Proceedings of the International Conference on Security of Information and Networks* (pp. 12–18). ACM.

Elahi, G., Yu, E., & Zannone, N. (2009). A vulnerability-centric requirements engineering framework: Analyzing security attacks, countermeasures, and requirements based on vulnerabilities. *Requirements Engineering*, 15(1), 41–62.

Freeman, E. H. (2007). Holistic information security: ISO 27001 and due care. *Information Systems Security*, 16(5), 291–294.

Ghanavati, S., Amyot, D., & Peyton, L. (2009). Compliance analysis based on a goal-oriented requirement language evaluation methodology. RE (pp. 133–142).

ISO/IEC. (2005). Information technology—Security techniques—Information security management systems—Requirements (ISO/IEC 27001). Geneva, Switzerland: International Organization for Standardization (ISO) and International Electrotechnical Commission (IEC).

ISO/IEC. (2013). Information technology—Security techniques—Information security management systems—Requirements (ISO/IEC 27001). Geneva, Switzerland: International Organization for Standardization (ISO) and International Electrotechnical Commission (IEC).

Jayawickrama, W. (2006). Managing critical information infrastructure security compliance: A standard based approach using ISO/IEC 17799 and 27001. In *Proceedings of the 2006 International Conference on on the Move to Meaningful Internet Systems: AWeSOMe, CAMS, COMINF, IS, KSinBIT, MIOS-CIAO, MONET–Volume Part i* (pp. 565–574). Springer.

Kersten, H., Reuter, J., & Schröder, K. -W. (2011). IT-Sicherheitsmanagement nach ISO 27001 und Grundschutz. Vieweg+Teubner.

Klipper, S. (2010). Information Security Risk Management mit ISO/IEC 27005: Risikomanagement mit ISO/IEC 27001, 27005 und 31010. Vieweg+Teubner.

Lin, H., & Fang, Y. (2013). Privacy-aware profiling and statistical data extraction for smart sustainable energy systems. *IEEE Transactions on Smart Grid*, 4(1), 332–340.

Liu, L., Yu, E., & Mylopoulos, J. (2003). Security and privacy requirements analysis within a social setting. In *Proceedings of the Requirements Engineering Conference (RE)* (pp. 151–161). IEEE Computer Society.

Li, N., Mitchell, J. C., & Winsborough, W. H. (2005). Beyond proof-of-compliance: Security analysis in trust management. *Journal of the ACM*, 52(3), 474–514.

Li, T., Liu, L., & Bryant, B. R. (2010). Service security analysis based on i*: An approach from the attacker viewpoint. In *Security, Trust, and Privacy for Software Applications (STPSA 2010)* (pp. 127–133). Seoul. IEEE Computer Society.

Lyubimov, A., Cheremushkin, D., Andreeva, N., & Shustikov, S. (2011). Information security integral engineering technique and its application in ISMS design. In *Proceedings of the International Conference on Availability, Reliability and Security (ARES)* (pp. 585–590). IEEE Computer Society.

Matulevičius, R., Mouratidis, H., Mayer, N., Dubois, E., & Heymans, P. (2012). Syntactic and semantic extensions to secure tropos to support security risk management. *Journal of Universal Computer Science*, 18(6), 816–844.

Mead, N. R., & Stehney, T. (2005). Security quality requirements engineering (square) methodology. *SIGSOFT Software Engineering Notes*, 30(4), 1–7.

Montesino, R., & Fenz, S. (2011). Information security automation: How far can we go? In *Proceedings of the International Conference on Availability, Reliability and Security (ARES)* (pp. 280–285). IEEE Computer Society.

Mouratidis, H., & Giorgini, P. (2007). Secure tropos: A security-oriented extension of the tropos methodology. *International Journal of Software Engineering and Knowledge Engineering, 17*(2), 285–309.

Stoneburner, G., Goguen, A., & Feringa, A. (2002). *Risk management guide for information technology systems* (NIST Special Publication No. 800-30). Gaithersburg, U.S.: National Institute of Standards and Technology (NIST).

Van Lamsweerde, A. (2004). Elaborating security requirements by construction of intentional anti-models. In *Proceedings of the 26th International Conference on Software Engineering* (pp. 148–157). IEEE Computer Society.

Van Lamsweerde, A., & Letier, E. (2000). Handling obstacles in goal-oriented requirements engineering. *IEEE Transactions on Software Engineering, 26*(10), 978–1005.

Watson, D., & Jones, A. (2013). *Digital forensics processing and procedures: Meeting the requirements of ISO 17020, ISO 17025, ISO 27001 and best practice requirements*. Amsterdam: Syngress Publishing.

Yu, E. (1995). *Modelling strategic relationships for process reengineering*. Unpublished doctoral dissertation, University of Toronto, Canada.

Chapter 7
Supporting ISO 27001 Establishment with CORAS

Abstract Establishing an information security management system (ISMS) compliant to the ISO 27001 standard is a way for companies to gain their customers trust with regard to information security. Key challenges of establishing an ISO 27001 compliant ISMS are removing the standards' ambiguities and providing an acceptable risk management approach. Risk management is vital to an ISMS establishment, because the aim of an ISMS is to manage security threats based on risk assessment. The security requirements engineering approach CORAS provides a structured way to implement risk management for a given company. We present an extension to this method called ISMS-CORAS, which enables security engineers to create an ISO 27001 compliant ISMS including the needed documentation. ISMS-CORAS uses another CORAS extension called Legal CORAS, which helps to be compliant to legal demands as well. The method is applied to a smart grid scenario provided by the industrial partners of the NESSoS project.

7.1 Introduction

Chapter 5 showed the problems with establishing the ISO 27001 (ISO/IEC 2005) standard, which are caused by ambiguous descriptions in the standard that may pose challenges during the establishment of an ISMS. Some of the ambiguity is deliberate to ensure that the standard can serve a multitude of different domains and stakeholders. This is nevertheless a problem for security experts, who have to choose a method for security analysis that is compliant with the standard. They moreover need to decide, e.g., on the abstraction level for the required documentation without any support from the standard. For example, security experts need to describe the business, organization, etc., and decide on their own which are the most relevant scope elements to consider. They also have to find a method that allows them to achieve completeness of identifying stakeholders, assets, security goals, and so forth. Moreover, the standard does not provide a method for assembling the necessary information or a pattern on how to structure that information. The importance of

© Springer International Publishing Switzerland 2015
K. Beckers, *Pattern and Security Requirements*,
DOI 10.1007/978-3-319-16664-3_7

these steps becomes apparent when one realizes that essential further steps of the ISO 27001 depend upon such information, e.g., the identification of *threats*, *controls*, and *vulnerabilities*.

In this chapter, we present an extension of the CORAS risk analysis method (Lund et al. 2010; Solhaug and Stølen 2013) to support the establishment of an ISO 27001 compliant ISMS. In previous work we analyzed the relations between different security requirements engineering and risk analysis methods (Chap. 5), and our results showed that the ISO 27001 standard has a significant focus on risk analysis. The standard describes how to build an ISMS, and CORAS (Sect. 2.5.2) already supports many of these steps due to its focus on risk management. A further motivation for building on CORAS is that it is based on the ISO 31000 (ISO 2009) standard, which is also the basis for the risk management process of ISO 27005 (ISO/IEC 2008). The latter standard refines the risk management process described in ISO 27001.

Additionally, the ISO 27001 standard demands legal aspects (such as laws, regulations, contracts and legally binding agreements) to be considered. CORAS provides support for the consideration of legal aspects during the risk analysis by the extension called *Legal CORAS* (Lund et al. 2010). CORAS moreover comes with tool support for all phases of the process, and the tool supports the specification of all CORAS diagrams, including Legal CORAS (Sect. 2.5.2). A further useful feature of CORAS is that it facilitates the reporting of the results by a formal mapping from its diagrams to English prose. This mapping enables ISO 27001 document generation.

In summary, we use CORAS as a basis because of its structured method for risk management, its compliance to ISO 31000, the consideration of legal concerns, the tool support and the support for document generation.

We refer to the CORAS extension presented in this chapter as *ISMS-CORAS*. In Sects. 2.2.2, and 2.5.2 we present the main background of this method, which are the ISO 27001 standard, CORAS, and Legal CORAS. We describe the ISMS-CORAS method in Sect. 7.2 including the steps of the ISMS-CORAS method, focusing on extensions to CORAS and demonstrating how it fulfills the demands of the standard, meaning how the security analysis and documentation demands are met.

We conducted the work in collaboration with some of the inventors of the CORAS method, namely Bjørnar Solhaug and Ketil Stølen. The author of this book is the main author of this work, but we relied upon their expertise in the form of discussions, feedback, and improvements of our work. This work would not have been possible without their guidance. The complete results of our collaboration is published in a technical report (Beckers et al. 2013c) and also as part of a book (Beckers et al. 2014) that illustrates the results of the NESSoS project,[1] on which this chapter is based.

The remainder of this chapter is organized as follows, Sect. 7.3 we show an example application of our method and the resulting artifacts. In particular, we apply ISMS-CORAS to a NESSoS industrial use case concerning a Smart Grid. In Sect. 7.4 we present related work and Sect. 7.5 concludes.

---

[1]The NESSoS project: http://www.nessos-project.eu.

7.2 The ISMS-CORAS Method

The method presented in this section extends CORAS in order to support security management compliant with ISO 27001. Our contribution, namely *ISMS-CORAS*, follows the steps depicted in Fig. 7.1. The figure also shows the resulting artifacts from applying our method. While keeping the names of the method steps, we focus in our description on the difference to CORAS, and we explain how our changes to CORAS are related to ISO 27001 and its documentation requirements as described above. The steps and artifacts of CORAS that we extended are marked in gray in the figure. Note that step 3 is white, even though its input changed with regard to original CORAS, but the changes are so minimal that we decided not to mark it gray. Note, importantly, that the ISO 27001 standard does not have specific demands on the form of the documentation, as "documents and records may be in any form or type of medium" (ISO/IEC 2005). Hence, we can use CORAS artifacts in the creation of our ISMS documentation. We illustrate a mapping between the ISMS-CORAS and the ISO 27001 documentation demands in Table 7.1. This includes the specification of the characteristics of the business, and information about technology relevant to the target description. We describe the activities of each step, the contribution to the ISMS documentation, and the novelty with regard to original CORAS.

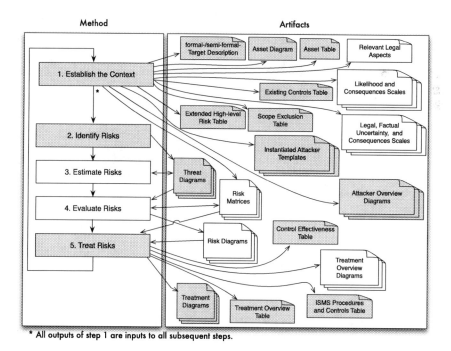

* All outputs of step 1 are inputs to all subsequent steps.

Fig. 7.1 The ISMS-CORAS method (our Contributions are marked in *Gray*)

Table 7.1 Mapping ISMS-CORAS artifacts to ISO 27001 documentation demands

No.	ISO 27001 document	ISMS-CORAS artifacts	Step
1	Scope of the ISMS	(Semi-) formal target description; scope exclusion table	1
2	ISMS policy statements	Extended high-level risk tables	1
3	Procedures and controls in support of the ISMS	ISMS procedure table	1, 5
4	Risk assessment methodology	Description of the CORAS method	1–5
5	Risk assessment report	Asset diagrams; asset tables; risk evaluation criteria (risk matrix); threat diagrams; risk diagrams; Legal CORAS diagrams	1–4
6	Risk treatment plan	Treatment diagrams; treatment overview diagrams	5
7	Procedures to the effective planning, operation and control of the ISMS	Treatment diagrams; treatment overview table; control effectiveness table; written documentation	5
8	Statement of applicability	Treatment diagrams; treatment overview table	5
9	Management decision	Written documentation	1, 5

Step 1: Establish the Context

The main objective of this step remains the same, but is extended to fulfill the demands of the standard. We divided Step 1 into sub-steps to reduce its complexity, which we describe in the following.

Sub-step 1.1—Target description including Locations	
Description	The target description begins with the scope and focus of the analysis, modeling the target of analysis at an adequate level of abstraction, identifying stakeholders and assets and defining the risk evaluation criteria. The target is modeled using a (semi-) formal language, such as UML (UML Revision Task Force 2010). These initial tasks remain unchanged from the original CORAS method (Sect. 2.5.2). An ISMS-CORAS specific subtask concerns the documentation of scope exclusions in a *scope exclusion table*. The table refers to elements in the target description and states reasons for excluding this particular element from the scope of the analysis. ISMS-CORAS requires also an explicit description of the location of each element in the target description due to demands of ISO 27001. Moreover, the location information is also essential for the consideration of legal aspects. For example, according to the German Federal Data Protection Act (BDSG) Section 4b, it is not allowed to store personal information outside of the European Union
Contribution to ISMS document(s)	Scope of the ISMS, ISMS Policy Statements
ISMS-CORAS Novelty	Explicit reasoning about scope exclusions, Explicit documenting of the location of scope elements

---

Changes for ISO 27001:2013 Compliance

The scope reasoning requires to be adapted to the new concept of external and internal context of the ISMS demanded in ISO 27001:2013. The concept of scope exclusion is replaced by the external and internal context of the ISMS. This is the basis for the *boundaries and applicability of the ISMS* and finally to establish the scope of the ISMS (ISO/IEC 2013, p. 1–2).
Furthermore, the context establishment of ISO 27001:2013 is based on ISO 31000. CORAS is based on the ISO 31000 standard as well. Hence, usage of CORAS for an ISO 27001:2013 establishment is a good fit.

Sub-step 1.2—Asset Identification and Documentation including existing Controls

Description	This sub-step describes how to use the target description to identify and document assets. ISMS-CORAS focuses in particular on *information assets* since an ISMS shall be built for their protection. A further task, as demanded by the standard, is to rate all assets according to their importance in order to prioritize the risk assessment. The rating and priority are documented in ISMS-CORAS specific *asset tables*. ISMS-CORAS requires also the definition of asset owners in these tables. An *asset owner* is an "individual or entity that has approved management responsibility for controlling the production, development, maintenance, use and security of the assets. The term 'owner' does not mean that the person actually has any property rights to the asset" (ISO/IEC 2005). Moreover, the first step of ISMS-CORAS involves the documentation of existing security controls. These shall be discussed when refining the target description and are documented in an *existing controls table*, which lists the controls and the assets these controls protect
Contribution to ISMS document(s)	Risk Assessment Report, Procedures and Controls in Support of the ISMS
ISMS-CORAS Novelty	Detailed description of assets in an asset table Descriptions of existing controls in an existing controls table

---

Changes for ISO 27001:2013 Compliance

ISO27001:2013 does not require an asset identification and asset owners are also not required. In contrast, ISO27001:2013 requires a risk identification for which an asset identification can help. Moreover, ISO27001:2013 demands to specify *risk owners*, which are persons responsible for managing a particular risk (ISO/IEC 2013, p. 4).

Sub-step 1.3—Attacker Identification and Documentation	
Description	ISMS-CORAS aims to identify relevant vulnerabilities of the systems in the ISMS scope, and how attackers may exploit the vulnerabilities. The ISO 27001 standard states that the possible impact an exploitation of each vulnerability has on the information asset must be estimated. The documentation of this impact shall consider confidentiality, integrity, and availability. We use *attacker templates* (see Table 7.7) to reason about attacker types and attacker motivations in relation to assets and the target description. The instantiation of these templates not only facilitates the security risk identification and in particular identifying attackers, but also results in documenting attackers that are out of scope and the assumptions that lead to scope exclusions. Their documentation is vital in order for other security experts to follow the reasoning of the threat model, e.g., in an audit of the ISMS. The attackers that are not excluded, their entry points in the target description, and the threatened assets are documented in *attacker overview diagrams*. These specify also the elements of the target descriptions and assets that are out of reach of a particular attacker, and therefore can be excluded from further analysis. Attacker templates and attacker overview diagrams are contributions of ISMS-CORAS This sub-step also involves the creation of a *high-level risk table* that defines who or what may cause incidents, how threats harm assets, and the vulnerabilities that the threat potentially exploits. ISMS-CORAS fulfills the ISO 27001 demands for a specific consideration of availability, confidentiality, and integrity for information assets as well. These are documented as high-level security objectives that mitigate the threats in an *extended high-level risk table*. This table refers to the attackers documented in the attacker templates and attacker overview diagrams
Contribution to ISMS document(s):	Risk Assessment Report
ISMS-CORAS Novelty	Reasoning about attacker in attacker templates and attacker overview diagrams. Extended high-level risk table use attacker templates and attacker overview diagrams as input. However, the structure is the same as the high-level risk table from the original CORAS method

---

Changes for ISO 27001:2013 Compliance

Threat analysis and attacker modeling is not a mandatory step in ISO27001:2013 ((ISO/IEC 2013, p. 4). However, the risk identification step in the standard could be supported by identifying vulnerabilities and threats, as proposed in the attacker identification step of ISMS-CORAS.

Sub-step 1.4—Identification of Risk Criteria

Description	This sub-step concerns the identification of risk criteria. In particular, the definition of scales of likelihoods, consequences, and risk evaluation criteria
	This sub-step also concludes with a review of all the documents created so far from the management and a written management commitment
Contribution to ISMS document(s):	Risk Assessment Report, Management Decisions
ISMS-CORAS Novelty	–

Sub-step 1.5—Considering Legal Aspects

Description	Step 1 also involves the creation of a *high-level risk table* that defines who or what may cause incidents, how threats harm assets, and the vulnerabilities that the threat potentially exploits. ISMS-CORAS fulfills the ISO 27001 demands for a specific consideration of availability, confidentiality, and integrity for information assets as well. These are documented as high-level security objectives that mitigate the threats in an *extended high-level risk table*. This table refers to the attackers documented in the attacker templates and attacker overview diagrams
	A further important aspect of the context establishment step is the definition of scales of likelihoods, consequences, and risk evaluation criteria
	Due to the standard it is mandatory to consider legal aspects in ISMS-CORAS. The identification of relevant legal aspects can be achieved, for example, by using our law patterns method (Beckers et al. 2012; Faßbender and Heisel 2013) or by involving domain experts and lawyers. Laws are considered in the Legal CORAS extension (Sect. 2.5.2), but Legal CORAS does not concern the identification of relevant laws. It concerns the risk assessment regarding likelihoods and consequences for getting convicted. Scales of likelihoods, consequences, and risk evaluation criteria for legal risks have to be described as well
Contribution to ISMS document(s):	Risk Assessment Report
ISMS-CORAS Novelty	Law identification and likelihoods, consequences, and risk evaluation criteria for legal risks

Step 2: Risk Identification

Description	This step is conducted similarly to CORAS, but includes also the identification of legal risks and creates threat diagrams that refer to the attacker types described in the attacker templates. The original CORAS risk identification uses threat diagrams that specify in more detail how the attackers identified in the previous step may cause security risks by exploiting vulnerabilities. Legal CORAS supports the identification of legal risks and the modeling of relevant legal norms and their sources, the legal and factual uncertainties, as well as the risks that are caused by legal norms
Contribution to ISMS document(s):	Risk Assessment Report
ISMS-CORAS Novelty	Threat diagrams refer to instantiated attacker templates

---

Changes for ISO 27001:2013 Compliance

Similar to Step 1.4 the threat analysis and the usage of the attacker template are not a mandatory step for ISO27001:2013 compliance ((ISO/IEC 2013, p. 4). However, the risk identification step in the standard should be supported by threat and attacker modeling, as proposed in ISMS-CORAS. Otherwise, a structured elicitation of risks is challenging. A similar comment can be made for the following steps, but for simplicity's sake we do not mention this change again.

Step 3: Risk Estimation

Description	This step is also similar to the original CORAS. However, ISMS-CORAS focuses additionally on the likelihood estimations of accidental misuses or exploits of the identified vulnerabilities. A specific task is to derive attacker types with a certain skill set, similar to the descriptions proposed in the Common Criteria (ISO/IEC 2012). The results of this step are documented in threat diagrams. A further task in ISMS-CORAS is to estimate the legal and factual uncertainty of the identified legal norms according to the description of Legal CORAS
Contribution to ISMS document(s):	Risk Assessment Report
ISMS-CORAS Novelty	Threat diagrams refer to exploits caused by the attackers described in the instantiated attacker templates and refer to the vulnerabilities these attackers exploit

Step 4: Risk Evaluation

Description	This step is performed according to original CORAS, but ISMS-CORAS considers also attacker types and their possible exploits of vulnerabilities for deciding whether a risk requires treatment
Contribution to ISMS document(s)	Risk Assessment Report
ISMS-CORAS Novelty	Risk treatments considers the attackers and vulnerabilities documented in the threat diagrams

Step 5: Risk Treatment

Description	Also this step follows CORAS, but ISMS-CORAS restricts the identification of risk treatments to the normative controls defined in Appendix A of the ISO 27001 standard. The treatments have to consider existing controls, and the asset owner is responsible for the controls protecting the asset. This information has to be included in the treatment diagrams. The residual risk has to be documented and the management has to approve it. As in original CORAS, the treatment plans should consider cost-benefit reasoning, for example, by using the CORAS extension proposed in Tran et al. (2013a, b)
	Step 5 further requires a reasoning why a particular Appendix A control is considered or left out. For this purpose we propose to use *treatment overview tables* that refer to an asset, its security objective, and relevant treatment or treatment overview diagrams, and a reasoning of why the treatment is sufficient. We also have to document how the effectiveness of each control can be measured in a *control effectiveness table* that defines measures to assess the effectiveness of each control. The procedures and controls that are part of the ISMS have to be documented, and a further subtask is to document each procedure that is part of a selected control
Contribution to ISMS document(s)	Risk Treatment Plan, Statement of Applicability, Procedures and Controls in Support of the ISMS, Risk Treatment Plan, Procedures to the effective planning, operation and control of the ISMS, Management Decisions
ISMS-CORAS Novelty	It is mandatory to consider the controls specified in ANNEX A of the ISO 27001 standard for risk treatment
	Risk treatment considers the attackers and vulnerabilities documented in the threat diagrams

ISMS-CORAS Support for the ISO 27001 Documentation Demands

In Table 7.1 we give an overview of how ISMS-CORAS fulfills the ISO 27001 documentation demands as listed in Sect. 2.2.2. Recall from that section that we do not address the documentation of the (8) ISMS records. The first column contains the document number, the second contains the name we assigned to the document, the third contains the ISMS-CORAS artifacts that support the documentation, and the last contains the method steps in which the artifacts are created.

> **Changes for ISO 27001:2013 Compliance**
> The controls in ANNEX A differ slightly in ISO 27001 and ISO 27001:2013.
> We provide a mapping between the controls of ISO 27001 and ISO
> 27001:2013 in Appendix C to support the efforts of supporting ISO
> 27001:2013 or updating an ISO 27001 ISMS to an ISO27001:2013 ISMS.
> Furthermore, we show the fulfillment of ISMS-CORAS of the ISO27001:2013
> documented information demands in Table 7.2.

Mapping to the Conceptual Framework for Security Standards—We present a mapping in Table 7.3 from this method to the conceptual framework for security standards, which is the foundation for our PEERESS framework (Chap. 3). The table lists the activities of the conceptual framework for security standards on the left column and the steps of the ISMS-CORAS method that concern these activities in the right column.

Table 7.2 Mapping ISMS-CORAS artifacts to ISO 27001:2013 documented information

No.	ISO 27001 documented information	ISMS-CORAS artifacts	Steps
1	Scope of the ISMS	(Semi-) formal target description	1
2	Information security policy	Extended high-level risk tables	1
3	Information security objectives	Extended high-level risk tables	1
4	Risk assessment and the risk treatment methodology	Description of the CORAS method	1–5
5	Risk assessment results	Asset diagrams; asset tables; risk evaluation criteria (risk matrix); threat diagrams; risk diagrams; Legal CORAS diagrams	1–4
6	Risk treatment results	Treatment diagrams; treatment overview diagrams	5
7	Statement of applicability	Treatment diagrams; treatment overview table	5
8	Competence records		5
9	Monitoring and measuring results	Documentation of security processes and their monitoring results	5
10	Audit programme and results	Documentation of the audits and their results	5
11	Management review results	Documentation of the management review activities and their results	5
12	Evidence of corrective actions	Documentation of selected measures of corrective actions	5
13	Procedures to the effective planning, operation and control of the ISMS	Treatment diagrams; treatment overview table; control effectiveness table; written documentation	5

Table 7.3 A mapping between the standard activities in the conceptual framework for security standards and the steps of the ISMS-CORAS method

Activities in the conceptual framework for security standards	Steps in the ISMS-CORAS method
Environment description	Sub-step 1.1—Target description including locations
Stakeholder description	Sub-step 1.1—Target description including locations
Asset identification	Sub-step 1.2—Asset identification and documentation including existing controls
Risk level description	Sub-step 1.4—Identification of risk criteria
Security property description	Sub-step 1.3—Attacker identification and documentation
Control assessment	Sub-step 1.2—Asset identification and documentation including existing controls
Vulnerability and threat analysis	Sub-step 1.3—Attacker identification and documentation
Risk determination	Sub-step 1.4—Identification of risk criteria Sub-step 1.5—Considering legal aspects Step 2: Risk identification Step 3: Risk estimation
Security assessment	Step 4: Risk evaluation
Security measures	Step 5: Risk treatment
Risk acceptance	Step 5: Risk treatment
Documentation	Steps 1–5 (cf. Table 7.1)

7.3 Application of Our Method

We apply our method in the following and focus on the most relevant artifacts.

Step 1: Establish the Context

Sub-step 1.1—Target description including Legal Norms and Locations
The smart home scenario concerns a house that is connected to the smart grid with two parties living in it. The scenario is provided by the industrial partners of the NESSoS[2] project. We show the target description in an UML (UML Revision Task Force 2010) class diagram (see Fig. 7.2). The associations in the diagram represent communication connections. All elements in the scope are the elements of the target description that are inside the smart home.

We do not consider the transport and production of energy in this scenario, because we aim to illustrate how ISMS-CORAS works for information and communication technologies. In part, because the ISO 27001 ISMS is only concerned with information assets (Sect. 7.2).

---

[2]http://www.nessos-project.eu.

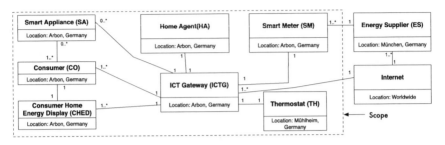

Fig. 7.2   The semiformal target description of the NESSoS smart home scenario

The *ICT Gateway (ICTG)* is the connection between the smart home and the information systems of the *Energy Supplier (ES)*. Every party in the smart home consists of *Consumers (CO)*, who use *Smart Appliances (SA)*. SAs are connected to the internet via the ICTG. For example, a fridge that can be remotely configured to cool down to a specific temperature in the evening is an SA. The parties can use services offered by the energy providers via a *Consumer Home Energy Display (CHED)*. A *Thermostat (TH)* measures the temperature of the home or of SAs. The temperature information is used for safety purposes, e.g., to prevent a stove from overheating. They are also used by applications that control SAs. In addition, customers can use THs to configure SAs, for example, to configure a heater to warm the smart home to a specific temperature during daytime. The *Home Agent (HA)* offers the CO a selection of different energy rates from different ES (Rodden et al. 2013). The *Smart Meter (SM)* transfers the energy consumption/production data to the ES.

The client, i.e., the commissioning party for this risk analysis, is the energy supplier, who conducts the study with the viewpoint of the consumers living in a smart home. The energy supplier is interested in analyzing privacy, availability, integrity, and confidentiality concerns of consumers and how these can be assured via an ISMS.

The energy supplier stated the following high-level security objectives:

- The integrity, confidentiality, and availability of consumer's configuration data of their home agents shall be preserved
- The privacy of the consumer's energy consumption data shall be preserved
- The integrity, confidentiality, and availability of the customer's configuration data for their smart appliances shall be ensured.

We state the exclusions from the scope in Table 7.4, based on the target description in Fig. 7.2.

Table 7.4   ISMS-CORAS scope exclusion table

Element of the target description	Reason for scope exclusion
Energy supplier	The energy supplier aims to create an ISMS to protect the consumers' security and privacy needs; hence, the energy supplier has no harmful intentions toward the consumers

Sub-step 1.2—Asset Identification and Documentation including existing Controls

The assets in this scenario are depicted in Fig. 7.3. This CORAS asset diagram shows three assets in the middle of the diagram. The *Consumers' Energy Consumption Data* shall be protected from attackers that use this data for creating behavioral profiles based on the consumption data. The value of the *Smart Appliances' Configuration* to the consumer is essential, because without it the consumer loses control of the appliances in his/her home. For example, a stove could heat up during the night and cause a fire. The *Home Agent's Configuration* states from which energy supplier the consumer buys and sells energy. An unauthorized change in the configuration of the home agent could result in, e.g., the purchase of electricity at a very expensive price. The arrows in the CORAS asset diagrams mean *harm to*. Hence harm to the assets *Consumers' Energy Consumption Data*, the *Smart Appliances' Configuration*, the *Home Agent's Configuration* also causes harm to the overall *Consumers' Security and Privacy*. CORAS defines the notion of an indirect asset. These are assets to which we cannot measure harm directly. This is just possible via the measurements of harm to other direct assets. The asset *Public's Trust in Smart Home* is such a case.

Due to limited resources the customer has to define some acceptable risks. The customer starts with a ranking of the assets according to their importance, shown

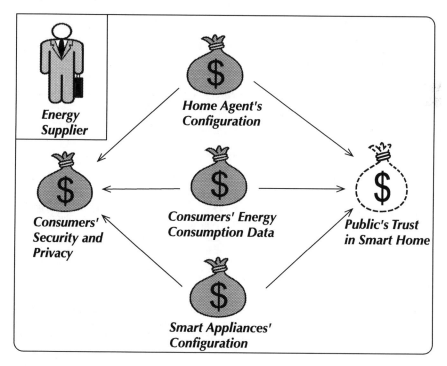

Fig. 7.3 Smart home asset diagram

Table 7.5 Asset table with asset owners

Asset	Importance	Type	Owner
Consumers' energy consumption data	1	Direct asset	Mr. Jones
Smart appliances' configuration	1	Direct asset	Ms. Smith
Home agent's configuration	3	Direct asset	Mr. Jones
Consumers' security and privacy	2	Direct asset	Ms. Jackson
Public's trust in smart home	1	Indirect asset	–

in Table 7.5. In the relative importance ranking, 1 stands for very important and 5 stands for minor importance. The most important asset is the Consumers' Energy Consumption Data. The damage that a leakage of this information can cause to the consumer is significant, in particular because these reveal behavior of persons in their homes. The smart appliances configurations ranks also one, because a not correct configuration of an oven or a stove could cause a fire in the smart home. Less significant damages would be fridges that turn off, which leads to spoilage of food. When the configuration of the home agent is incorrect, this can cause financial loss. The reason is that energy is bought at a too high price. The asset table has an extra column for stating the asset owner. We refer to the meaning of owner as defined in ISO 27001 Section 4.2.1d (Sect. 12.4), meaning persons employed by the energy supplier that have management responsibly for considering the security concerns of a particular asset. Indirect assets do not have owners, because these cannot be protected by direct security measures. The protection of these assets is assured via protecting the direct assets that have a *harm to* relation to that indirect asset.

We list existing controls for assets in Table 7.6. These refer to the controls implemented by the energy supplier. The supplier recommends also to choose security controls based on security standards. ISMS-CORAS supports this demand by using the ISO 27001 standard. In Step 5 of ISMS-CORAS the existing controls are mapped to ISO 27001 Appendix A controls.

Table 7.6 Existing controls table

Asset	Existing control
Consumers' energy consumption data	Secure communications between the Smart Meter and the energy supplier's systems: encrypted data communication and encryption of all data on removable devices like SD-cards. In addition, the data integrity is checked using certificates and hash values
Smart appliances' configuration	None
Home agent's configuration	Access control: The prices and tariffs are stored in the Smart Meter and can only be read by the consumer. Only the energy supplier is allowed to update prices and tariffs
Consumers' security and privacy	All of the control listed above
Public's trust in smart home	–

Sub-step 1.3—Attacker Identification and Documentation
This sub-step of the context establishment is concerned with narrowing down the number of possible attacker types for our target description (see Fig. 7.2). We reason about the unwanted incidents attacker types can cause, to what assets, and we also document assumptions that rule out specific threats.[3] Moreover, we consider attacker motivations to sharpen the description of the proposed attackers. This task is based on three kinds of artifacts. We contribute attacker templates that present a structured way to describe attacker types, motivations, assumptions, and resulting threats via instantiation. Attacker overview diagrams show vital information of an instantiated attacker template.

Our template (see Table 7.7) consists of three parts: a *basic attacker description*, a *refined attacker description*, and a *results* part.

Under the *basic attacker description*, the *attacker type* specifies the kind of attacker. A specific attacker can combine several types, so it is possible to tick several boxes. The classification of attackers into these types is based on our previous work shown in Chap. 8 and in Beckers et al. (2013a, b). For a specific analysis *threatened assets* are instantiated with the assets documented in the asset diagram, and the assets that may be threatened by the attacker type in question are marked. This work classifies *Attackers* into the following categories. *Physical Attackers* threaten the physical elements of the system, e.g., hardware or buildings that host computers. *Network Attackers* threaten *network connections* within the target of analysis. *Software Attackers* threaten software components of the system, e.g., the *smart meter*. *Social Engineering Attackers* threaten humans, e.g., consumers. We reason about these types of attackers and choose if they are relevant for our target of analysis, given its scope, and assets. A reason for the exclusion of an attacker is that the attacker has nothing to threaten, e.g., if we analyze an autonomous system that has no humans in its scope, *Social Engineering Attackers* do not need to be considered during the remaining analysis. The reasons for exclusion of an attacker from the scope of the analysis have to be documented. The specification of the *threatened security goals* is demanded by the ISO 27001 standard. The *entry points* are specified by marking the relevant references to the target description elements. In an instantiated attack template the elements are extracted from the target models. The entry points and the specification of the *attack paths* are based on Microsoft Threat Modeling (Swiderski and Snyder 2004). An attack path is a description of an attack from an entry point to an asset, within the scope of the analysis. The *attacker motivation* is specified based on criteria from a SANS institute white paper (Allen 2006).

We propose this process of elimination of attacker types to allow the security and risk experts to focus on relevant threats of the target analysis, rather than considering every attacker type in the entire method. Each of the attacker types introduced previously (physical attacker, network attacker, software attacker, and social engineering

---

[3]Note that in the CORAS terminology threats are attackers, persons, or other elements that cause unwanted incidents. This is different from other terminologies in which threats are actual exploits of vulnerabilities. In this we mean that by the word threatened that an attacker causes an unwanted incident.

Table 7.7 Attacker template

Basic attacker description	
Attacker type	☐Physical Attacker ☐Network Attacker ☐Software Attacker ☐Social Engineering Attacker
Threatened assets	☐Asset 1
	☐Asset 2
	☐…
Threatened security goals	☐Availability
	☐Confidentiality
	☐Integrity
	Reasoning
	• Explain why the selected security goals of an asset are threatened
	• Reason also why the remaining security goals are excluded
Entry points	☐ Target Description Element 1 ☐Target Description Element 2 ☐…
	Reasoning
	• State why the selected elements are entry points for this attacker
	• Reason why the remaining entry points are not relevant
Attack paths (possible vulnerabilities)	Describe all attack paths from the entry points to the assets
Assumptions of the Target description	☐Target Description Element 1 ☐Target Description Element 2 ☐…
	Describe all assumptions about the target description
Refined attacker description	
Required attack skills	State which kind of skills the attacker needs to succeed
Attacker motivation	☐financial gain ☐self-interest ☐revenge ☐external pressure ☐curiosity
	Reasoning
	• Describe why the selected attacker motivations are relevant
	• Explain also all exclusions of attacker motivations
Required resources	Describe the resources required for the attacker to conduct the attack
Assumptions about the attacker	What are the assumptions about the motivation, skills, and resources of the attacker
Insider/Outsider	Describe the difference if persons that are inside the scope and persons that are outside are the attacker
Results	
Threats	Describe the high-level threats the attacker presents
Reasons for scope exclusion	Describe the reasons for excluding the attacker or variants of the attacker from the scope of the threat analysis

attacker) should be considered in at least one instantiated attacker template and it should be reasoned in the template to what extent the attacker should be considered in the scenario. This shall achieve a sense of completeness of the threat analysis, because even if an attacker is considered not relevant for a specific scenario it is documented in the template why this reasoning was done. The documented attacker reasoning can be used during audits or reviews of the analysis.

The usage of the template requires a statement, which assets are threatened by the attacker documented in the template. Afterwards, a task is to state which of the security goals confidentiality, integrity, and availability is threatened and the task also requires a reasoning of *why* assets and security goals are selected or ruled out. The reasoning should be based on the attacker type, e.g., a network that is limited to the physical boundaries of a building cannot be threatened by a network attacker outside the building. Another example is that a physical attacker can threaten the availability of a digital file, but the attacker cannot threaten its integrity if the file exists in digital form and not in physical form.

We based the fields *entry points* and *attack paths* on the work done at Microsoft (Swiderski and Snyder 2004). Their methodology focuses on analyzing all interfaces of the target description elements with the outside world, so-called *entry points* and afterwards analyzing how an attacker can reach a particular asset from this entry point. A sequence of actions of an attacker leading him/her to the asset is a so-called *attack path*. An attack path without mitigating controls is a vulnerability. Our attacker template has to be instantiated with the elements of the target description for each analysis. Afterwards, a task is to reason about why an attacker can use an entry point or not. The subsequent task involves a description of the resulting attack paths. The last task for instantiating the attacker template is to identify assumptions about elements of the target description that reduce the number of entry points or attack paths. For example, if a network connection is embedded into a layer of concrete, an assumption is that a physical attacker cannot reach this connection, due to the significant effort required for penetrating the concrete.

The *refined attacker description* requires a description of the skills an attacker requires to succeed in harming the assets. For example, a network attacker might require skills to tamper with the network addresses of messages send over a network. The field *attacker motivation* is based on a study from the SANS Institute mentioned previously that revealed four fundamental motivations of social engineering attackers: *Financial gain, self-interest, revenge*, and *external pressure*. We believe these motivations are generic enough to serve as basic types of attackers in the information system domain. We also added the motivation curiosity, which we identified in discussions with the industrial partners of the NESSoS project. Curiosity motivates an attacker wanting to know the content of a certain data set or simply to find out if his/her skills are sufficient to penetrate a system. A subsequent task is to reason about why an attacker can have a certain motivation. This reasoning shall be done in relation to the attacker type and the threatened assets. For example, the motivation *financial gain* does not make sense in regard to a physical attacker who only threatens the availability of information assets by destroying them.

The instantiation of the template also involves the elicitation of *assumptions about the attacker*. For example, a consumer can state the attackers motivated by curiosity do not concern him/her, because this motivation may not lead to intentional damage of assets. The *insider/outsider* field shall invoke a reasoning of attackers that are part of the target description (insiders) and those that are not (outsiders). The *results* part of the template sums up the information collected about an attacker. A task is to define the *threats* an attacker causes and also the *Reasons for Scope Exclusions* of attackers.

Filling in the attacker template using as input the target description and the identified assets serves as a means to ensure completeness of the security risk analysis with respect to the target of analysis and the ISO 27001 demands. It furthermore serves as a basis for the subsequent and more detailed security risk identification, and it documents what may be excluded from the analysis and why. We propose attacker overview diagrams (see Figs. 7.4, 7.5, 7.6 and 7.7), which have two parts. The top part of the diagram shows the assets that the attacker threatens. The lower part shows the elements of the target description that an attacker can use to enter the system. The diagram also shows what assets are not threatened by the attacker and which elements of the target description are not entry points for the attacker. These assets and target description elements are positioned outside the frame of the attacker overview diagrams. Attacker overview diagrams always refer to a specific instantiation of an attacker template and represent some vital information of the template for security reasoning. We show instantiated attacker templates and attacker overview diagrams for physical attackers (see Table 7.8 and Fig. 7.4), for network attackers (see Table 7.9 and Fig. 7.5), for software attackers (see Table 7.10 and Fig. 7.6), and for social engineering attackers (see Table 7.11 and Fig. 7.7).

In Table 7.12 we present an extended CORAS high-level risk table. The purpose of this table is to get an overview of all identified unwanted incidents of all attackers documented in attacker templates and attacker overview diagrams. This overview servers as a means to identify unwanted incidents that are missing. This table is part of the original CORAS method as well. We add in ISMS-CORAS the instantiated attack templates and template overview diagrams as a mandatory input for the high-level risk table. The column *what makes it possible* refers only to security incidents, because security is the focus of this work. Please note that CORAS concerns also other unwanted incidents, e.g., in relation to safety, which normally appear also in this column. We focus on security issues in this work, because this is the primary concern of the ISO 27001 standard and leave out other unwanted incidents.

Validation Conditions—We propose to check the instantiations of the attacker template via several validation conditions. We have identified the following conditions so far:

- Check if all assets marked are threatened at least once by the attacker.
- Check if the security goals marked are used at least once in relation to an asset.
- Check if each template contains a reason why a motivation is selected.
- Are all not marked security objectives, motivations, attacker types, etc., explained?

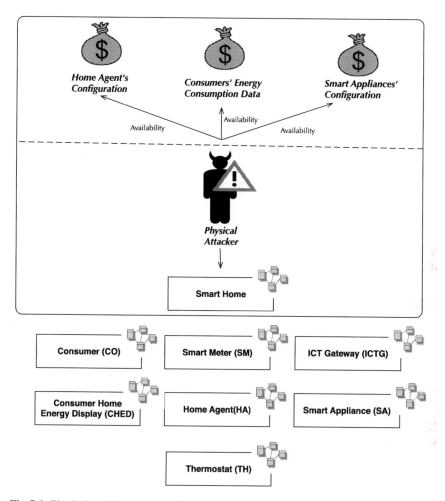

Fig. 7.4 Physical attacker overview diagram

- Are the resources and the skills of the attacker correct? Is something missing to achieve the unwanted incidents?
- Are all target description elements in the templates?
- Are all assets in the template?
- Does the high-level risk table refer to all threats stated in the instantiated templates?

Conduct Customer Verification Review—The instantiated attacker templates and attacker overview diagrams have to be verified by the customer, which is the energy supplier in our example. A meeting with the customer should check for completeness of considered entry points and attack paths. Workshops should be conducted that verify the correctness and completeness of the instantiated attacker templates, attacker overview diagrams, and the extended high-level risk table. The

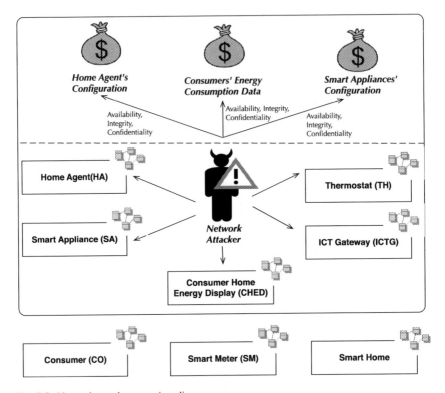

Fig. 7.5 Network attacker overview diagram

domain knowledge of the customer can help to identify gaps in the analysis conducted until this point. We also exclude attackers that collect energy consumption information to sell these. The assumption is that illegal obtained energy consumption information has little monetary value at this point in time.

Sub-step 1.4—Identification of Risk Criteria
We define scales of likelihoods, consequences, and risk evaluation criteria in this sub-step. Risk assessment can be conducted either quantitatively or qualitatively. Quantitative risk assessment demands that the likelihood and consequences scales contain numeric values. These have to express in which time frame a risk is likely and what the consequences are, e.g., the number of affected persons. The system in our example has not fully being build and deployed in a large scale yet. Meaning that the functionality is only partially available at this point and only implemented in small testbeds. This is the reason why we express the likelihood and consequences tables using qualitative scales that do not contain numbers. These are a starting point for risk assessment, and should the numeric values become available, a quantitative risk assessment should be done. We illustrate our likelihood scale in Table 7.13 and the consequences in Table 7.14.

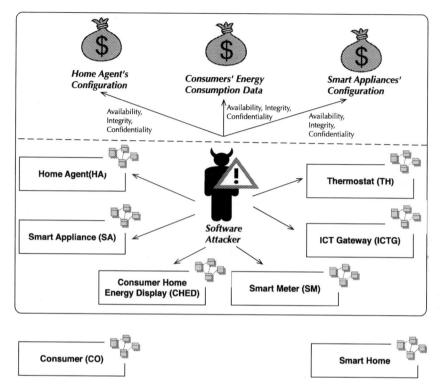

Fig. 7.6 Software attacker overview diagram

We use these scales for all assets, and we use the risk evaluation criteria specified by the matrix in Table 7.15. The matrix shows the acceptable combinations of like-lihoods and consequences in light shading, and unacceptable combinations in dark shading.

Sub-step 1.5—Considering Legal Aspects
The smart home scenario involves certain legal issues. We consider the German law, because our consumer and energy supplier are located in Germany. Hence, this law is applicable and we provide some insight on the EU norms regarding smart grid.

Germany's Energy Industry Act (EnWG) Section 21b Paragraph 3a states that all new buildings, as well as newly renovated ones, have to use smart meters. In addition, the network operators have to provide a smart meter to all consumers that request one. The network operators can provide these themselves or hire a third party to do so. Moreover, Section 40 of the EnWG states that energy suppliers have to offer energy tariffs that provide incentives for guiding or reducing energy consumption. These are so-called *load variable and daytime dependent tariffs*.

The principles of data avoidance and data minimisation in German Federal Data Protection Act (BDSG) Section 3a prevent the collection of energy consumption

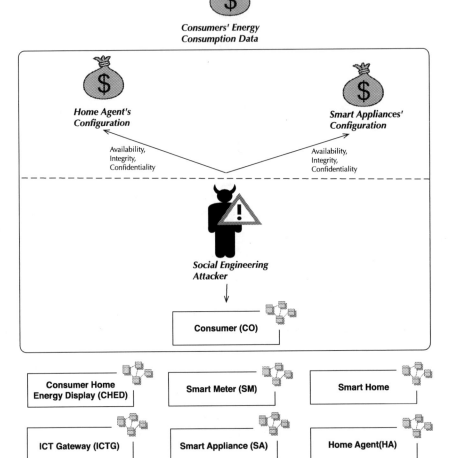

Fig. 7.7 Social engineering attacker overview diagram

Table 7.8 Attacker template instantiated for a physical attacker

Basic attacker description	
Attacker type	⊠ Physical attacker ☐Network attacker ☐Software attacker ☐Social Engineering Attacker
Threatened assets	⊠ Home agent's configuration
	⊠ Consumers' energy consumption data
	⊠ Smart appliances' configuration
Threatened security goals	⊠ Availability
	☐Confidentiality
	☐Integrity
	Reasoning
	• The physical attacker threatens only the availability of the assets, because all of them are in electronic form
	• The assets would need to have physical form for the attacker to read (threats to confidentiality) or change their content (threats to integrity)
	• The availability is threatened, because the attacker can destroy the SM, HA, and SA
Entry points	☐SM ☐HA ☐SA ☐TH ☐ICTG ☐CHED ☐CO ⊠ Smart Home
	Reasoning
	The physical attacker has to enter the smart home in order to threaten the availability
Attack paths (possible vulnerabilities)	The physical attacker enters the smart home in order to destroy the SM, HA, and SA
Assumptions of the target description	⊠ SM ☐HA ☐SA ☐TH ☐ICTG ☐CHED ☐CO ⊠ Smart Home
	The smart home is protected with at least two locks on the front door, and when the consumers are not home all windows and doors are locked. In addition, it is envisioned that the smart home has an alarm system connected to the SM that reports unauthorized entries to the police. Moreover, the cables connecting the smart home are under ground and advanced attackers that are willing to dig up the ground are considered unrealistic
Refined attacker description	
Required attack skills	Basic burglary skills
Attacker motivation	☐financial gain ☐self-interest ⊠ revenge ⊠ external pressure ⊠ curiosity
	Reasoning
	• The physical attacker cannot breach the confidentiality or integrity of the assets. This makes financial gain, self-interest, and curiosity unlikely motivations, because the attacker cannot sell information or change it in order to have a benefit, e.g., less payment for electricity
	• The physical attackers are motivated by revenge or external pressure, which could motivate him/her to threaten the availability of the assets

(continued)

Table 7.8 (continued)

Required resources	Basic burglary tools
Assumptions about the attacker	We do not assume that an attacker is motivated by self-interest, or external pressure to enter the home specifically for harming the smart home assets. We also assume that armed physical attackers that could force their way into the home at gunpoint do not specifically target the smart home assets. The reason is that these attackers would not risk getting caught just to harm the smart home assets. These would likely conduct kidnapping or burglary crimes
Insider/Outsider	If the attacker is an insider, he/she does not have to breach the perimeter of the house. However, an attack would raise immediate suspicion if there are no signs of an external attacker
Results	
Threats	A physical attacker can break into the smart home and destroy elements of the target description
Reasons for scope exclusion	A highly skilled physical attacker might be able to manipulate the sensors of the SM to the extent that these measure wrong values for the energy consumption. However, the skills of such an attacker make it unlikely to happen in a normal usage scenario. Hence, we exclude this attacker from the scenario. We also exclude attackers that cause integrity problems of hard drives by physically attacking them via, e.g., bit flipping. Attackers with these skills are very unlikely for this scenario

data without a valid reason.[4] The energy provider has such reasons, which are billing purposes. The consumer and the energy provider in our example have a tariff that requires the meter reading every day.

The EU and Germany envision that meter readings occur in very small time intervals. The readings are transmitted to the energy provider. The smallest transmission interval is assumed to be 15 min (Knyrim and Trieb 2011; Raabe et al. 2011; Karg 2009). These transmissions sum up to 35 040 transmissions of energy values a year per household. In our example, we assume transmissions of meter readings once a day, which sums up to 365 data transmissions a year. Before the usage of smart meters the energy data was transmitted only once a year. In addition, the transmission contained only one value, which was the sum of the overall energy consumption at the time of the meter reading. The usage of smart meters allows to measure energy consumption every second and store it in a separate value. Thus, the transmission of energy data every 15 min can contain up to 900 different values. To sum up, the intervals in between energy data transmissions and the values this data contains increases significantly with the introduction of smart meters from previously one transmission with exactly one value once a year to hundreds or thousands of transmissions a year with at least that many values (if not more) in each transmission. This means 35 040

---

[4]The BDSG refers to personal information and according to Knyrim and Trieb (2011), Raabe et al. (2011), Karg (2009) energy consumption data is personal information.

Table 7.9   Attacker template instantiated for a network attacker

Basic Attacker Description	
Attacker type	☐Physical Attacker ☒ Network Attacker ☐Software Attacker ☐Social Engineering Attacker
Threatened assets	☒ Home agent's configuration
	☒ Consumers' Energy Consumption Data
	☒ Smart Appliances' Configuration
Threatened security goals	☒ Availability
	☒ Confidentiality
	☒ Integrity
	Reasoning
	• The network attacker can use DoS attacks to reduce the availability of all network assets, e.g., send more message to the HA than it can process
	• The network attacker can also read the network messages regarding energy consumption (breach of confidentiality) and even change them (threat to integrity)
	• The same can be said for the other assets if they are transferred over the network
Entry points	☐SM ☒ HA ☒ SA ☒ TH ☒ ICTG ☒ CHED ☐CO ☐Smart Home
	Reasoning
	• The attacker can pretend to be a HA, TH, ICTG, or CHED in the network
	• The SM cannot be impersonated, because of strong authentication mechanism
	• The CO cannot be spoofed in the network, because it is a person. A similar argument goes for the smart home
Attack paths (possible vulnerabilities)	The network attacker can enter the network and pretend to be one of the network devices listed in the entry points and afterwards open a communication with any other device in the smart home
Assumptions of the target description	☒ SM ☐HA ☐SA ☐TH ☒ ICTG ☒ CHED ☐CO ☐Smart Home
	We assume the consumer has a firewall installed that protects the ICTG from uninitiated connections with the internet.
	In addition, the sensor for measuring energy consumption is part of the SM. Hence, information for measuring energy consumption is not transmitted over the internal network of the smart home. Only the already measured energy consumption is transmitted between the SM and the CHED via ICTG

(continued)

Table 7.9 (continued)

Refined attacker description	
Required attack skills	Network hacking skills
Attacker motivation	☐financial gain ☐self-interest ☒revenge ☒external pressure ☒curiosity
	Reasoning
	• The attacker can conduct a revenge crime by making the energy consumption profile public or change the configuration of an SA that is a heater
	• The attacker could turn the heater off in winter as a result of external pressure
	• The attacker could also be curious which SA he/she can remote control, e.g., increase the temperature of a fridge to its maximum and freeze all the food
Required resources	The attacker requires a computer with network access and probably a number of network attack tools
Assumptions about the attacker	We assume that the attacker in this scenario has just a basic skill set and that a correctly configured firewall will protect the system from an outside attacker. The inside attacker presents the important threat in this scenario
Insider/Outsider	Outsiders have to pass the firewall of the consumer first. Thus, these have to have a greater skill set than internal attackers
Results	
Threats	Spoofing and Tampering with the network messages. This can occur when the assets are transmitted over or via the network
Reasons for scope exclusion	The network attacker from the outside with mediocre skill is excluded from the threat analysis

measuring values a year instead of one as is the case today, in our example these are still 356 more than usual. In addition, these values do not detail the degree of measurements that are transmitted, e.g., if the smart meter measured the energy values every minute.

Metering data are personal data according to BDSG Section 1 Paragraph 2 and Section 3 Paragraph 1, because these provide information about the personal and factual living conditions of the consumers (Raabe et al. 2011; Karg 2009). Moreover, the BDSG Section 4 Paragraph 2 states that personal information has to be collected with the *involvement of the concerned person*, which is the consumer in our example. Hence, the consumer has to provide the energy consumption data. Mechanical meters have to be read by a person, who is often the consumer or a technician of the energy supplier. Either the consumer reads the value and transmits it to the energy supplier and, thus, the involvement of the concerned person (the consumer) can be assumed. Or the concerned person grants the technician access to the meter, which also implies the involvement of the concerned person. In both cases BDSG Section 4 Paragraph 2 is satisfied. However, the situation changes if a smart meter transmits meter readings

Table 7.10 Attacker template instantiated for a software attackers

Basic attacker description	
Attacker type	☐Physical Attacker ☐Network Attacker ☒ Software Attacker ☐Social Engineering Attacker
Threatened assets	☒ Home agent's configuration
	☒ Consumers' energy consumption data
	☒ Smart appliances' configuration
Threatened security goals	☒ Availability
	☒ Confidentiality
	☒ Integrity
	Reasoning
	• A software attacker could bypass the access control mechanism of the SM threaten the asset Consumers' Energy Consumption Data
	• The attacker could delete the software that runs on the meter (threats to availability), change the meter readings (threats to integrity), or read the meter reading (threats to confidentiality)
	• The attacker present similar threats to the other assets on the SA and HA
Entry points	☒ SM ☒ HA ☒ SA ☒ TH ☒ ICTG ☒ CHED ☐CO ☐Smart Home
	Reasoning
	• A software attacker could exploit vulnerabilities in the SM software, e.g., buffer overflow attacks to gain control of the SM
	• The HA, SA, TH, ICTG use web interfaces for their configuration
	• The software attacker could use, e.g., cross side scripting attacks to reduce the availability of these software systems
	• The software attacker could also use, e.g., SQL injections to read and change the assets Home Agent's Configuration, and Smart Appliances' Configuration
Attack paths (possible vulnerabilities)	The attacker could compromise ICTG, TH first and afterwards use their connection to the HA or SA to read (threats to confidentiality) or change (threats to integrity) their configuration. Afterwards it could use the connection from HA or ICTG to attack the SM
Assumptions of the target description	☒ SM ☒ HA ☐SA ☐TH ☒ ICTG ☐CHED ☒ CO ☐Smart Home
	We assume the SM does not have a direct communication/web interface for the CO. The communication is guided via the HA or ICTG, so the attacker has to consider their controls

(continued)

Table 7.10 (continued)

Refined attacker description	
Required attack skills	Basic/advanced knowledge of software attacks
Attacker motivation	☒ financial gain ☒ self-interest ☒ revenge ☒ external pressure ☒ curiosity
	Reasoning
	• The change in the energy consumption data could lead to financial gain for the attacker
	• Self-interest could also protect other COs from spending money
	• Revenge could be the intend to harm the CO via increasing the values for energy consumption in the smart grid. Revenge can also be the motivation to configure SAs to conduct attacks like creating a fire using a misconfiguration of the oven
	• All of these threads could be motivated by external pressure as well
	• The reading and changing of energy consumption data could be motivated by curiosity, as well as the reconfiguration of SA and HA
Required resources	Computer with an interface to the SA, HA, ICTG, TH, and probably software hacking tools
Assumptions about the Attacker	We assume that SA, HA, ICTG, TH have basic protection against attacks and that a certain skill level is required to attack them
Insider/outsider	Financial gain is unlikely for an outsider, because she/he does not participate in the energy consumption of the smart home. An exception of this could be that the attacker is collaborating with a physical attacker and the electricity line is compromised and energy is redirected to the outside attacker. The software attacker modifies the energy consumption in order for the CO not to recognize the attack. However, the physical connections that transport the electricity are unlikely to remain undetected in the smart home. Hence, we exclude this attack
Results	
Threats	Software attacker can exploit the software of the SA, HA, ICTG, TH, and manipulate their configuration and energy consumption. This could also lead to the use of SAs for attacks, reduce the availability of SM and cause the HA to negotiate a tariff that causes financial harm to the CO
Reasons for scope exclusion	Outside attacker with a financial gain motivation or self-interest

automatically to the energy supplier and the consumer is not aware of it. In this case, the involvement of the concerned person (the consumer) cannot be assumed. This is a compliance problem with BDSG Section 4 Paragraph 2, but legal experts (see Raabe et al. 2011; Karg 2009) argue that a large number of data transmissions causes an unacceptable effort for the concerned person. In our example this means 365 data transmissions a year (and up to 35 040 transmissions in other scenarios). Hence, the legal experts argue that if the concerned person (the consumer) provides an *informed*

Table 7.11 Attacker template instantiated for a social engineering attacker

Basic attacker description	
Attacker type	☐Physical attacker ☐Network attacker ☐Software attacker ☒ Social engineering attacker
Threatened assets	☒ Home agent's configuration ☐Consumers' energy consumption data ☒ Smart appliances' configuration
Threatened security goals	☒ Availability ☒ Confidentiality ☒ Integrity
	Reasoning
	• The social engineering attacker can manipulate humans in deleting (threats to availability), modifying (threats to integrity), or telling (threat to confidentiality) the configuration of SA and HA. However, the energy consumption data is stored in the SM and the CO does not have the access rights to change these
	• They could tell the overall consumption, but since the CO does not tell all the details of the measurements, we consider the energy consumption secure from confidentiality, as well as availability threats
Entry points	☐ SM ☐HA ☐ SA ☐TH ☐ICTG ☐CHED ☒ CO ☐Smart home
	Reasoning
	• The only human the social engineering attacker can manipulate in this scenario is the CO
Attack paths (possible vulnerabilities)	The social engineering attacker contacts the CO and manipulates the person into deleting (threats to availability), modifying (threats to integrity), or telling (threat to confidentiality) the content of all the information assets
Assumptions of the target description	☐SM ☐HA ☐SA ☐TH ☐ICTG ☐CHED ☒ CO ☐Smart home
	We assume the CO has not been trained to detect social engineering attackers. In addition, the CO has no strong mechanisms implemented to authenticate persons contacting him/her
Refined attacker description	
Required skills	The attacker needs to be able to communicate with the CO and be able to pretend to be a person the CO trusts
Attacker motivation	☐financial gain ☐self-interest ☒ revenge ☒ external pressure ☒ curiosity
	Reasoning
	• The social engineering attacker is not motivated by financial gain, because the CO is not able to change the metering data in the SM
	• The social engineering attacker is not motivated by self-interest, because the manipulation of the configuration of SA and HA can only support other parties in the smart home marginally
	• The social engineering attacker can be motivated by revenge and cause financial harm via misconfiguration by the HA to the CO or physical harm by misconfiguration of the SA ,e.g., a fire caused by an overheating stove
	• These acts can be motivated by external pressure as well
	• The social engineering can also be curious how the SAs and the HA are configured

(continued)

Table 7.11 (continued)

Refined attacker description	
Required resources	The social engineering attacker requires skills to pretend to be another person that the CO trusts. The attacker requires impersonation skills, e.g., the attacker can impersonate the technical support of the energy supplier
Attacker assumptions	We assume the CO only responds to reasonable requests of social engineering attackers. An example for an unreasonable effort would be to physical destroy the smart meter or an SA
Insider/outsider	Insiders are familiar with the behavior of the persons living in the smart home and might be able to conduct the attack with less effort
Results	
Threats	The social engineering attacker can impersonate persons the CO trusts, e.g., support employees of the energy supplier. This way the attacker can manipulate the CO to misconfigure SA or HA
Exclusion	We exclude social engineering attackers motivated by financial gain. A social engineering attacker could pressure the consumer to switch to an energy provider that pays the social engineering attacker. At this point in time we assume that the energy providers do not use these kind of attacks toward the consumers

consent as described in BDSG Section 4 Paragraph 1, the energy provider is allowed to initiate the transmissions of the energy consumption data in compliance with the BDSG (Raabe et al. 2011; Karg 2009).

Moreover, the collection of energy consumption data requires a legal contract between the energy supplier and the consumer to be in compliance with BDSG Section 28 Paragraph 1 Nr. 1. The contract has to specify in which intervals data is collected, the type of data collected, the time frames when the data is collected, and how the stored data is documented. The data also can only be used for the purpose it is collected (Karg 2009).

The collection, processing, or distribution of energy consumption data without a tariff as defined in EnWG Section 40 requires an informed consent in compliance with BDSG Section 4 Paragraph 1 (see above). If energy data is collected from an energy provider (or any other stakeholder) without an informed consent, even though the technical means in smart meters exist, this is a misdemeanor according to BDSG Section 43 Paragraph 2 Nr. 1. This misdemeanor can result in a maximum fee of 300.000 € in Germany (see Karg 2009). In addition, BDSG Section 7 provides the basis for the consumer to claim compensation for damages and defects (Karg 2009).

In our example, a default configuration of a smart meter that collects energy data in 15 min intervals would be a violation of the BDSG, because the energy provider has only the informed consent of the consumer to collect the energy consumption once every day, and the transmitted value is supposed to be just the sum of the energy consumption of the consumer of each day. Every other data collected is a violation of the BDSG and is punishable by the fines stated above.

Table 7.12 High-level risk table with security objectives

Who or what causes it?	How? What is the incident? What does it harm?	What makes it possible?	What are the security objectives?
Network attacker	System break-in and theft of energy consumption data	Insufficient protection of connection	Confidentiality of energy consumption data
Network attacker	System break-in and manipulation of smart appliances configuration data	Insufficient protection of connection	Availability of smart appliances, Confidentiality and Integrity of smart appliances' configuration data
Network attacker	System break-in and manipulation of the home agents configuration	Insufficient protection of connection	Availability of the home agent, confidentiality and integrity of the home agent configuration data
Software attacker	System break-in and manipulation of energy consumption data	Insufficient security	Confidentiality, integrity, and availability of energy consumption data
Software attacker	System break-in and theft of energy consumption data	Insufficient security	Integrity of energy consumption data
Software attacker	System break-in and deletion of the smart meter software	Insufficient security	Availability of the smart meter
Software attacker	System break-in and deletion of the smart appliances software	Immature technology	Availability of smart appliances
Software attacker	System break-in and deletion of the home agent software	Immature technology	Availability of the home agent
Software attacker	System break-in and misconfiguring of the home agent to increase the price of energy	Immature technology	Integrity of home agent's configuration
Software attacker	System break-in and configuring the CHED to display wrong energy consumption data	Insufficient security	Integrity of energy consumption data
Software attacker	System break-in and configuring smart meter to delete the metering data	Insufficient security	Integrity of energy consumption data

(continued)

Table 7.12 (continued)

Who or what causes it?	How? What is the incident? What does it harm?	What makes it possible?	What are the security objectives?
Software attacker	System break-in and configure smart appliance to raise a burglary alarm	Insufficient security	Integrity of smart appliances' configuration
Social engineering attacker	Manipulation of the consumer to provide access to smart appliances	Insufficient protection	Integrity of smart appliances' configuration
Social engineering attacker	Manipulation of the consumer to change the home agents configuration to buy only expensive energy	Insufficient protection	Integrity of home agent's configuration
Social engineering attacker	Manipulation of the consumer to remove the energy of the smart meter	Insufficient security	Availability of energy consumption data
Physical attacker	System break-in and destruction of the smart meter	Insufficient security	Availability of energy consumption data
Physical attacker	System break-in and destruction of the home agent	Insufficient security	Integrity of the home agent's configuration
…		…	…

Table 7.13 Qualitative likelihood scale for the smart home

Likelihood value	Description
Certain	A high number of similar incidents have been recorded; has been experienced a very high number of times by several consumers
Likely	A significant number of similar incidents have been recorded; has been experienced a significant number of times by several consumers
Possible	Several similar incidents on record; has been experienced more than once by the same consumer
Unlikely	Only very few similar events on record; has been experienced by few consumers
Rare	Never experienced by most consumers throughout the total lifetime of the Smart Home

Table 7.14 Qualitative consequence scale for the smart home

Consequence	Generic interpretation
Catastrophic	Recovery from failure is not possible; Has a significant potential to put the energy supplier out of business
Major	Failure to recover can potentially put the energy supplier out of business
Moderate	Several occurrences over time can potentially put the energy supplier out of business
Minor	Tolerable if easy to recover from and if very rare
Insignificant	Generally tolerable and easy to manage to recover from

Table 7.15 Risk evaluation matrix for the smart home scenario (Color table online)

		Consequence				
		Insignificant	Minor	Moderate	Major	Catastrophic
Frequency	Rare					
	Unlikely					
	Possible					
	Likely					
	Certain					

A further important task when considering legal aspects is the definition of scales of likelihoods, consequences, and risk evaluation criteria with respect to the identified legal norms. The likelihoods and risk evaluation matrix are similar to the ones shown previously. However, we describe a different consequence scale for the legal aspects of the smart home scenario, which ranks from a minor breach that the supplier detects and fixes him-/herself to an ordered cease of personal information. We describe a legal consequence scale for the smart home scenario in Table 7.16. The scale is concerned

Table 7.16 Qualitative legal consequence scale for the smart home

Consequence	Generic interpretation
Catastrophic	Processing of personal data ordered to cease
Major	Civil law liability and fine; Criminal law liability and prison sentence
Moderate	Enforcement notice
Minor	Information notice
Insignificant	Minor breach of consumer's privacy discovered and corrected

with compliance with data protection laws and regulations, which in our scenario is particularly related to the Consumers' Security and Privacy. We introduce the asset of compliance with governmental laws and regulations, using the short name *Legal Compliance* for the asset for the remainder of the work. We do not introduce it as a separate asset in the asset diagram, because it refers also to the asset Consumers' Security and Privacy with the difference that these refer to a legal norm.

Given our target of analysis, legal compliance issues are mostly relevant to incidents related to consumers' security and privacy. Nevertheless, by introducing legal compliance as an asset of its own, we make the legal risks explicit in the analysis and in the documentation of the results. Note that for the consequence scale in Table 7.16, each consequence typically includes all lower consequences, as the legal consequences with respect to this asset usually escalate. We also use the risk matrix in Table 7.15 as the evaluation criteria for compliance.

Step 2: Identify Risk

The risk identification refines the attacker descriptions in the high-level risk table into threat diagrams. These show the detailed attack paths of attackers into the system, and how an unwanted incident may be caused. CORAS makes use of workshops,

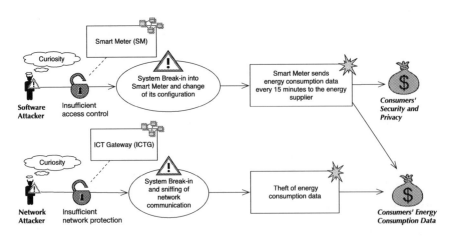

Fig. 7.8 Extended threat diagram for the software and network attackers

structured brainstorming, and other techniques to elicit unwanted incidents and describe the scenarios that may lead to them. We show a small example threat diagram in Fig. 7.8 to illustrate how ISMS-CORAS extends the CORAS threat diagram notation. One extension is the attacker motivation, depicted as clouds over the attacker symbol. We consider a software attacker and a network attacker in our example. Another extension is the relation from a vulnerability to the element of the target description that contains the vulnerability. In our example, a software attacker exploits the vulnerability "insufficient access control," which is contained in the smart meter. The threat scenario is that the attacker breaks into the smart meter and changes the configuration. The software attacker is not trying to enrich her-/himself, because the motivation is "curiosity." The attacker changes the smart meter configuration in such a way that the meter sends the energy consumption data to the energy supplier every 15 min.

The network attacker has the same motivation and exploits the "insufficient network protection" vulnerability, which is also contained in the smart meter. The unwanted incident is a theft of the energy consumption data. We show the extended threat diagram notation that uses Legal CORAS in Fig. 7.9. In our example scenario the German Federal Data Act (BDSG) applies, because the energy consumption data of the consumer is considered personal data, and because the sending of energy consumption data in shorter intervals than the tariff requires without the consumers informed consent is a violation of BDSG Section 4. The energy supplier is therefore subject to the risk of getting fined or sued, due to liability. This can cause the unwanted event of prosecution of the energy supplier for storing and processing of personal information without an informed consent. The law suit can result in a fine of 300.000 € , as explained previously (Karg 2009).

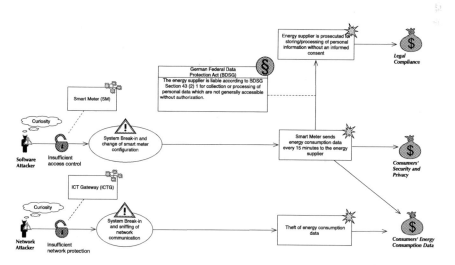

Fig. 7.9 Extended threat diagram for the software and network attackers including legal concerns

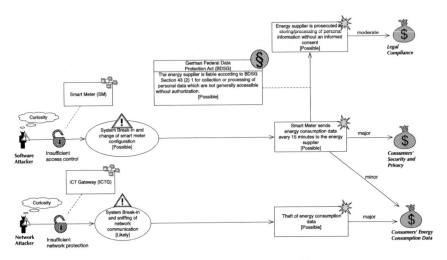

Fig. 7.10 Extended threat diagram for the software and network attackers including legal concerns and likelihoods

Step 3: Estimate Risk

Steps 3 and 4 remain almost unchanged from CORAS, and we therefore describe this part more briefly. In Step 3 the likelihoods and consequences are estimated and discussed with the customer, and the results are annotated in the threat diagram using the scales introduced previously. The annotated threat diagram is depicted in Fig. 7.10.

The system break-in of the software attacker has the likelihood "possible", due to the existing vulnerability "insufficient access control". The resulting unwanted incident is assigned a major consequence for the asset *Consumers' Security and Privacy*, because the energy provider gains more details about the consumers' energy consumption, and could also derive behavioral profiles of the consumers. This is moreover in violation of the BDSG. As depicted by the legal norm in Fig. 7.10, it is held as "possible" that that this norm applies to these circumstances. Consequently, it is held as "possible" that the incident of legal prosecution occurs. The legal consequence with respect to compliance is estimated to be "moderate".

Similar reasoning is conducted to estimate the risks with respect to the consumers' energy consumption data. The consequence of consumption data readings every 15 min is held as "minor" since misuse by the energy supplier is not assumed. The consequence of theft of energy consumption data is, however, held as "major", because the data could be used to analyze the behavioral profiles of the customer and used, for example, to commit burglaries when the consumer is not likely to be at home.

Step 4: Evaluate Risk

The values for consequences of unwanted incidents and likelihoods estimated in the previous step are combined into risks and drawn in risk diagrams (see Fig. 7.11). The

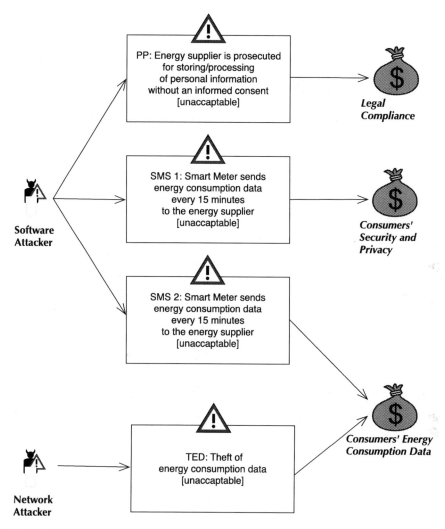

Fig. 7.11   Risk diagram

values for likelihoods and consequences are plotted into the risk matrix, depicted in Table 7.17. The matrix determines whether a risk is acceptable or should be considered for treatment, which is conducted in the next step.

As depicted in Fig. 7.11, the software attacker gives rise to three risks. The risk *PP* states that the energy supplier can be prosecuted for storing or processing of personal information without an informed consent, and is assessed as unacceptable (see Table 7.17). The situation is similar for the *SMS 1* risk, which consists of sending energy consumption data every 15 min to the energy supplier. The risk *SMS 2* on the other hand is acceptable due to the low consequence. Similarly, the *TED* risk is also categorized as unacceptable (see Table 7.17).

Table 7.17 Risk evaluation using the risk matrix (Color table online)

		Consequence				
		Insignificant	Minor	Moderate	Major	Catastrophic
Frequency	Rare					
	Unlikely					
	Possible		SMS2	PP	SMS1 TED	
	Likely					
	Certain					

Step 5: Treat Risk

The unacceptable risks have to be evaluated for possible treatment. Appendix A of the ISO 27001 standard describes the normative controls of the standard, and ISMS-CORAS requires these to be considered. We present a short overview of these controls in Table 7.18. The numbering of the controls starts with A.5 and ends with A.15 because this is the numbering used in the standard. The standard provides guidelines on how to implement these and other controls, but the implementation is not normative.

We support the selection of controls with a mapping of controls to attacker types in Table 7.19. The table lists the controls and the attacker types whose threats can be mitigated by these controls. In addition, we list the control objectives and the types of target description elements that can be protected by these controls. Considering the information in our enhanced threat diagrams in combination with this table, the control selection should become more time efficient. In addition, ruling out a particular control for a specific attacker based on this table narrows down the choices for relevant controls to treat the risk. For example, the software attacker in Fig. 7.10 has the motivation *curiosity*, exploits the vulnerability *insufficient access control* and the concerned target description element is the *smart meter*. Using Table 7.19 we realize that the control *A.11* concerns *access control* and the attacker type *software attacker*. Thus, this control is of relevance for mitigating the particular threat described in Fig. 7.10, because the threat is caused by the software attacker. We select relevant sub-controls of *A.11* as risk treatments, and proceed for the network attacker in a similar manner.

We illustrate the selection of controls using our extended CORAS treatment diagram notation. These diagrams are used for identifying and documenting risk treatments, and the novelty of ISMS-CORAS is that the risk treatment and the diagrams have to consider ISO 27001 controls. Additionally, the attacker types and related target elements are specified as for threat diagrams.

In the example depicted in Fig. 7.12, we have identified treatments for the risks PP and SMS, which are caused by a software attacker. We can reduce a risk with controls to reduce its likelihood or consequence (or both). First, we select a control to reduce the likelihood of both risks. The risks PP and SM S are caused by the vulnerability *insufficient access control*. We select the control *A 11.2.4 Review of*

Table 7.18 Controls of the ISO 27001 standard

Control name	Control objective	Important demands
A.5 Security policy	Provide directions for information security	Documentation and review requirements
A.6 Organization of information security	Manage security within the organization and with external parties	Clear management commitment, responsibilities, coordination, and independent consultation and review
A.7 Asset management	Achieve and ensure appropriate protection levels for assets	Identify assets, assign responsibilities for assets, classify assets, define and document rules for treatment of assets
A.8 Human resources security	Provide security training for employees, communicate responsibilities, provide structured exit procedures	Specify role and terms of employment, define responsibilities and provide security education and training, define disciplinary process, define termination responsibilities, return of assets and removal of rights
A.9 Physical and environmental security	Prevent unauthorised physical access, damage and interference to secure areas and equipment	Establish security perimeter, physical controls for access to secure rooms. Equipment shall be protected, e.g., from power failure and the support for the equipment shall be ensured, e.g., protect cable connection from interference
A.10 Communications and operations management	Ensure secure operations of information processing, especially for service delivery from third parties, ensure availability, integrity, and confidentiality of information processing	Guidelines for processes, e.g., segregation of duties, and specific demand that ensure the goals e.g. backup and monitoring of processes
A.11 Access control	Control the access to information	Ensure access control on information systems, networks, operating systems, etc.
A.12 Information systems acquisition, development and maintenance	Embed security in information systems and prevent misuse of information	Specific measure are demanded, e.g., security requirements analysis, input/output data validation, use of cryptography, prevent information leakage, etc.
A.13 Information security incident management	Identify security events and weaknesses associated with information security and provide timely corrective action, ensure a consistent and effective approach	Ensure a reporting for security events and security weaknesses, learn from information security incidents

(continued)

Table 7.18 (continued)

Control name	Control objective	Important demands
A.14 Business continuity management	Protect critical business processes from effects of information system failures and ensure their timely resumption	Include security and risk management in the business continuity management process, reassess and test the business continuity plans
A.15 Compliance	Ensure compliance with laws, regulations, contractual obligations, security requirements, organizational security policies, and standards, consider system audits	Identify relevant laws, regulations, contractual obligations, etc., and also data and privacy protection measures, check the compliance to these laws, regulations, contractual obligations, etc., and use also audits to check compliance

Table 7.19 Mapping controls of the ISO 27001 standard to our attacker types

Control name	Attacker types	Control objective	Relevant target elements
A.5 Security policy	All	Provide directions for information security	All
A.5.1.1 Information security policy document	All	Get approval by management, and publish and communicate to all relevant parties	All
A.5.1.2 Review of the information security policy	All	Review and improve the policy continuously	All
A.6 Organization of information security	All	Security management activity, e.g., clear management commitment	All
A.6.1 Internal organization	All	Manage information security within the organization	All
A.6.2 External parties	All	Maintain the security of the organizations information and information processing facilities	All
A.7 Asset management	All	Activities regarding identify, classify and protect assets	All
A.7.1 Responsibility for assets	All	Achieve and maintain appropriate protection of organizational assets	All

(continued)

Table 7.19 (continued)

Control name	Attacker types	Control objective	Relevant target elements
A.7.2 Information classification	All	Ensure that information receives an appropriate level of protection	All
A.8 Human resources security	Social engineering attacker	Activities regarding training, responsibility assignment, designing and implementing exit procedures, etc.	All that are humans
A.8.1 Prior to employment	Social engineering attacker	Ensure that employees, contractors and third party users understand their responsibilities	All that are humans
A.8.2 During employment	Social engineering attacker	Ensure that all employees, contractors and third party users are aware of information security threats and concerns, their responsibilities	All that are humans
A.8.3 Termination or change of employment	Social engineering attacker	Ensure that employees, contractors and third party users exit an organization or change employment	All that are humans
A.9 Physical and environmental security	Physical attacker	Activities regarding concerning physical access and prevention of damage/interference of hardware	All that are physical e.g. hardware
A.9.1 Secure areas	Physical attacker	Prevent unauthorised physical access, damage and interference	All that are physical e.g. hardware
A.9.2 Equipment security	Physical attacker	Prevent loss, damage, theft or compromise of assets	All that are physical e.g. hardware
A.10 Communications and operations management	All	Activities regarding guidelines for processes, e.g., segregation of duties	All
A.10.1 Operational procedures and responsibilities	All	Ensure the correct and secure operation of information processing facilities	All

(continued)

Table 7.19 (continued)

Control name	Attacker types	Control objective	Relevant target elements
A.10.2 Third party service delivery management	All	Implement and maintain the appropriate level of information security	All
A.10.3 System planning and acceptance	All	Minimize the risk of systems failure	All
A.10.4 Protection against malicious and mobile code	Software attacker	Protect the integrity of software and information	All that are software
A.10.5 Backup	All	Maintain the integrity and availability of information and information processing facilities	All that are information or software
A.10.6 Network security management	Network attacker	Ensure the protection of information in networks	All that are part of the network
A.10.7 Media handling	All	Prevent unauthorised disclosure, modification, removal or destruction of assets	All that are media
A.10.8 Exchange of information	All	Maintain the security of information and software exchanged within an organization	All
A.10.9 Electronic commerce services		Maintain the security of information and software exchanged within an organization	
A.10.10 Monitoring	Software attacker, network attacker	Ensure the security of electronic commerce services	All that are electronic commerce services
A.11 Access Control	Software attacker/network attacker/physical attacker	Activities regarding implement and monitor access to information	All that are software
A.11.1 Business requirement for access control	All	Control access to information	All
A.11.2 User access management	All	Ensure authorised user access	All
A.11.3 User responsibilities	All	Prevent unauthorised user access	All

(continued)

Table 7.19 (continued)

Control name	Attacker types	Control objective	Relevant target elements
A.11.4 Network access control	Network attacker, software attacker	Prevent unauthorised access to networked services	All that are part of the network or a networked service
A.11.5 Operating system access control	Software attacker, social engineering attacker	Prevent unauthorised access to operating systems	All operating systems or in relation to operating system security
A.11.6 Application and information access control	Software attackers	Prevent unauthorised access to information held in application systems	Applications or in relation to application security
A.11.7 Mobile computing and teleworking	Software attacker, network attacker	Ensure information security when using mobile computing	All that are mobile computing or teleworking
A.12 Information systems acquisition, development and maintenance	Software attacker/network attacker	Activities regarding eliciting of security requirements and vulnerability detection, e.g., penetration testing and specific measures, e.g., cryptography	All that are software or network components
A.12.1 Security requirements of information systems	All	Ensure that security is an integral part of information systems	All
A.12.2 Correct processing in applications	Software attacker	Prevent errors, loss, unauthorized modification	All applications
A.12.3 Cryptographic controls	Software and network attacker	Protect the confidentiality, authenticity or integrity of information	All applications and networks
A.12.4 Security of system files	Software attacker, social engineering attacker	Ensure the security of system files	All that are applications
A.12.5 Security in development and support processes	All	Maintain the security of application system software and information	All
A.12.6 Technical Vulnerability Management	Software attacker, network attacker	Reduce risks resulting from exploitation of published technical vulnerabilities	All that are technical

(continued)

Table 7.19 (continued)

Control name	Attacker types	Control objective	Relevant target elements
A.13 Information security incident management	All	Activities regarding reporting security events and issues, ensuring a consistent and effective response, learning from security incidents,	All
A.13.1 Reporting information security events and weaknesses	All	Ensure information security events and weaknesses associated with information systems are communicated	All
A.13.2 Management of information security incidents and improvements	All	Ensure a consistent and effective approach is applied to the management of information security incidents	All
A.14 Business continuity management	All	Activities regarding business continuity management for business processes, e.g., security and risk management	All
A.14.1 Information security aspects of business continuity management	All	Counteract interruptions to business activities	All
A.15 Compliance	All	Activities regarding identifying laws, regulations and contractual obligations. privacy protection, monitor compliance to the laws regulations and contractual obligations, compliance audits	All
A.15.1 Compliance with legal requirements	All	Avoid breaches of any law, statutory, regulatory or contractual obligations, and of any security requirements	All
A.15.2 Compliance with security policies and standards, and technical compliance	All	Ensure compliance of systems with organizational security policies and standards	All

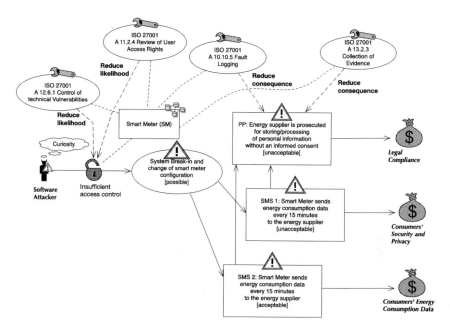

Fig. 7.12 Treatment diagram I

User Access Rights, because an analysis of this issue resulted in the conclusion that the smart meter did not restrict access to the configuration for any user. In addition, we select *A 12.6.1 Control of technical Vulnerabilities*, which reduces the likelihood of the existence of technical vulnerabilities that allow software attackers to change the access control rules established with the control *A 11.2.4*.

Second, we select controls to reduce the risk PP. This is of particular relevance, because in case the unwanted incident of sending energy consumption data in short intervals happens, a law suit is a possibility. The likelihood of this event cannot be influenced further with reasonable effort. Hence, it is important to reduce the possible consequence of such a lawsuit. The controls *A 10.10.5 Fault Logging* and *A 13.2.3 Collection of Evidence* are selected to reduce the consequences of the risk PP. PP is the result of the likelihoods and consequences shown in Fig. 7.10. Both controls aim to document the occurrences of the incident with the purpose of proving that the violation of the privacy of the consumer was not intended by the energy supplier. Hence, the legal sentence (the consequence) should be reduced. The risks SMS 1 and SMS 2 are only reduced via their likelihoods. A focused view on only the risks PP, SMS 1, SMS 2, and the selected controls is provided in a treatment overview diagram, as shown in Fig. 7.13. We illustrate the selected controls for the TED risk in Fig. 7.14, which concerns the theft of energy consumption data caused by a network attacker. In this diagram, one element of the target description has to be protected, namely the ICT gateway. The vulnerability that has to be addressed by the controls is insufficient network protection. We identified the controls *A 11.4.6 Network Control Connection*

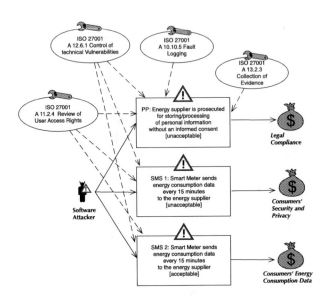

Fig. 7.13 Treatment overview diagram I

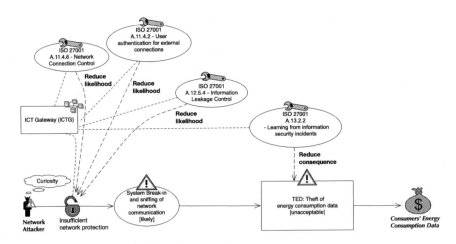

Fig. 7.14 Treatment diagram II

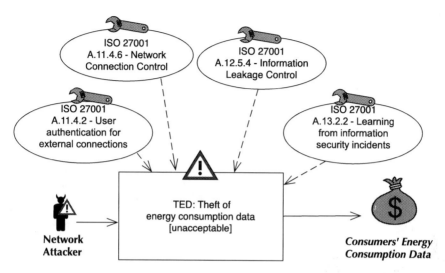

Fig. 7.15   Treatment overview diagram II

and *A 11.4.2 User Authentication for External Connections* as relevant, because the controls restrict the access to certain network devices and implement strong authentication mechanisms. The control restrictions to the network refer to the ICT Gateway, while the authentication mechanisms refer to the Smart Meter. Moreover, the control *A 12.5.4 Information Leakage Control* shall reduce the likelihood of loss of energy consumption data from the ICT Gateway. The consequences of the TED shall be reduced with the control *A 13.2.2 Learning from Information Security incidents*. The control shall show that the energy provider will investigate every incident and learn from his/her mistakes. This shall result in improved controls and prevent the same exploit to happen twice. A focused view on only the risk TED and the selected controls is provided in a treatment overview diagram, shown in Fig. 7.15. We present all selected treatments in Table 7.20, which is a so-called treatment overview table. The first column shows the assets that shall be protected by the control. The second column shows the asset owner, who is responsible for implementing the control. The column is filled with made up names for illustrative purposes. The following columns state the addressed security objective, the selected treatment or control and the reason for selecting the control.

We list all the controls that are not selected for an asset in the control exclusion table, which is depicted in Table 7.21. The table lists the asset in the first column, the control in question in the second, and the reason for not selecting the control in the last column. The control overview table and the control exclusion table form the statement of applicability for the ISO 27001 documentation. The statement of applicability provides a reasoning of the controls selected for the ISMS.

The control effectiveness measure table shown in Table 7.22 lists all controls and describes how to measure them. The ISMS Procedure and Control Table is depicted in Table 7.23. The table lists the procedures and the controls necessary to ensure the protection of each asset. The table states the asset in the first column, the treatment or control in the second table, the target description in the third, and a description of how the procedure or control is applied in the last column.

Table 7.20 Treatment overview table

Asset	Asset owner	Security objective	Treatment	Reasoning
Consumers' Security and Privacy	Mrs. Jackson	Confidentiality	A 12.6.1—Control of technical vulnerabilities	The novelty of the technology makes undetected vulnerabilities likely and effort should be spent in detecting those
Consumers' Security and Privacy	Mrs. Jackson	Confidentiality	A 10.10.5—Fault Logging	The logging of all events regarding the smart meter supports the analysis of what may lead to a data leakage of energy consumption data. The novelty of the technology and lack of experience with attackers in this domain makes it essential to be able to retrace steps of an attacker via logs and to detect the vulnerability that caused this problem
Consumers' Security and Privacy	Mrs. Jackson	Confidentiality	A 13.2.3— Collection of Evidence	An attacker may enter the smart meter without authorisation and change the configuration to send energy consumption data every 15 min. The energy supplier is violating legal norms like the German BDSG if this happens, because the data are send without an informed consent of the consumer. If a detailed log can prove that the energy supplier was not the one that initiated this configuration, the penalty after prosecution is likely to be lower

(continued)

Table 7.20 (continued)

Asset	Asset owner	Security objective	Treatment	Reasoning
Consumers' Energy Consumption Data	Mr. Jones	Confidentiality	A.12.5.4—Information Leakage Control	The new technology of the smart meters can lead to a loss of the energy consumption data. It should be checked what can be done to prevent information from leaking, either on the design side of the smart meters via separation of data in the devices or ensure that the data can only be read by the energy supplier, e.g., via encryption mechanism
Consumers' Energy Consumption Data	Mr. Jones	Confidentiality	A.11.4.6—Network Connection Control	Network connection has to be authenticated properly and it also has to be assured that transmissions cannot lead to a leakage of information. For example, even if the data is encrypted, a flaw in the protocol (e.g., that energy consumption data is only send to the energy supplier if energy is actually consumed) might cause information leakage. In this case, missing transmissions between the consumer and the energy supplier could indicate that the consumer is not at home (and not consuming energy), and in turn trigger burglary
Consumers' Energy Consumption Data	Mr. Jones	Confidentiality	A.11.4.2—User authentication for external connections	All external users that connect to the smart meter have to be authenticated in order to avoid the unauthorized change of its configuration
Consumers' Energy Consumption Data	Mr. Jones	Confidentiality	A.13.2.2—Learning from information security incidents	The new system depends upon a detailed recording of security events to facilitate later analyses. These can also lead to the discovery of new vulnerabilities
...	...	...	...	...

Table 7.21 Control exclusion table

Asset	Control	Reason for control exclusion
Home Agent's Configuration	A.11—Access Control	The distribution and organization of the home agents are not part of the scope of the ISMS. In addition, the consumer is acquiring and configuring the home agent, and the energy supplier has no influence on the configuration of these devices
Smart Appliances' Configuration	A.12—Information systems acquisition, development and maintenance	The consumer is acquiring and maintaining the smart appliances. Hence, the energy supplier has no influence on which smart appliances are part of the smart home. In addition, the energy supplier does not develop or maintain the appliances, and these activities therefore cannot be influenced
…	…	…

Table 7.22 Control effectiveness measure table

Treatment	Effectiveness measure
A 12.6.1—Control of technical vulnerabilities	Check if new vulnerabilities are found
Check if found vulnerabilities are fixed
A 13.1.2—Reporting security weakness	Check if weaknesses are reported
Conduct interviews and check if weaknesses are known that are not reported
A 10.10.5—Fault Logging	Check if a logging system is working properly via, e.g., functional testing
…	…

7.4 Related Work

To the best of our knowledge no specific methods for security requirements engineering or security risk analysis exist that support the establishment of an ISO/IEC 27001 compliant ISMS, and that satisfies the standard's documentation demands as is the goal of ISMS-CORAS.

Looking at established standards and methods for security risk analysis, several alternatives could be considered for facilitating the establishment of an ISMS, but none of them provide systematic support for ISO/IEC 27001 compliance. OCTAVE (Alberts and Dorofee 2001) is a suite of tools, techniques and methods for risk-based information security assessment and planning. Although the security risk analysis process is similar to ISMS-CORAS, the aim of OCTAVE is not to create and document an ISMS. The same is the case for CRAMM (Siemens 2003). Both CRAMM and OCTAVE are compliant with the BS 7799 information security standard, which was adopted by ISO/IEC 27001. However, the focus is still on the security risk analysis, and less on systematically fulfilling the standard's requirements to ISMS

Table 7.23 ISMS procedure and control table

Asset	Treatment	Target Element(s)	Procedure or Control
Consumers' Security and Privacy	A. 12.6.1—Control of technical vulnerabilities	Smart Meter (SM)	Technical vulnerabilities have to be identified, e.g., via penetration testing. These vulnerabilities have to be patched. We define a security black box penetration testing once a month for the first twelve month the device is operational and afterwards every two months
Consumers' Security and Privacy	A. 10.10.5—Fault Logging	Smart Meter (SM)	In case the smart meter has a malfunction or is attacked, it is important to document the event to prevent future occurrences. A check has to verify that the logging functionality is activated and that the information in the log is sufficient to trace unwanted events. These checks should be conducted once a month
Consumers' Security and Privacy	A. 13.2.3—Collection of Evidence	Smart Meter (SM)	The collection of evidence regarding unwanted events is important for a new technology like smart meters. This can happen via logging (also Control A. 10.10.5—Fault Logging) or via external observation of the smart using, e.g., network monitoring tools that check the traffic to and from the device. The chosen mechanisms should be checked once every two month to ensure their viability
Consumers' Energy Consumption Data	A. 12.5.4—Information Leakage Control	Smart Meter (SM)	The SM shall be configured in such a way that network connections from any device in the smart home have to be initiated from the CHED if the SM shall respond with energy consumption data. The settings have to be tested every 6 months

(continued)

Table 7.23 (continued)

Asset	Treatment	Target Element(s)	Procedure or Control
Consumers' Energy Consumption Data	A.11.4.6—Network Connection Control	ICT Gateway (ICTG)	The ICTG has to control that only devices from inside the smart home can connect to the SM. The configuration of the ICTG has to be checked every six months
Consumers' Energy Consumption Data	A.11.4.2—User authentication for external connections	Smart Meter (SM)	If external parties connect to the SM the connection has to be routed via the SSN. All users that access the SM have to be authenticated. This setting of the SM has to be checked every six months
Consumers' Energy Consumption Data	A.13.2.2—Learning from information security incidents	Smart Meter (SM)	When the logs or other sources report on security incidents involving the SM, these incidents shall be analyzed. Every six months this information of SM has to be checked and a meeting has to take place that documents these events and the actions that must be taken by the energy supplier to prevent further occurrences
...	...	...	...

establishment and documentation. The CRAMM repositories of assets, threats and countermeasures could, however, support the ISMS-CORAS process.

EBIOS (ANSSI 2010) is a method for assessing and treating risks related to information systems security, and is consistent with the ISO 31000, ISO/IEC 27001 and ISO/IEC 27005 standards. While consistent with these standards, the method is designed for security risk identification and mitigation and provides therefore only partial support for establishing an ISO/IEC 27001 ISMS. The Microsoft Security Risk Management Guide (Microsoft 2006) is developed to support organizations in the overall security management and risk assessment. The fulfillment of ISO/IEC 27001 is beyond the scope, although there are many overlaps. The same applies to FRAAP (Peltier 2010), which is a method for the analysis of information security related issues, focusing on protection of data confidentiality, integrity and availability.

Other existing works provide some guidance in interpreting the demands of the ISO/IEC 27001 standard. Calder (2009) and Kersten et al. (2011) provide advice for an ISO/IEC 27001 realization. In addition, Klipper (2010) focuses on risk management according to ISO 27005. The author also includes an overview of the ISO 27000 series of standards. However, none of these works consider using structured methods to fully support the standard, as is it the aim of ISMS-CORAS.

Other authors try to capture the most important relations presented in the standard by using models. Cheremushkin and Lyubimov (2010) present a UML-based meta-model for several terms of the ISO 27000. These meta-models can be instantiated and, thus, support the refinement process (Lyubimov et al. 2011). However, the authors do not present a holistic method to information security.

Works also exist that aim at improving the establishment of an ISMS via automation. Mondetino et al. investigate possible automation of controls that are listed in the ISO/IEC 27001 and ISO 27002 (Montesino and Fenz 2011). Their work can complement our own by providing some automation, but does not provide a complete method for establishing and documenting an ISMS.

For the Common Criteria (CC) standard (ISO/IEC 2012) there exists a security requirements engineering approach that uses the standard as a baseline for a method. Mellado et al. (2006a) created the Security Requirements Engineering Process (SREP), which is an iterative and incremental security requirements engineering process. In addition, SREP is asset-based, risk driven, and follows the structure of the Common Criteria (Mellado et al. 2006b). The work differs from ours, because the authors do not support the ISO 27001 standard and also do not aim at security standard compliance or satisfying the Common Criteria documentation demands. In addition, Ardi and Shahmehri (2009) extend the CC Security Target document with a section that considers knowledge of existing vulnerabilities. The authors aim at improving the CC and not at supporting its establishment.

7.5 Summary

In this chapter we have presented ISMS-CORAS, which is an extension of the CORAS model-driven approach to risk analysis. The method has been designed to support the establishment of an ISO 27001 compliant Information Security Management System, and to produce all the documentation that is demanded by the standard.

In summary, the benefits of our method are:

- A structured method for establishing an ISMS compliant to ISO 2700 that focuses on risk management and structured elicitation of attackers.
- The method is aligned with the ISO 31000 risk management standard, as well, because CORAS already builds on the former standard.
- ISMS-CORAS fulfills the demands of the standards ISO 27001 for security management and the ISO 31000 for risk management. Hence, the ISMS-CORAS method supports an overall process for a risk and cost[5] aware software

---

[5]Note that we provide in this work the relation to ISO 27001 and CORAS. The treatment plans consider cost-benefit reasoning by using the CORAS extension proposed in (Tran et al. 2013a).

development life cycle (SDLC). This SDLC is one of the main goals of the NESSoS project.

- Detailed steps for asset identification, threat analysis, risk management and security reasoning based on a detailed attacker modeling with templates.
- Consideration of legal compliance via steps for identifying laws and regulations.
- A systematic support to generate the required ISMS documentation in compliance to the standard, based on the proven existing method CORAS.

ISO 27001 defines the so-called Plan-Do-Check-Act (PDCA) model that specifies how to establish, implement, monitor, and maintain an ISMS. ISMS-CORAS is developed to support the plan phase, and therefore focuses on the establishment and documentation of an ISMS.

Establishing an ISMS involves conducting a security risk analysis following a process similar to those defined by ISO 31000 and ISO 27005. Because CORAS is based on the former standard, CORAS already fulfills many of the ISO 27001 requirements to risk analysis and documentation. CORAS moreover comes with techniques, guidelines, modeling support and tool support that facilitate several parts of the ISO 27001 tasks. A further useful feature of CORAS in the ISMS context is the support for modeling and analyzing legal aspects.

ISMS-CORAS extends CORAS with the features, artifacts, and techniques that are needed to provide complete support for establishing and documenting an ISMS. Some of the main novelties of ISMS-CORAS are the following. The method comes with detailed steps for asset identification, threat analysis, risk management and security reasoning; it is supported by attacker templates, classification of attacker types and attacker overview diagrams to facilitate and ensure completeness of attacker identification; it is supported by several kinds of diagrams for threat and risk modeling with attacker types, modeling of vulnerabilities and attacker entry points, as well as legal aspects; it provides a mapping between attacker types and ISO 27001 controls to facilitate treatment identification. These and other novelties in combination provide a systematic support for generating the required ISMS documentation in compliance with the standard.

References

Alberts, C. J., & Dorofee, A. J. (2001, December). OCTAVE Criteria. Technical Report No. CMU/SEI-2001-TR-016. Washington, USA: CERT.

Allen, M. (2006). *Social engineering: A means to violate a computer system.* SANS Institute White Paper.

ANSSI. (2010). *EBIOS 2010—Expression of needs and identification of security objectives.* Paris, France: Agence nationale de la sécurité des systémes d'information (ANSSI).

Ardi, S., & Shahmehri, N. (2009). Introducing vulnerability awareness to common criteria's security targets. In *Proceedings of the Fourth International Conference on Software Engineering Advances ICSEA,* (pp. 419–424). IEEE Computer Society.

Beckers, K., Fasbender, S., Küster, J.-C., & Schmidt, H. (2012). A pattern-based method for identifying and analyzing laws. In *Proceedings of the International Working Conference on Requirements Engineering: Foundation for Software Quality (REFSQ)* (pp. 256–262). Springer.

Beckers, K., Côté, I., Hatebur, D., Fasbender, S., & Heisel, M. (2013a). Common criteria compliant software development (CC-CASD). In *Proceedings 28th Symposium on Applied Computing* (pp. 937–943). ACM.

Beckers, K., Hatebur, D., & Heisel, M. (2013b). A problem-based threat analysis in compliance with common criteria. In *Proceedings of the International Conference on Availability, Reliability and Security (ARES)* (pp. 111–120). IEEE Computer Society.

Beckers, K., Heisel, M., Solhaug, B., & Stolen, K. (2013c). ISMS-CORAS: A Structured Method for Establishing an ISO 27001 Compliant Information Security Management Standard. Technical Report. Oslo, Norway: SINTEF ICT.

Beckers, K., Heisel, M., Solhaug, B., & Stolen, K. (2014). ISMS-CORAS: A structured method for establishing an ISO 27001 compliant information security management system. *Advances in engineering secure future internet services and systems* (pp. 315–344). Springer.

Calder, A. (2009). Implementing information security based on ISO 27001/ISO 27002: A management guide. Van Haren Publishing.

Cheremushkin, D. V., & Lyubimov, A. V. (2010). An application of integral engineering technique to information security standards analysis and refinement. In *Proceedings of the International Conference on Security of Information and Networks* (pp. 12–18). ACM.

DCSSI. (2004, February). Expression des Besoins et Identification des Objectifs de Sécurité (EBIOS)—Section 2—Approach. General Secretariat of National Defence Central Information Systems Security Division (DCSSI).

Faßbender, S., & Heisel, M. (2013). From problems to laws in requirements engineering using model-transformation. In *ICSOFT 2013—Proceedings of the 8th International Conference on Software Paradigm Trends* (pp. 447–458). SciTePress.

ISO. (2009). ISO 31000 risk management—Principles and guidelines Geneva. Switzerland: International Organization for Standardization (ISO).

ISO/IEC. (2005). Information technology—Security techniques—Information security management systems—Requirements (ISO/IEC 27001). Geneva, Switzerland: International Organization for Standardization (ISO) and International Electrotechnical Commission (IEC).

ISO/IEC. (2008). Information technology—security techniques—information security risk management (ISO/IEC 27005). Geneva, Switzerland: International Organization for Standardization (ISO) and International Electrotechnical Commission (IEC).

ISO/IEC. (2012). Common Criteria for Information Technology Security Evaluation (ISO/IEC 15408). Geneva, Switzerland: International Organization for Standardization (ISO) and International Electrotechnical Commission (IEC).

ISO/IEC. (2013). Information technology—Security techniques—Information security management systems—Requirements (ISO/IEC 27001). Geneva, Switzerland: International Organization for Standardization (ISO) and International Electrotechnical Commission (IEC).

Karg, M. (2009). Datenschutzrechtliche Bewertung des Einsatzes von intelligenten Messeinrichtungen für die Messung von gelieferter Energie (Smart Meter) Technical Report Kiel, Germany: ULD. (https://www.datenschutzzentrum.de/smartmeter/20090925-smartmeter.html).

Kersten, H., Reuter, J., & Schröder, K.-W. (2011). It-sicherheitsmanagement nach ISO 27001 und grundschutz. Wiesbaden: Vieweg+Teubner.

Klipper, S. (2010). Information security risk management mit ISO/IEC 27005: Risikomanagement mit ISO/IEC 27001, 27005 und 31010. Wiesbaden: Vieweg+Teubner.

Knyrim, R., & Trieb, G. (2011). Smart metering under eu data protection law. *International Data Privacy Law, 1*, 121–128.

Lund, M. S., Solhaug, B., & Stolen, K. (2010). *Model-driven risk analysis: The CORAS approach* (Vol. 1). Berlin: Springer.

Lyubimov, A., Cheremushkin, D., Andreeva, N., & Shustikov, S. (2011). Information security integral engineering technique and its application in isms design. In *Proceedings of the international conference on availability, reliability and security (ARES)* (p. 585–590). IEEE Computer Society.

Mellado, D., Fernandez-Medina, E., & Piattini, M. (2006a). A comparison of the common criteria with proposals of information systems security requirements. In *The first International Conference on Availability, Reliability and Security, ARES* (pp. 654–661). IEEE Computer Society.

Mellado, D., Fernández-Medina, E., & Piattini, M. (2006b). Applying a security requirements engineering process. In *Proceedings of Computer Security—ES-ORICS 2006*. LNCS (Vol. 4189, pp. 192–206). Springer.

Microsoft. (2006). The Security Risk Management Guide. http://technet.microsoft.com/en-us/library/cc163143.aspx.

Montesino, R., & Fenz, S. (2011). Information security automation: How far can we go? In *Proceedings of the International Conference on Availability, Reliability and Security (ARES)* (pp. 280–285). IEEE Computer Society.

Peltier, T. R. (2010). *Information security risk analysis* (Vol. 3). Boca Raton: Auerbach Publications.

Raabe, O., Lorenz, M., Pallas, F., Weis, E. (2011). Datenschutz im smart grid und in der elektromobilität Technical Report Karslruhe, Germany: KIT. (http://compliance.zar.kit.edu/21438.php).

Rodden, T. A., Fischer, J. E., Pantidi, N., Bachour, K., & Moran, S. (2013). At home with agents: Exploring attitudes towards future smart energy infrastructures. In *Proceedings of the SIGCHI Conference on Human Factors in Computing Systems* (pp. 1173–1182). ACM.

Siemens. (2003). *CRAMM—The total information security toolkit*. http://www.cramm.com/.

Solhaug, B., & Stolen, K. (2013). The CORAS language—Why it is designed the way it is. In *Safety, Reliability, Risk and Life-Cycle Performance of Structures and Infrastructures, Proceedings of 11th International Conference on Structural Safety & Reliability (ICOSSAR'13)*. CRC Press.

Swiderski, F., & Snyder, W. (2004). *Threat modeling*. Redmond: Microsoft Press.

Tran, L. M. S., Solhaug, B., & Stolen, K. (2013a). An Approach to select cost-effective risk countermeasures. In *Proceeding of the Conference on Data and Application Security and Privacy*. LNCS (Vol. 7964, pp. 266–273). Springer.

Tran, L. M. S., Solhaug, B., & Stolen, K. (2013b). An approach to select cost-effective risk countermeasures exemplified in CORAS Technical Report No. A24343. Oslo, Norway: SINTEF ICT.

UML Revision Task Force. (2010, May). OMG unified modeling language: Superstructure.

Chapter 8
Supporting Common Criteria Security Analysis with Problem Frames

Abstract Software vendors have to build their customers' trust through appropriate security functionalities of their products. The Common Criteria (ISO 15408) security standard provides an evaluation process for a software product, the application of which results in a set of documents that can be reviewed by a certification body. Creating this comprehensible set of documents is difficult, due to a detailed threat analysis, security objectives elicitation, and a selection and implementation of appropriate security measures. Moreover, the descriptions of what to do in the document are given in ambiguous natural language. We propose a model-driven approach for Common Criteria threat analysis and the subsequent security analysis based on the problem frames security requirements engineering method. Our method contains a UML profile that aligns the problem frames and Common Criteria concepts and terminology. Furthermore, we provide OCL checks for these models for consistency and reasoning support. In addition, our tool support contains a functionality to transform the information stored in UML models to natural language texts in LaTeX and HTML format. We illustrate the application of our approach for a smart grid example based on a published Common Criteria protection profile.

8.1 Introduction

We introduced the concept of supporting the security analysis and documentation demands of security standards in Chap. 5. In this chapter, we present a concrete example of this research, which leads to the publications (Beckers et al. 2013a,b, 2014). We proposed to extend the problem frame method (Sect. 2.5.3) to support the security reasoning and documentation demands of the Common Criteria (CC) (Sect. 2.2.4). The CC demand several steps, such as a description of the software and hardware in its environment, asset identification, and threat analysis, but does not provide a description of when these steps are complete and all threats are elicited. A detailed description on how to conduct these steps is also not available. We provide a method that extends the UML4PF profile (Sect. 2.5.3) with CC-specific terminology, e.g., the so-called *ToE* is the system containing hardware and software to be certified. In addition, we introduce an attacker classification and define specific kinds

© Springer International Publishing Switzerland 2015
K. Beckers, *Pattern and Security Requirements*,
DOI 10.1007/978-3-319-16664-3_8

of Jackson's domains that these attackers threaten, e.g., a software attacker threatens only causal domains. Our method is supported by a modeling tool that contains OCL expressions (UML Revision Task Force 2010a), which support security reasoning by checking for completeness of the threat analysis expressed in UML models (UML Revision Task Force 2010b). For example, it checks whatever all network connections are threatened by a network attacker. These expressions also check for consistency problems in the resulting model, e.g., if assets are also attackers.

Our method uses models and inserts all relevant information and texts in these models. We support the creation of security target (ST) and protection profile (PP) documents. We support the creation of all sections from the *introduction* section to the *security objectives* section (Sect. 2.2.4). This requires a transformation from our UML models into texts and tables. We provide tool support that supports an automatic transformation from our models to tables and texts for ST and PP documents. We limit our work to security objectives, because these correspond to requirements in the conceptual framework (Sect. 2.4), and the problem frames method focuses on the requirements phase of software engineering. The support for the sections in ST and PP documents, e.g., security functional requirements that refer to software design are part of our future work. We validated our work by applying our method to an existing CC protection profile for a smart metering gateway (BSI 2011).

This chapter is structured as follows: We introduce our threat analysis and security reasoning techniques in Sect. 8.2 and present arguments why we choose to base our work on problems frames and not other security requirements engineering methods. Section 8.3 shows our CC-specific extension of the UML4PF profile, Sect. 8.4 contains our method, and Sect. 8.5 presents an example application of our method. We illustrate our tool support in Sect. 8.6. We report the results of discussions with practitioners, who evaluated our method, in Sect. 8.7. Section 8.8 presents related work, and Sect. 8.9 concludes.

8.2 Supporting Common Criteria Using Problem Frames

The ISO 15408 Standard—Common Criteria for Information Technology Security Evaluation (short CC)—(ISO/IEC 2012) demands a detailed documentation of the software system that should be evaluated. This software system is the so-called *Target of Evaluation (ToE)* and consists of hardware and software. A *ToE* has to be described in detail, including its environment.

In this work, we focus on the threat analysis of the CC and on the description of the *ToE* in its environment, which is the input for this threat analysis. This considers *assets, attackers,*[1] *threats, assumptions, security objectives,* and *security functional requirements* of the *ToE*. The challenge of any threat analysis is to achieve a complete coverage of all possible threats. Security requirements engineering (SRE) methods

---

[1] The CC uses the term *threat agent* for attacker. However, we use attacker as a synonym for threat agent in this work.

exist, which provide structured threat analysis on an abstraction of the system. However, these abstractions often only consider parts of the system-to-be (Fabian et al. 2010), whereas for a CC-compliant threat analysis we require a complete model of the *ToE*.

Goal-based methods, e.g., SI* (Massacci et al. 2010) and KAOS (van Lamsweerde 2009), investigate the goals and views of all stakeholders of the system. These methods model threats based upon structured goal models. Hence, they consider all goals and relevant software artifacts to these goals. However, they do not consider a complete view of the system-to-be. Other SRE methods follow similar steps, e.g., the asset-driven risk management method CORAS (Lund et al. 2010) identifies assets and determines threats to these assets. CORAS models the system-to-be in artifacts that have a relation to an asset and also does not represent the complete system-to-be. Therefore, we do not use any of these methods for our CC-compliant threat analysis.

The Problem Frames method (Jackson 2001) uses an abstraction of the system-to-be and models the environment of the system around it. Hence, this method is our choice for satisfying the CC's demand to model the *ToE* in its environment. The method models the *ToE* and its environment in domains with certain characteristics, and we propose a threat analysis that uses these characteristics to determine assets, possible attackers, and subsequent threats for these domains. We introduce a structured method that elicits attackers and threats for each domain. We also provide computer-aided support for consistency, document creation, and security reasoning for this method by using OCL queries on the problem frame models. Hence, we benefit from having a complete model of the system-to-be and its environment to conduct a threat analysis. Our method iterates over all domains of the system and reasons if these are threatened by an attacker. Hence, our method achieves completeness when all parts of the model are considered during security reasoning and the system model is complete.

8.3 UML Profile for Problem-Based and Common Criteria-Compliant Security Analysis

Our Common Criteria extension for the UML4PF profile is shown in Figs. 8.1 and 8.2. We split the profile into two figures in order to improve readability. All parts of the Common Criteria extension are marked in gray.

We contribute relations between problem frames and common criteria elements like *countermeasures*, which are now a kind of domain (see Fig. 8.1). The profile considers the *machine* to be evaluated, which is the *ToE*, and the *ToEOwner*, the person using the ToE. Assets are domains in Jackson's sense (Sect. 2.5.3) or part of a domain. We use an OCL expression to enforce this condition (see expression AE02CON in Table 8.1). *Assets* have a *description* and a *need for protection* attribute. The CC has also the concept of *SecondaryAssets*. Harm to *SecondaryAssets* do not cause a loss to the *ToEOwner* directly, but the harm can cause harm to an *Asset*.

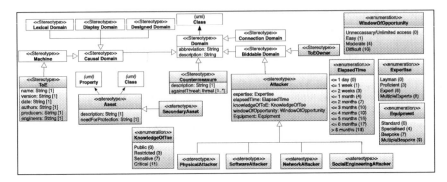

Fig. 8.1 A common criteria extension of the UML4PF profile (1/2)

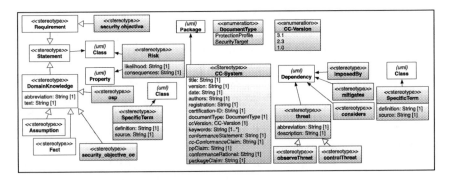

Fig. 8.2 A common criteria extension of the UML4PF profile (2/2)

This in turn can cause a loss to a *ToEOwner*. *Threat*s can harm assets and have an abbreviation and a description. A *Risk* caused by threats has a likelihood and a consequence. An *osp* is an organizational security policy, which states rules that have to be followed when using the *ToE*. A *SpecificTerm* is a term that is not common knowledge and is defined for the ST/PP document. The specific term has a definition and a source of the definition.

Threats are caused by *Attackers* that we classify into the following categories. *Physical Attackers* threaten the physical elements of the system, e.g., hardware or buildings that host computers. *NetworkAttacker*s threaten *Network connections* or *ConnectionDomains* in our models. *SoftwareAttacker*s threaten the parts of the system that are software, e.g., the *ToE* or other *CausalDomains* that are a software. *SocialEngineeringAttacker*s threaten biddable domains, e.g., users of the system.

The UML profile also contains ≪enumerations≫ to represent the attributes required to describe an *Attacker* according to the CC. The attributes, e.g., *ElapsedTime* or *WindowOfOpportunity* have numeric values attached in brackets. These values are defined by the CC and are used to determine the EAL (evaluation assurance level) for a system. For example, an attacker with a combined score between 10 and 13 results in a recommendation to implement the security assurance classes

Table 8.1 OCL expressions for ensuring model consistency

OCL-EXPR-ID	Referenced class	Expression
Domain knowledge		
FA01CON	Fact	– Refers to at least one domain
AS01CON	Assumption	– Refers to at least one domain
AE01CON	Asset	– Has a relation to the ToE domain (e.g., composition) and is not an attacker
AE02CON	Asset	– Is a domain or part of a domain
ST01CON	Secondary asset	– Has a relation to an asset and is not an attacker
Attackers		
AT01CON	Attacker	– Presents at least one threat and is not an asset
NA01CON	Network attacker	– Threatens only connection domains, connections, or subtypes
PA01CON	Physical attacker	– Threatens a domain
SA01CON	Software attacker	– Threatens only causal domains
SE01CON	Social engineering attacker	– Threatens only biddable domains
Threats		
TH01CON	Threat	– Threatens only assets
OT01CON	observeThreat	– Window of opportunity of the attacker is greater than 0
CT01CON	controlThreat	– Window of opportunity of the attacker is greater than 0

AVA_VAN.1 and *AVA_VAN.2*. This recommendation results in an *EAL* requirement of at least *EAL 2*. This classification of attackers is used in the CC during the evaluation of existing implementations. We propose to use it already during the requirements stage. Some information might not be available in the requirements phase, but the information that is already present can be included in the model and used for the security reasoning.

We extended the CC basic security model in order to distinguish different kinds of threats. *Threats* can be further divided into *controlThreats* and *observeThreats* (see Fig. 8.2). *ControlThreats* take control of a domain, while *observeThreats* only observe information about the behavior of a domain. For example, an *observeThreat* is the eavesdropping of confidential information, whereas the manipulation of a key exchange is a *controlThreat*. The distinction between observe and control threats helps to determine the security objectives. For example, observe threats are likely to cause confidentiality problems. We propose a threat analysis during requirements engineering where the exact flow of information is only partially known.

Hence, we assume that informational assets can be reached by all domains or interfaces in the model. This is why we allow threats not only to be attached directly to assets, but to all domains or interfaces in our model. We explain how we model the ToE in its environment in Sect. 8.4.

Each *security objective mitigates* at least one threat and concern the ToE. *Security_objective_oe* state demands for the environment of the ToE. A security objective can also ≪referTo≫ a domain and ≪consider≫ an organizational security policy.

8.4 A Method for a Systematic Security Analysis and Documentation

Figure 8.3 shows our security analysis method, which we explain in the following.

1. Define scope To perform the security analysis systematically, we start with creating a context diagram that contains the scope of our analysis. The context diagram contains all domains (e.g., persons and technical systems) in the environment of the machine that are referred to by the functional requirements. For an example of a context diagram, (see Fig. 8.5).

We defined several consistency checks for checking that a context diagram is correct. These are described in detail in (Hatebur 2012) and are already part of UML4PF.

2. Asset identification For all domains in the context diagram, we check if the domain contains an *asset* or is an asset. Assets are documented in domain knowledge diagrams (Sect. 2.5.3) as classes with the stereotype ≪Asset≫ as introduced in the UML profile in Fig. 8.1. If the entire domain is an asset, we add the stereotype ≪Asset≫ to that class. In the case that an asset is only part of a domain, we use UML aggregation or composition relations between the asset and the domain it belongs to. We also identify secondary assets, which cause harm to other assets. We use OCL to check model consistency and completeness, which we also use for security reasoning. We state examples for each step of the method and how this step benefits from our OCL expressions. We provide an overview of OCL expressions for model consistency in Table 8.1. These expressions query the entire model, meaning context diagram and all domain knowledge diagrams. The table has a unique ID for each expression in the first column, the referenced class of the UML profile shown in Figs. 8.1 and 8.2 in the second column, and the consistency and expression checks in the third column.

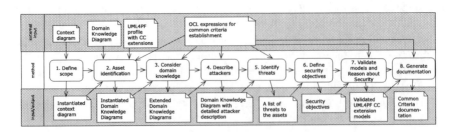

Fig. 8.3 A method for common criteria-compliant security analysis

```
let                                                                          1
  stereotype :  String = 'Attacker'                                          2
in                                                                           3
let                                                                          4
  attackers :  Set (Class) =                                                 5
  Class. allInstances ()–>select(                                            6
  let                                                                        7
      first :  Set(Stereotype) = getAppliedStereotypes ()–>asSet ()          8
  in                                                                         9
  first –>union(first –>closure(general. oclAsType(Stereotype ))).          10
      name–>includes(stereotype))
in                                                                          11
attackers –>forAll( a | a–>asSequence ()–>first (). clientDependency–>size () >=   12
    1 )
  and                                                                       13
attackers. clientDependency–>forAll(                                        14
        getAppliedStereotypes (). name–>includes('controlThreat')           15
        or                                                                  16
        getAppliedStereotypes (). name–>includes('observeThreat')           17
        or                                                                  18
        getAppliedStereotypes (). name–>includes('threat')                  19
        )                                                                   20
  and                                                                       21
  let  s :  Set(Class) =                                                     22
      Class. allInstances ()–>select(getAppliedStereotypes (). name         23
        –>includes('Attacker') and
        getAppliedStereotypes (). name–>includes('Asset'))
  in                                                                        24
s–>isEmpty ()                                                               25
```

Fig. 8.4 *AT01CON* checks that all attackers present at least one threat and are not assets

As an example, we discuss the OCL expression *AT01CON* in more detail (see Fig. 8.4). The expression selects all attackers (lines 1–11) and checks that the attackers have at least one dependency (line 12), all their dependencies have a stereotype ≪threat≫ or a subtype (lines 14–20), and attackers are also not assets (lines 21–25).

We use *AE01CON* (see Table 8.1) to check if an *asset* has no relation to the *machine* (in the case of the Common Criteria this is the *ToE*). The reason is that the common criteria certifies products and assets have a relation to that machine. Otherwise, countermeasures in the machine could not protect the assets. If this is the case, this relation has to be added, or the class is not an asset and the stereotype should be removed. We use *ST01CON* (see Table 8.1) to check secondary assets in a similar manner. Moreover, we use OCL to check for missing assets by *AE01REA* (see Table 8.4). The OCL expression *AE01REA* (see Table 8.4) returns all classes that are not an asset and do not contain an asset. Security engineers can reason if this is correct for all the listed assets. This list helps to identify missing assets. For secondary assets, we proceed in a similar way, using *ST01REA* and further expressions in Table 8.1.

We introduce our identified OCL expressions that support security reasoning in Tables 8.3, 8.4, and 8.5, e.g., by checking for completeness of the threat analysis in domain knowledge diagrams.

3. Consider domain knowledge As a next step, for all assets, either an *assumption* or *fact* about its protection has to be described. Facts and assumptions help to estimate if an asset already has sufficient protection and no further security requirement is necessary. They also help to formulate focused requirements that do only address security issues that are not already addressed. In addition, relevant facts or assumptions about assets, which can be exploited by an attacker, have to be documented. Facts and assumptions are documented in domain knowledge diagrams with classes and the stereotypes ≪Fact≫ or ≪Assumption≫ using the UML profile in Fig. 8.2. The relation between facts and/or assumptions and assets can be documented with dependencies and the stereotype ≪refersTo≫. ≪refersTo≫ states that a statement refers to some domains. It extends the UML meta-class *Dependency*. We use the OCL expressions *FA01CON* and *AS01CON* (see Table 8.1) to check that each fact and assumption ≪refersTo≫ at least one domain. In addition, we use *FA01REA* and *AS01REA* (see Table 8.4) to list all domains having no facts or assumptions (considering all domain knowledge diagrams). For these domains, one should make sure that the most obvious facts have been considered and that facts and assumptions have been distinguished correctly.

4. Describe attackers All attackers have to be described using the attributes shown in Fig. 8.1. We iterate through all *assets* and check if they have *assumptions* or *facts* that prevent them from being threatened by a specific kind of attacker. For example, a piece of software that has no connection with the *ToE* provides no attack vector for a network attacker. Otherwise, an *attacker* has to be introduced that threatens the *asset*. Moreover, the introduced attackers also have assumptions and facts. These have to be modeled explicitly, as well, to support a correct threat assessment.

We use OCL expressions to query our model for getting an overview of all existing threat analysis elements, e.g., assets. These expressions end with the letters "DOC" that stands for documentation. We list all DOC expressions in Table 8.2. For example, expression *AE01DOC* lists all *assets* that can be threatened by an *attacker*. We consider for each asset if an attacker can cause harm to it. Afterward, we use the expressions *FA01DOC* and *AS01DOC* to check for each assumption and the fact if these can be used to cause harm to an asset. If this is the case, another attacker has to be introduced.

5. Identify threats Threats are a relation between an attacker and an asset. This relation can be modeled with dependencies and the stereotypes ≪threat≫, ≪observeThreat≫, or ≪controlThreat≫. In this step we iterate over all the attackers and introduce threats. *Assumptions* or *facts* have to be considered or introduced when deciding if the attacker represents an ≪observeThreat≫, or ≪controlThreat≫. We use the *AT01DOC* (see Table 8.2) to list all attackers to start our iteration. Afterwards, we introduce *threats* for each attacker. The OCL expressions *AE01DOC*, *ST01DOC*, *FA01DOC*, *AS01DOC* (see Table 8.2) provide us with lists of domain knowledge artifacts, e.g., facts and assets. After we have introduced threats for the attackers under the consideration of domain knowledge, we check if all attackers represent at least one threat using the expression *AT01CON* (see Table 8.1). If an attacker does not represent a threat, the attacker should either be removed, or a threat should be added. We execute the OCL consistency expressions

Table 8.2 OCL expressions of document generation

OCL-EXPR-ID	Referenced class	Expression
Domain knowledge		
DO01DOC	Domain	– List all facts and assumptions for each domain
FA01DOC	Fact	– List all facts in the domain knowledge diagrams
AS01DOC	Assumption	– List all assumptions in the domain knowledge diagrams
AE01DOC	Asset	– List all considered assets
ST01DOC	Secondary asset	– List all considered secondary assets
Attackers		
AT01DOC	Attacker	– List all considered attackers including all attributes
NA01DOC	Network attacker	– List all considered network attackers
PA01DOC	Physical attacker	– List all considered physical attackers
SA01DOC	Software attacker	– List all considered software attackers
SE01DOC	Social engineering attacker	– List all considered social engineering attackers
Threats		
TH01DOC	Threat	– List all considered threats and threatened assets
OT01DOC	observeThreat	– List all considered observe threats and threatened assets
CT01DOC	controlThreat	– List all considered control threats and threatened assets

for network, physical, software, and social engineering attacker in a similar fashion (see Table 8.1).

6. Define security objectives Each of the threats leads to the formulation of a security objective. The Common Criteria distinguishes between security objectives (SO), which concern the ToE, and the ones concerning the environment. The latter ones are so-called *security objectives for the environment (SO-OE)*. We model SOs in problem diagrams, because these are directed towards the ToE. SO-OEs are modeled in domain knowledge diagrams, because these concern the environment.

7. Validate models and Reason about Security We use OCL to check model consistency of the various diagrams. We defined several validation conditions as OCL expressions, which are listed in Table 8.1. These check, e.g., if facts and assumptions refer to at least one domain. These conditions have to be executed via our support tool in order to validate the models. All existing inconsistencies have to be removed in this step. We have developed debug expressions for all OCL expressions, which state precisely which domain(s) caused the model inconsistency.

In this step we also check for completeness of attackers in the model. For example, the expressions *SA01REA* (see Table 8.3) checks for all causal domains if these are threatened by a software attacker. If this is not the case, we should check for existing assumptions using *AS01CON* (see Table 8.3) for these domains. The resulting information of these expression should serve as a basis for security reasoning for these

Table 8.3 OCL expressions that support security reasoning—attackers

OCL-EXPR-ID	Referenced class	Expression	Reasoning support for security experts
Attackers			
AT01REA	Attacker	– List all attackers that have only observe threats or only controls threats	– Is the attacker's potential modeled correctly?
NA01REA	Network attacker	– List all connection domains and connections that are not threatened by a network attacker	– Are threats to all relevant domains from that attacker considered?
NA02REA	Network attacker	– List all connection domains and connections that are not threatened by a network attacker and do not have an assumption	– Are threats to all relevant domains from that attacker considered or do we need to add an assumption?
PA01REA	Physical attacker	– List all biddable domains that are not threatened by a physical attacker	– Are all humans considered that a physical attacker can threaten?
PA02REA	Physical attacker	– List all causal domains that are not threatened by a physical attacker	– Are all physical devices considered that a physical attacker can threaten?
SA01REA	Software attacker	– List all causal domains that are not threatened by a software attacker	– Is every software considered in the threat analysis?
SA02REA	Software attacker	– List all causal domains that are not threatened by a software attacker and that do not have an assumption	– If a software attacker is not considered and we do not have an assumption, we should add an assumption or include further software attackers for the resulting domains in the threat analysis
SE01REA	Social engineering attacker	– List all biddable domains that are not threatened by a social engineering attacker.	– Are all possible threats by social engineering attackers considered?
SE02REA	Social engineering attacker	– List all biddable domains that are not threatened by a social engineering attacker and that do not have an assumption specified	– For each biddable domain that is not threatened by a social engineering attacker, we should provide at least an assumption why this is not necessary. If no valid assumption can be found, the threat analysis should be revised to include this attacker

domains. The question if we need to consider a software attacker for these domains should be answered in particular. The other attacker types are considered in a similar manner using the expressions in Table 8.3. We also use the security reasoning expressions in Tables 8.4 and 8.5 to reason about the completeness of domain knowledge and threats.

Table 8.4 OCL expressions that support security reasoning—domain knowledge

OCL-EXPR-ID	Referenced class	Expression	Reasoning support for security experts
Domain knowledge			
DO01REA	Domain	– List all domains that have no facts or assumptions	– Do we really have no domain knowledge at all about a domain?
FA01REA	Fact	– List all domains that have no facts	– Have at least the most obvious facts been considered?
AS01REA	Assumption	– List all domains that have no assumptions	– Have at least the most obvious assumptions been considered?
AE01REA	Asset	– List all classes that are not assets or secondary assets or attackers	– Is an asset still missing?
AE02REA	Asset	– List all assets that have no need-for-protection property	– Is that asset really an asset if it has no need for protection?
AE03REA	Asset	– List all connections or connection domains that do not transmit assets	– Does a connection between domains really transport no assets?
ST01REA	Secondary asset	– List all secondary assets	– Are these all really not assets?

Table 8.5 OCL expressions that support security reasoning—threats

OCL-EXPR-ID	Referenced class	Expression	Reasoning support for security experts
Threats			
TH01REA	Threat	– List all assets that are not threatened	– Is an asset not threatened at all?
OT01REA	observeThreat	– List all assets that have no observe Threats	– Has an asset only control threats?
CT01REA	controlThreat	– List all assets that have no control Threats	– Has an asset only observe threats?

8. Generate documentation Finally, we generate textual documents from the information in the models. Our support tool provides functionalities to query the models and select all relevant information and transform it into text and tables. The output is possible in LATEX or HTML documents to form the basis for a PP or ST document.

Mapping to the Conceptual Framework for Security Standards—We present a mapping in Table 8.6 from this method to the conceptual framework for security standards, which is the foundation for our PEERESS framework (Chap. 3). The table lists the activities of the conceptual framework for security standards on the left

Table 8.6 A mapping between the standard activities in the conceptual framework for security standards and the steps of the UML4PF-CC method

Activity in the conceptual framework for security standards	Steps in the UML4PF-CC method
Environment description	Step 1. Define scope
Stakeholder description	Step 1. Define scope
Asset identification	Step 2. Asset identification
Risk level description	Step 3. Consider domain knowledge
	Step 4. Describe attackers
Security property description	Step 6. Define security objectives
Control assessment	Step 3. Consider domain knowledge
	Step 4. Describe attackers
Vulnerability and threat analysis	Step 5. Identify threats
Risk determination	Step 5. Identify threats
Security assessment	Step 6. Define security objectives
Security measures	Step 7. Validate models and reason about security
Risk acceptance	Step 7. Validate models and reason about security
Documentation	Step 8. Generate documentation

column and the steps of the UML4PF-CC method that concern these activities in the right column. Note that the Common Criteria focuses on documenting vulnerabilities and security controls of the ToE. It does not consider risk management per se, but rather provides the information about threats and countermeasures to stakeholders. Afterwards the stakeholders can use this information to conduct a risk analysis (cf. Annex 4). Hence, all activities of the conceptual framework for security standards map to steps concerning describing threats and attackers.

8.5 Application of Our Method

Application scenario We use the protection profile for the smart metering gateway as an example for our method (BSI 2011). We apply our method to the creation of a security target and base it on this protection profile. The gateway is a part of the smart grid. The smart grid is a commodity network that intelligently manages the behavior and actions of its participants. The commodity consists of electricity, gas, water, or heat that is distributed via a grid (or network). The benefit of this network is envisioned to be a more economic, sustainable, and secure supply of commodities. Smart metering systems meter the consumption or production of energy and forward the data to external entities. This data can be used for billing and steering the energy production. The "Protection Profile defines the security objectives and corresponding requirements for a Gateway which is the central communication component of such a Smart Metering System" (BSI 2011, p.16).

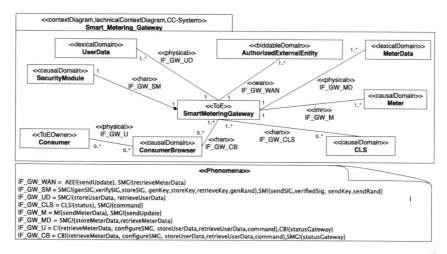

Fig. 8.5 The context diagram of the smart metering gateway

1. Define scope The context diagram shown in Fig. 8.5 describes the machine to be built in its environment. It is part of the overview description of the security target. The ≪ToE≫ is the SmartMeteringGateway, which serves as a bridge between the Wide Area Network ≪wan≫ and the Local Network ≪physical≫ of the Consumer, the ≪ToE Owner≫. The Meter is connected to the ToE via a Local Metrological Network ≪lmn≫. This is an in-house equipment that can be used for energy management. The Controllable Local System CLS can be, for example, an air conditioning unit or an intelligent refrigerator. The Consumer can also access the ToE (BSI 2011) via a ConsumerBrowser. We extended the description for our specification with the following phenomena. The Meter sends meter data to the SmartMeteringGateway. The SmartMeteringGateway stores this data. The Meter can also receive updates from the AuthorizedExternalEntity forwarded via the SmartMeteringGateway. The AuthorizedExternalEntity gets receives meter data in fixed intervals from the SmartMeteringGateway. The SecurityModule provides cryptographic functionalities for the SmartMeteringGateway such as key generation and random number generation. The Consumer can retrieve meter data via the SmartMeteringGateway. The Consumer can also configure the Smart-MeteringGateway, send commands to the CLS, receive status messages from the SmartMeteringGateway, and store UserData in it.

2. Asset identification We iterate over the domains in Fig. 8.5 and identify the MeterData as an ≪asset≫. Figure 8.6 presents a domain knowledge diagram that contains the description of this asset. The meter data has value for the Consumer, because his/her billing depends upon it and a behavior profile about the Customer can be created from it. Integrity, authenticity, and confidentiality of this data need to be protected. Another asset of the SmartMeteringGateway is the Gateway-Time (see Fig. 8.6). The asset is revealed via investigating assumptions about the

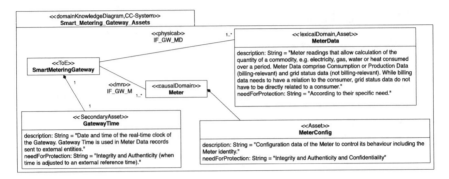

Fig. 8.6 An example for a common criteria-compliant asset description

SmartMeteringGateway, namely that the meter data is recorded with a correct
time stamp. The time is used in MeterData records that are sent to AuthorizedExternalEntity, e.g., for billing. Its integrity and authenticity have to be protected and
especially the time adjustment using an externally referenced time is critical.

We use *AE01REA* (see Table 8.4), which returns all classes that are not an asset
and do not contain assets. For the smart metering gateway running example, we have
so far only identified the assets MeterConfig, MeterData, and GatewayTime (see
Fig. 8.6). The expression *AE01REA* returns: UserData, AuthorizedExternalEntity,
CLS, Consumer, ConsumerBrowser, and SecurityModule. For these domains, a
good rationale has to be given why they are not assets, or they are have to be marked
as assets. For example, the SecurityModule and the connection IF_GW_WAN are
indeed assets.

3. Consider domain knowledge The Common Criteria demands that assumptions
about domains and connections are made explicit. We choose the assumptions about
the AuthorizedExternalEntity, the IF_GW_WAN, and the SmartMeteringGateway as examples (taken from BSI 2011). The assumptions document the assumed
behavior of the authorized external entities, reliability and bandwidth of the connection, and the installation location of the SmartMeteringGateway. Assumptions
refer to domains in the context diagram. Additionally, facts can be included, e.g., that
the SmartMeteringGateway needs electricity to operate. These facts are stated in
a domain knowledge diagram, depicted in Fig. 8.7.

4. Describe attackers We introduce a ≪NetworkAttacker≫, who threatens the
WAN connection, depicted in Fig. 8.7. We have also assumptions regarding this
attacker. AssumptionWLANAttacker states that the attacker is located in the WAN
and that he/she has the capability to threaten the smart grid, e.g., via sending forged
meter data into the grid. This assumption ≪refersTo≫ the WANAttacker. Based
upon the AssumptionWLANAttacker we instantiate the attacker with the following
attributes: the attacker has the Expertise = Expert (6) and the ElapsedTime = ≤ 1
day (0), the KnowledgeOfToE = Restricted (3), and the WindowOfOpportunity =
Unnecessary/Unlimited access (0). We can calculate the value for these attributes

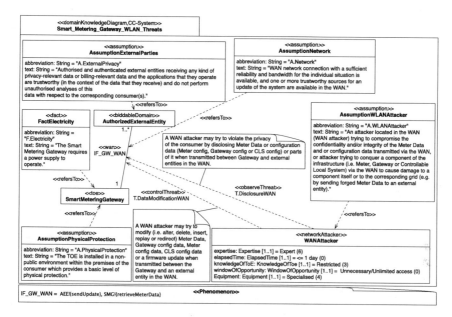

Fig. 8.7 A domain knowledge diagram for common criteria-compliant threat description

for addressing the assurance component AVA_VAN for vulnerability assessment of the CC (ISO/IEC 2012, part3,p.16). The results demand at least an EAL 2 of the CC.

We also know that the Meter depends upon electricity and we introduce the Fact-Electricity. This can lead to the introduction of a ≪PhysicalAttacker≫. However, the AssumptionPhysicalProtection states that a basic level of physical protection exists. Hence, the introduction of a ≪PhysicalAttacker≫ is not required for this scenario if we install the Meter in locked box. An alternative is to remove the AssumptionPhysicalProtection and consider a sophisticated ≪PhysicalAttacker≫, who can penetrate the physical barriers of the *ToE*.

5. Identify threats The WANAttacker gives rise to the ≪observeThreat≫ T.Disclosure WAN (see Fig. 8.7), which states that the WAN attacker can disclose meter data or meter configuration data. The WANAttacker causes also the ≪control Threat≫ T.DataModificationWAN, which allows the attacker to modify several different data, e.g., meter data, and meter config data.

We list all threats stated in the protection profile (BSI 2011) in Fig. 8.8. The *WANAttacker* causes the following threats related to modification and observation of data:

T.DataModificationWan   the unauthorized modification of configuration parameters via using the IF_GW_WAN connection of to the gateway.

T.DisclosureWan   disclosing   of   meter   data   from   the   IF_GW_WAN connection.

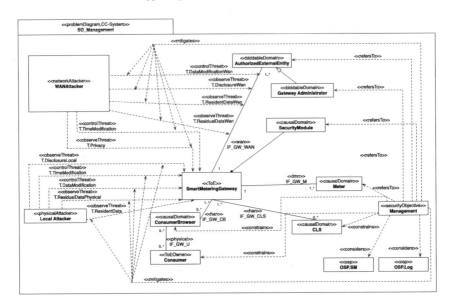

Fig. 8.8 Problem diagram for the security objective "Management"

T.ResidentDataWan   the unauthorized reading of data on the gateway that is no longer required and originated from the IF_GW_WAN connection.

T.TimeModification   the unauthorized modification of the gateway time.

T.Privacy   the IF_GW_WAN connection transports data that is not required by the *AuthorizedExternalEntity*.

The *Local Attacker* threatens the *SmartMeteringGateway* with regards to modification and observation of data:

T.DisclosureLocal   disclosing of meter data from the IF_GW_CB, IF_GW_M, and IF_GW_CLS connections.

T.TimeModification   the unauthorized modification of the gateway time.

T.DataModification   the unauthorized modification of configuration parameters using the connections IF_GW_CB, IF_GW_M, and IF_GW_CLS connection.

T.ResidualDataPhysical   the unauthorized reading of data on the gateway that is no longer required and originated from the IF_GW_M, and IF_GW_CLS connections.

T.ResidentData   the unauthorized modification of configuration parameters via physical access to the gateway.

We use *AT01DOC* (see Table 8.2) to check all considered attackers, which so far only returns the WAN attacker. In order to reason that all relevant attackers have been considered, we execute the following OCL expressions from the security

reasoning about attackers listed in Table 8.3. We consider only the domains shown in
Fig. 8.7. *NA01REA* does not list any domain, because the connection IF_GW_WAN is
already threatened by a network attacker. *PA01REA* returns the *SmartMeteringGate-*
way, because a threat caused by physical attacker should be considered. We do not
follow this suggestion, because we introduced AssumptionPhysicalProtection in
the previous step of our method. *SA01REA* returns that a software attacker should be
considered for the SmartMeteringGateway. The software attacker may penetrate
the SmartMeterGateway and present a ≪controlThreat≫ towards it. An assumption about the software attacker is that she/he can control the SmartMeterGateway
and modify all meter data the gateway has access to. In addition, the OCL expressions
SE01REA states that a social engineering attacker can threaten AuthorizedExternalEntity. The social engineering attacker presents a ≪controlThreat≫ towards the
AuthorizedExternalEntity, and an assumption is introduced that the attacker can
control the AuthorizedExternalEntity in such a way that the attacker gains access to
the meter data and to the keys and certificates necessary to access the SmartMeter-
Gateway. Hence, the new assumption states that the social engineering attacker can
access and configure the SmartMeterGateway. For example, the attacker could use
the gateway to control a CLS, e.g., a heater or a refrigerator. This assumption has to
be included in Fig. 8.7, as well as the subsequent threats.

6. Define security objectives We define security objectives in problem diagrams.
An example is the security objective *Management* taken from (BSI 2011, p.39).

The protection profiles formulates the security objective *Management* as follows:

- The ToE shall only provide authorized Gateway Administrators with functions for
  the management of the security features.
- The ToE shall ensure that any change in the behavior of the security functions can
  only be achieved from the WAN side interface. Any management activity from a
  local interface may only be read.
- Further, the ToE shall implement a secure mechanism to update the firmware of
  the ToE that ensures that only authorized entities are able to provide updates for
  the ToE and that only authentic and integrity protected updates are applied.

We depict the security objective *Management* in Fig. 8.8. The security objective *Management* ≪refersTo≫ *Gateway Administrators*, a specific kind of *AuthorizedExternalEntity*. The administrators are the only domain allowed to access the
management functionality of the *Smart Metering Gateway*. The functionality for
authentication is provided by the *Security Module*, which is described in a separate protection profile (BSI 2013). The mandatory usage of the *Security Module* is
also defined in the organizational security policy *OSP.SM*, which ≪considers≫ the
security objective *Management*. The objective considers the policy *OSP.Log*, as well.
This policy states that a set of log files have to be compiled, e.g., for the information
flow between the WAN and the smart metering gateway. In addition, all accesses to the
configuration data of the *SmartMeteringGateway* are recorded. The security objective
Management ≪constraints≫ the *ConsumerBrowser*. The *ConsumerBrowser* is not
allowed to access the management configuration of the system. The same constraint

has to hold for the *Consumer* and the *CLS*. The restriction of access to the management functions of the *SmartMeteringGateway* shall ≪mitigate≫ threats related to modification and observation of data from a *WANAttacker* introduced in the previous step. These are *T.DataModificationWan*, *T.DisclosureWan*, *T.ResidentDataWan*, *T.ResidualDataWan*, *T.TimeModification*, and *T.Privacy*.

The restriction of access to the management functions of the *SmartMeteringGateway* shall also ≪mitigate≫ threats related to modification and observation of data from a *Local Attacker* introduced in the previous step, as well. These are *T.DisclosureLocal*, *T.TimeModification*, *T.DataModification*, *T.ResidualDataPhysical*, *T.ResidentData*.

7. Validate Models and Reason about Security Table 8.1 states several conditions that check for consistency problems in the entire model (which contains all context, domain knowledge and problem diagrams). The expression TH01CON checks if all threatened domains and connections are assets. This check fails on our model. One of the issues is that the *WANAttacker* threatens the connection IF_GW_WAN Fig. 8.7, which is not identified as an asset in the protection profile. This is an issue for discussion when improving the protection profile. The connection might not be an asset, because it is not a part of the ToE, but including the connection at least as secondary asset would probably be a good solution.

Table 8.7 shows the use of our OCL expressions for security reasoning. The first column of the table states the expression used. We used the expressions only on a few domains. The second column on the table states the domains considered by the OCL expression. The third column states the results of the query and the last column the resulting security reasoning based on the results of the query.

FA01REA (see Table 8.4) checks if we have modeled facts about domains. This is not the case for the WANAttacker and the AuthorizedExternalEntity. Hence, we have to reason why our assumptions are sufficient and should get feedback on these assumptions from further security experts.

NA01REA (see Table 8.3) queries the model if we have considered a NetworkAttacker for all network connections or connection domains. This is not the case for the network connection IF_GW_CB, which connects the ConsumerBrowser and the SmartMeteringGateway (see Fig. 8.5). During security reasoning either an assumptions is added to the model instead of a network attacker. We assume that there are no malicious insiders in the ≪han≫, who misuse network traffic. Otherwise, a security objective has to be added that states the SmartMeteringGateway has to protect the IF_GW_CB connection by encryption (see Table 8.7).

SE02REA (see Table 8.3) checks if we considered for all biddable domains SocialEngineeringAttackers or have modeled assumptions. For the Consumer and the AuthorizedExternalEntity we have modeled neither. Hence, the result of SE02REA should be discussed in an expert workshop. The experts decide if SocialEngineeringAttackers have to be included into the threat analysis. Alternatively, they have to add assumptions, which explain why the consideration of SocialEngineeringAttackers is not needed.

Table 8.7 An example for OCL-based security reasoning

OCL-EXPR-ID	Class or relation	Result	Reasoning
FA01REA (see Table 8.4)	SmartMetering Gateway, WANAttacker, AuthorizedExternalEntity	WANAttacker, AuthorizedExternalEntity	We do not have facts about the WANAttacker and the AuthorizedExternalEntity. Hence, we have to check if the assumptions documented are valid, e.g., by discussing the results with an independent security expert
NA01REA (see Table 8.3)	Consumer, ConsumerBrowser, SmartMeteringGateway, Meter	IF_GW_CB	We have to add a security objective that the communication on the IF_GW_CB connection is encrypted by the SmartMeteringGateway or an assumption that there are no malicious insiders in the ≪han≫, who misuse network traffic
SE02REA (see Table 8.3)	All domains	Consumer, AuthorizedExternalEntity	We have not considered SocialEngineeringAttackers and have no assumptions specified why SocialEngineeringAttackers do not need to be considered. The result of this OCL expression should trigger a threat analysis regarding if SocialEngineeringAttackers are relevant for the Consumer and theAuthorizedExternalEntity

8. Generate documentation The CC demands a particular description of the *ToE*, which has to follow a specific structure. For example, it starts with an introduction that has to contain a description of the *ToE*, its interfaces, and the operational environment, e.g., the operating system the *ToE* runs on. We propose to use the requirements-driven software engineering process *ADIT* (Sect. 2.5.3), which also relies on UML4PF. The application of ADIT results in a detailed documentation of the software. The application of the UML4PF CC extension to ADIT results in Common Criteria-compliant documentation of the software. The ADIT process demands to generate texts, figures, or tables during software development. All of these artifacts can be reused to create a CC documentation. Furthermore, we reuse ADIT's traceability and consistency checks between different models of the software engineering process. For example, ADIT provides checks that can analyze if all model elements, which are used in the architecture, refine a model element of the context description in the analysis phase. These mechanisms can be combined with the ones proposed in the previous steps of this method.

Table 8.8 Mapping ADIT phases to PP and ST documents of the common criteria

Common criteria—PP and ST		ADIT	ADIT artifact type
Introduction	PP/ST reference	A1	Context diagram
	ToE Reference (ST)	A1	Context diagram
	ToE-Overview	A1	Context diagram
		A2[a]	Problem diagram[a]
		A3[a]	Sequence diagram[a]
		A4	Technical context diagram
	ToE Description (ST)	A1	Context diagram
		A2	Problem diagram
		A3[a]	Sequence diagram[a]
		A4	Technical context diagram
Conformance claims	CC-Conformance	A1	Context diagram
	PP-Conformance or security requirements package	A1	Context diagram
	Explanation for conformance	A1	Context diagram
	Conformance definition (PP)	A1	Context diagram
Security problem definition	External entities	A2	Domain knowledge diagram
			Problem diagram
		A4[a]	Technical context diagram[a]
	Assets	A2	Domain knowledge diagram
	Assumptions	A2	Domain knowledge diagram
	Threats	A2	Domain knowledge diagram
	Organizational security policy (OSP)	A2	Domain knowledge diagram
Security objectives	ToE security objectives	A2	Domain knowledge diagram
		A2 (PP)	Problem diagram
		A3 (ST)	Sequence diagram
	Security objectives for the environment	A2 (PP)	Domain knowledge diagram
		A3 (ST)	Sequence diagram
	Objectives rational (OR)	A2	Domain knowledge diagram
			Problem diagram
		A3	Sequence diagram

The entries with [a]are optional for common criteria documentation (not for ADIT). *ST* Security target and *PP* Protection profile

We analyzed the documents resulting from the analysis phase of the *ADIT* software development process (Sect. 2.5.3) and the document requirements of the *Common Criteria*, (see Table 8.8). The table lists the chapters and sections of *Security Target*s and *Protection Profiles* in the first two columns. The sections that are only relevant for a *Security Target* are annotated with an "(ST)", the sections only relevant for a *Protection Profile* are annotated with a "(PP)". The following columns state the related ADIT phase and the output document types of this phase. We introduced specific classes to represent *CC-Systems* and specific CC artifacts, e.g., *organizational security policies*, in the *UML4Pf* profile (see Figs. 8.1 and 8.2). The *Introduction*, *Conformance Classes*, and *Security Problem Definition* chapters of the *ST* and *PP* are CC-specific. Hence, these require numerous extensions to *ADIT* as extension to the UML4PF profile and in the form of OCL expressions. The *Security Objectives* chapter of *ST/PP* documents requires enhanced *DomainKnowledgeDiagrams* to represent *Attackers*, *Threats*, and *Assumptions*. The term *security objective* in the CC is similar to the term *security requirement* in the problem frame method (see Table 2.1) and ADIT uses the term *security requirement* in the phases A1, A2, and A3.

We provide only a short example of relevant ADIT artifacts we use for supporting the generation of a CC-compliant *ST/PP* document. We show an example how a *ToE* description can be generated using the *UML4PF CC-system extension* and a list of assets. We provide example artifacts for a *ST/PP*: the *ToE Reference* in the *Introduction* and the *Assets* in the *Security Problem Definition*. The *ToE Reference* for a *PP* can be generated from the information collected in the *context diagram* and *technical context diagram*. We use an OCL expression to extract the collected information. Table 8.9 presents the output of the example reference. We collected the information by an OCL expression that collects the information contained in the applied stereotypes ≪CC-system≫ and ≪ToE≫.

We can automatically generate a list of *Assets* using OCL on the information contained in the domain knowledge diagram (see Fig. 8.6). We use the OCL expression AE01DOC (see Table 8.2) and we show AE01DOC in Fig. 8.9. This expression selects all classes, whose name contains the String *Asset*. The expression further creates a sequence that contains the *name* of the classes and the values of the attributes *description* and *needForProtection*. These values are separated using a ";". We use our tool support to transform the resulting String to a LATEX table (see Sect. 8.6 for more details on the support tool). The resulting table is shown in Table 8.10.

Table 8.9 PP reference—generated by our tool support

Title	Protection profile for the gateway of a smart metering system (Gateway PP)
Version	01.01.01(Final draft)
Date	25.08.11
Authors	Dr. Helge Kreutzmann, Stefan Vollmer (BSI), Nils Tekampe and Arnold Abromeit (TÜV Informationstechnik GmbH)
Registration	Bundesamt für Sicherheit in der Informationstechnik (BSI) Federal office for information security Germany
Certification-ID	BSI-CC-PP-0073
CC-Version	3.1
Keywords	Smart metering, protection profile, meter, gateway, PP

```
 1  Property.allInstances()->select(
 2  getAppliedStereotypes().name->includes('Asset'))
 3  ->collect(c |
 4    let st: Stereotype =
 5  c.getAppliedStereotypes()->select(name='Asset')->
 6        asSequence()->first() in
 7      c.name.oclAsType(String).concat(';')
 8        .concat(c.getValue(st,'description').
 9        oclAsType(String)).oclAsType(String)
10        .concat(';')
11        .concat(c.getValue(st,'needForProtection')
12        .oclAsType(String))
13  )
```

Fig. 8.9 AE01DOC—collecting assets and their attributes

The Common Criteria demands a *cross-table* as part of the security objectives section of the ST and PP. The cross table presents the security objectives rational, which analyzes the relations between threats, assumptions, organizational security policies and security objectives, as well as security objectives for the environment. The cross-table for our example is depicted in Table 8.11. It shows all security objectives and security objectives for the environment states in the smart metering gateway PP on the horizontal axis and all threats the objectives mitigate on the vertical axis. If an element on the vertical axis is addressed by an objective on the horizontal axis, the box is marked by an "x". For the security rational it is essential that all threats are addressed by at least one security objective or security objective for the environment. Assumptions can only be addressed by security objective for the environment according to the Common Criteria. Organizational security policies have to be considered by at least one security objective or security objective for the environment. Whether the threats, assumptions, and organizational security policies are addressed reasonably is to be determined by a CC certification body. The cross-table is of utmost importance for a structured argumentation.

We use OCL expression to collect all threats, assumptions, and OSPs. In addition, we collect the relations between these threats, assumptions, and OSPs and the security objectives for the environment. We check if the objectives have dependencies with the stereotype ≪mitigates≫ to threats or dependencies with the stereotype ≪considers≫ to assumptions or OSPs. Afterwards, our support tool creates a LATEX table (Sect. 8.6 for more details on the support tool). Table 8.11 is generated for our example using our support tool.

Reusing and extending ADITs Traceability and Consistency checks For the ADIT development process, more than 70 traceability and consistency checks between the different diagrams have been defined (Côté 2012). For example, it can be checked that each element in a problem diagram has a relation to the context diagram. A possible relation is that an element in the problem diagram is part of an element in the context diagram. Such relations refine security properties. These are essential in order to protect the ToE against threats.

Table 8.10 Table for assets of the PP/ST—generated by our tool support

Asset	Description	Need for protection
MeterData	Meter readings that allow calculation of the quantity of a commodity, e.g., electricity, gas, water, or heat consumed over a period. Meter data comprise consumption or production data (billing-relevant) and grid status data (not billing-relevant). While billing data needs to have a relation to the consumer, grid status data do not have to be directly related to a consumer	According to their specific need
Consumption data	Billing-relevant part of meter data Please note that the term consumption data implicitly includes production data	Integrity and authenticity (comparable to the classical meter and its security requirements), Confidentiality (due to privacy concerns)
Data/user data	The terms data or user data are used as a hyperonyms for meter data and supplementary data	According to their specific need
Supplementary data	The Gateway may be used for communication purposes by devices in the LMN or HAN. It may be that the functionality of the Gateway, that is used by such a device, is limited to pure (but secure) communication services. Data that is transmitted via the Gateway but that does not belong to one of the aforementioned data types is named supplementary data	Integrity and authenticity (comparable to the classical meter and its security requirements), Confidentiality in the WAN (due to privacy concerns)
Status data	Grid status data, subset of meter data that is not billing-relevant	Integrity and authenticity (comparable to the classical meter and its security requirements), Confidentiality (due to privacy concerns)
Gateway config	Configuration data of the Gateway to control its behavior including the Gateway identity and the access control profiles	Integrity and authenticity, Confidentiality
Firmware update	Firmware update that is downloaded by the ToE to update the firmware of the ToE	Integrity and authenticity
Gateway time	Date and time of the real-time clock of the Gateway. Gateway time is used in meter data records sent to external entities	Integrity and Authenticity (when time is adjusted to an external reference time)
CLS config	Configuration data of a CLS to control its behavior	Integrity and authenticity, Confidentiality
Firmware	The firmware of the ToE	Integrity, Authenticity
MeterConfig	Configuration data of the meter to control its behavior including the meter identity	Integrity and authenticity, Confidentiality

Table 8.11 Common criteria cross table generated by our tool support

	SO.Access	SO.Conceal	SO.Crypt	SO.Firewall	SO.Log	SO.Management	SO.Meter	SO.Protection	SO.SeparateIF	SO.Time	SO_OE.ExternalPrivacy	SO_OE.Network	SO_OE.PhysicalProtection	SO_OE.Profile	SO_OE.SM	SO_OE.TrustedAdmins	SO_OE.Update
T.ResidentData	X			X				X								X	
T.InfrastructureMeter		X						X								X	
T.DisclosureWAN		X		X	X			X	X							X	
T.ResidualDataWAN																X	
T.InfrastructureGateway		X				X	X	X								X	
T.TimeModification						X		X		X			X			X	
T.ResidualDataWan								X									
T.TimeModificationPhysical						X		X								X	
T.InfrastructureCLS			X	X				X	X							X	
T.ResidualDataPhysical						X		X								X	
T.ResidentDataWan								X					X				
T.Privacy				X			X	X	X						X	X	
T.ResidentDataWAN	X																
T.DataModificationWAN		X		X				X								X	
T.DataModificationLocal		X				X	X	X					X			X	
T.DisclosureLocal		X				X	X	X					X			X	
A.AccessProfile														X			
A.ExternalPrivacy											X						
A.Network												X					
A.PhysicalAttacker													X				
A.PhysicalProtection													X				
A.TrustedAdmins																X	
A.Update																	X
A.WLANAttacker												X					
OSP.Log	X			X	X			X								X	
OSP.SM			X		X			X							X	X	

We wrote more than 50 OCL expressions for model consistency checks, security reasoning, and document generation. These expressions are based on the existing checks of UML4PF. We show the expressions for security reasoning and model validation in Appendix A.

8.6 Tool Support

Our method is based on tool support, otherwise the manual creation of all diagrams and manual mapping to textual documents would be very costly in terms of time. We based our tool on the UML4PF tool (Sect. 2.5.3) and named our tool

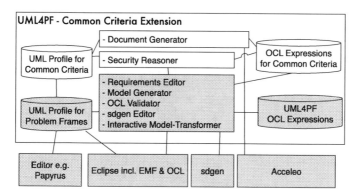

Fig. 8.10 UML4PF common criteria extension support tool

UML4PF-CC.[2] Figure 8.10 shows the architecture of UML4PF-CC. The white boxes in Fig. 8.10 state components that we implemented specifically for UML4PF-CC and the gray boxes are components that we reused from the UML4PF tool.

In the following, we list the functionalities and items of UML4PF-CC:

- The *UML Profile for Common Criteria* defines the relevant stereotypes for the Common Criteria, e.g., ≪ToE≫.
- The *Document Generator* uses the ACCELEO[3] model-to-text transformation tool, which is also an Eclipse plug-in. The *Document Generator* creates HTML and LATEX documents (see Fig. 8.13) from UML4PF-CC models.
- The *Security Reasoner* contains several OCL expressions that support security reasoning based on UML4PF-CC models, e.g., if all attackers are considered. The reasoner is based on the OCL validator. The security reasoner is depicted in Fig. 8.12.
- The *OCL Validator* checks if a model is valid and consistent. The *OCL Validator* has been extended to use specific *OCL expressions* for UML4PF-CC models (see Fig. 8.11). We contributed these OCL expressions, as well.

We illustrate an overview of the UML4PF GUI in Figs. 8.12 and 8.13. The graphical representation of our domain knowledge diagram that contains assets is shown on the top (see Fig. 8.14). The bottom shows the results of the *OCL Validator* and the *Security Reasoner* for the entire model. Results with a green background mean that the checks were successful, and red background means the expression has identified a problem (see Fig. 8.15[4]).

---

[2]This extension is available under the following homepage: http://www.uml4pf.org/ext-cc/.

[3]The ACCELEO homepage: http://www.acceleo.org/pages/home/en.

[4]Note that the in the upper left corner the tool states that the validation is suspended. The reason is that we paused the validation for taking this screenshot.

Fig. 8.11 UML4PF-CC screenshot: OCL validator menu

8.7 Discussion of Our Results with Practicioners

Our method was developed based on the experience from several security and especially Common Criteria projects. To illustrate the procedure, we used a case study that creates a Security Target for an existing Protection Profile. The method was discussed with two security consultants, who have already applied parts of the method in industrial projects. In Common Criteria projects, cross-tabulations are created for checking the consistency of documents. Especially, the effort for this task is significantly reduced by the presented method and tool. The security consultants also mentioned that this structured procedure

- helps to describe the attackers' abilities in more detail,
- supports to identify all threats to the given assets,
- helps not to forget relevant assumptions or facts, and
- supports to identify and classify assets.

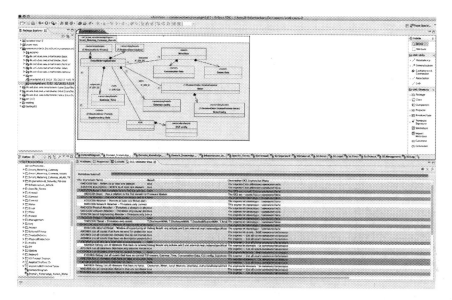

Fig. 8.12 UML4PF-CC screenshot: overview

Fig. 8.13 UML4PF-CC screenshot: generated LATEX documents

We also asked two evaluators, who check Common Criteria documentations. They responded that they prefer the graphical representation used in our method instead of the plain text and tables in current Common Criteria documents.

The evaluators mentioned the following limitations of our method:

- The amount of text in a class is sometimes distracting.
- The modeling is time consuming.
- The problem frame notation has to be learned beforehand, and
- Our method does not support the entire process of Common Criteria certification.

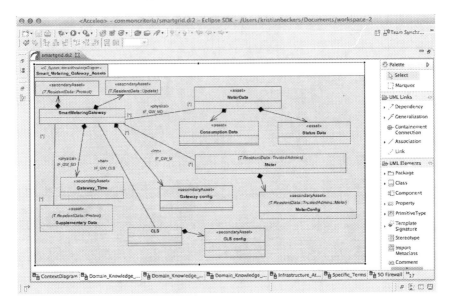

Fig. 8.14 UML4PF-CC screenshot: domain knowledge diagram with assets

Validation suspended!		

OCL Expression Name Result Description

FA01CON Fact – Re true	The expression collects all facts and checks if these have a dependency with a refersTo stereotype (line 1),
AS01CON Assumpt true	The expression collects all assumptions (line 1) and checks that the number of dependencies is greater than
AE01CON Asset – F false	The OCL expression collects all classes with the ToEs (line 10) and Asset (line 20) stereotypes and also all pr
AE01CON Asset Firmw	The OCL expression collects all classes with the ToEs (line 10) and Asset (line 20) stereotypes and also all pr
AT01CON Attacker false	The expression collects all attackers (line 1–15), collects attackers that have no dependencies (line 18–25),
AT01CON Attack	The expression collects all attackers (line 1–15), collects attackers that have no dependencies (line 18–25), ;
NA01CON Netwc	This expression collects all network attackers (line 1–15), checks if all dependencies are threats (line 18–20)
PA01CON Physical true	The expression collects all physical attackers (line 1–22), checks if these cause threats (line 25–45) and that
SA01CON Software true	The expression collects all software attackers (line 7–14), checks if these have dependencies (line 17), check
SE01CON Social Eng true	The expression collects all social engineering attackers (line 7–14), if these have dependencies (line 17),
TH01CON Threat – false	The expression collects all threats (lines 7–14) and checks if these threaten all assets (line 16).
TH01CON Threa T.Data	The expression collects all threats (lines 7–14) and checks if these threats threaten all assets (line 16).
OT01CON observe1 false	The expression collects all observe threats (line 1–11), and checks if any of the attackers causing the threats
OT01CON obser Debug	The expression collects all observe threats (line 1–11), and checks if any of the attackers causing the threats
AE02REA List all as false	The expression collects all assets (line 1–14) and checks if the 'needForProtection' attribute is empty (line 19
AE02REA Debug Meter	The expression collects all assets (line 1–14) and checks if the 'needForProtection' attribute is empty (line 19
AE03REA List all co true	The expression collects all classes with the stereotype <ConnectionDomain> (line 3–14),
AE04REA List all as false	The expression collects all classes with the stereotype <Asset> (line 1–14),
AE04REA Debug Meter	The expression collects all classes with the stereotype <Asset> (line 1–14),
AS01REA List all do false	The expression collects all domains (lines 3–14), and all assumptions and their dependencies (lines 16–19).
AS01REA Debug Debug	The expression collects all domains (lines 3–14), and all assumptions and their dependencies (lines 16–19).
AT01REA List all att true	The expression collects all attackers (lines 4–15)
CT01REA List all as false	The expression collects all assets (line 1–12),
CT01REA Debug Status	The expression collects all assets (line 1–12),

Fig. 8.15 UML4PF-CC screenshot: OCL reasoner overview

8.8 Related Work

We have analzyed related work in the knowledge area of security requirements engineering methods.

Schmidt proposes the problem frame-based method *Security Engineering Process using Patterns (SEPP)* (Schmidt 2010), which introduces security concerns into the problem frame notation, e.g., an attacker. The author decomposes the security issues of a system into security problem frames. Each security problem frame concerns a particular security concern of the system in relation to existing functional requirements. The author refines these into concretized security requirements, which also state possible solutions for the security concerns. The author maps the artifacts produced by the SEPP method to the Common Criteria. However, the mapping is proposed, but not shown in an example or integrated into the method itself. Moreover, the SEPP method does not use the CC terminology like ToE and does not support CC-specific document creation.

Mellado et al. (2006a) created the Security Requirements Engineering Process (SREP). SREP is an iterative and incremental security requirements engineering process. In addition, SREP is asset-based, risk driven, and follows the structure of the Common Criteria. The method uses use cases to model security objectives, and misuse cases to model threats. The authors also developed a template for ranking threats, attacks, and risks. They propose a Security Resources Repository (SRR) that can store elicited threats, attacks, and risks. The method differs from our work in the sense that SREP is a method that supports the security reasoning according to the CC. The authors use misuse cases for eliciting threats and their method does not provide clear criteria to decide when all threats are elicited.

Rottke et al. (2002) present a problem-driven requirements engineering method for CC-compliant systems. This high level method also considers problem frames. The method focuses on creating reliable models for context and problem descriptions. This work differs from ours, because we do not limit our method to context and problem descriptions.

Yin and Qiu (2010) model so-called *early-phase* security requirements with an extended i* model. The authors also describe security policies using a formal model and so-called *late-stage security requirements* in an extended UML model. The extended i* framework adds the modeling element *security flaw*, which can have a relation to goals and soft goals. The goals can be influenced by a threat and eliminated by a security goal, e.g., confidentiality. The policies provide three templates for so-called *stream control*, which specifies rules for network traffic. For example, allowed IP addresses. The extended UML model considers explicitly for each element if it belongs to the ToE, external entities, or communication entities. The method differs from ours, because it focuses on generating CC policies for *stream control*. The method does not aim at providing a holistic support for the generation of *ST/PP* documents.

Abuse Frames are a method for analyzing security issues and the corresponding threats and vulnerabilities by using problem frames (Lin et al. 2004). So-called anti-requirements and the corresponding abuse frames are defined. An anti-requirement expresses the intentions of a malicious user, and an abuse frame represents a security threat. In contrast to our method, abuse frames do not consider specific notions of the common criteria and do not support computer-aided security reasoning, e.g., for missing threats.

Mayer et al. present a conceptual model called *Information System Security Risk Management (ISSRM)* (Mayer et al. 2007; Mayer 2009). The model defines terms and notions of risk management for IT systems with regard to security and relates this conceptual model to definitions in standards like Common Criteria. ISSRM does not provide a structured method for creating Common Criteria documentation.

The Security Requirements Engineering Framework (SREF) (Haley et al. 2008; Moffett et al. 2004) considers security requirements elicitation and analysis with problem frames. The method is an iterative process consisting of four steps. Step one aims to identify business (or functional) requirements. Step two identifies assets and relevant security goals for these assets. Step three concerns to identify security requirements, which are refinements of the security goals to functionalities of the system-to-be. The last step is to reason about why the functionalities fulfill the security goals. In contrast to our work Haley et al. (2008), Moffett et al. (2004) do not support Common Criteria-compliant security analysis or document generation.

Several works focus on ontologies for the Common Criteria.

Bialas (2009) introduces an ontology that supports the CC security problem definition (SPD). The SPD contains threats, security policies, and assumptions concerning the ToE. The ontology provides relations between security-related elements, e.g., risks and threats. The relations can be used to create an SPD. For example, the ontology maps specific threats to specific risks. In addition, the ontology can be queried to find countermeasures for specific risks. The author extends the method to a IT security development framework, which is complaint with the CC (Białas 2009). The method differs from ours, because the author focuses on creating just the *SPD* and not a holistic support for generating *ST/PP* documents. Nevertheless, Bialas work can complement our own. The stored threats and their relation to the ToE could be implemented as a function to suggest threats in our method.

Chang and Fan (2010) design an ontology that is intended to decrease the time for CC certification. The ontology supports four different use cases. The first is to query content of the CC standard using a tree. The second use case considers a markup tool that allows the user to mark specific parts of the CC. These marks can contain a choice of predefined comments that can be used to ease the review of CC documents. The third use case considers a CC review tool that can provide a checklist of required documents for evaluating a ToE for a specific EAL. The CC review tool also contains information about required documents. This includes already written and approved documents and documents that have to be revised. The last use case concerns a review report tool. This tool provides an assessment of the review process using the data from the previous use case. This work can complement our own. We could use their ontology after generating documents with our method.

Automated Risk and Utility Management (AURUM) is a method for supporting the NIST SP 800-30 risk management standard (Ekelhart et al. 2009). The method is based upon an ontology that supports the elicitation of threats, choosing fitting countermeasures, and calculating risks. In contrast to our work, AURUM focuses exclusively on risk management.

Some works also exist with the aim to improve the Common Criteria or use it as an input for security analyzes.

Ardi and Shahmehri (2009) extend the CC Security Target document with the knowledge of existing vulnerabilities. In particular, the authors add threats from known vulnerabilities to the Security Problem Definition, security objectives from vulnerabilities, and information on how to consider these vulnerabilities in the Security Objectives section. The authors use vulnerability cause graphs and security activity graphs to refine the information from the vulnerabilities. This work can complement our own. We can use the information about existing vulnerabilities in our process as well.

Schneider et al. (2012) use organizational learning to check software documentation for relevant parts to elicit security requirements. The basis for the organizational learning software the authors use is the Common Criteria. This work differs from our own, because we aim to create Common Criteria documentation, while the work of Schneider et al. uses the content of the Common Criteria standard to identify relevant parts for security requirements elicitation in software documentation.

We looked in state-of-the-art threat research of security analysis based on the data flow diagrams as proposed by Microsoft (Howard and Lipner 2006).

Dhillon (2011) models the flow of information in a system and investigates possible interaction points of an attacker with the system. The author proposes to use annotations on the models for security relevant information, e.g., authentication data flows. These annotations are used to check a database for possible threats, but the work does not focus on supporting security standards.

8.9 Summary

We have extended the ADIT process (Sect. 2.5.3) that is based on Jackson's problem frame method (Sect. 2.5.3). Our extension is a threat analysis and security requirements elicitation compliant to Common Criteria. Our threat analysis considers attacker types that threaten specific kinds of Jackson's domains. Thereby, we built on the existing UML4PF tool, and its UML profile for dependability (Hatebur 2012). Our contribution is a structured method for problem-based threat analysis including several OCL expressions, which provide validation, security reasoning, and document generation support. Security reasoning is meant in the sense that we can check for completeness of the considered attackers during the threat analysis. Our method includes a structured elicitation, documentation, and validation of assets, assumptions, threats, attackers, and security requirements.

Our method offers the following main benefits:

- A structured process for elicitation of threat analysis elements for a Common Criteria certification
- A tool-supported identification of assets, assumptions, and threats
- Support for the reasoning of Common Criteria threats based upon attacker types for Jackson's domain types

- Explicit consideration of domain knowledge in terms of facts and assumptions about attackers, the environment, existing security controls.
- The method can also be used without the Common Criteria. The UML profile and the OCL constraints can be adapted with little effort to other security standards or methods.
- Computer-aided generation of tables and figures for each chapter of a Common Criteria *Protection Profile* or *Security Target*
- Support for the reasoning about Common Criteria *Security Objectives* and *Security Objectives for the Environment*
- Traceability from elements of the *ToE* description to threats to Common Criteria *Security Objectives*
- Consistency checks of all elements and diagram of the UML4PF-CC model

We created a UML-based method to support the security analysis and documentation demands of the Common Criteria. Our method is tool supported and has the ability to check models for completeness, validate the models, and generate textual documents from these models. Our method has the potential to introduce model-based analysis for the Common Criteria certification, which is currently based on textual documents and tables. These models can support the discussions about security issues and support a structured threat analysis.

References

Ardi, S., & Shahmehri, N. (2009). Introducing vulnerability awareness to common criteria's security targets. In *Proceedings of the Fourth International Conference on Software Engineering Advances. ICSEA* (pp. 419–424). IEEE Computer Society.

Beckers, K., Côté, I., Hatebur, D., Faßbender, S., & Heisel, M. (2013a). Common criteria compliant software development (CC-CASD). In *Proceedings 28th Symposium on Applied Computing* (pp. 937–943). ACM.

Beckers, K., Hatebur, D., & Heisel, M. (2013b). A problem-based threat analysis in compliance with common criteria. In *Proceedings of the International Conference on Availability, Reliability and Security (ARES)* (pp. 111–120). IEEE Computer Society.

Beckers, K., Hatebur, D., & Heisel, M. (2014). Supporting common criteria security analysis with problem frames. *Journal of Wireless Mobile Networks, Ubiquitous Computing, and Dependable Applications (JoWUA)*, 5(1), 37–63.

Bialas, A. (2009). Ontology-based security problem definition and solution for the common criteria compliant development process. In *Proceedings of the Fourth International Conference on Dependability of Computer Systems. DepCos-RELCOMEX* (pp. 3–10). IEEE Computer Society.

Bialas, A. (2009). Ontological approach to the it security development. In E. Tkacz & A. Kapczynski (Eds.), *Internet—technical development and applications* (Vol. 64, p. 261–269). Springer Berlin/Heidelberg.

BSI. (2011). Protection Profile for the Gateway of a Smart Metering System (Gateway PP) (Version 01.01.01(final draft)). Bonn, Germany: Bundesamt für Sicherheit in der Informationstechnik (BSI)—Federal Office for Information Security Germany. (https://www.bsi.bund.de/SharedDocs/Downloads/DE/BSI/SmartMeter/PP-SmartMeter.pdf?_blob=publicationFile).

BSI. (2013). Protection Profile for the Security Module of a Smart Meter Gateway (Security Module PP) (Version 1.0)). Bonn, Germany: Bundesamt für Sicherheit in der Informationstechnik (BSI)—Federal Office for Information Security Germany. https://www.commoncriteriaportal.org/files/ppfiles/pp0077b_pdf.pdf).

Chang, S.-C., & Fan, C.-F. (2010). Construction of an ontology-based common criteria review tool. In *Proceedings of the 2010 International Computer Symposium (ICS)* (pp. 907–912). IEEE Computer Society.

Côté, I. (2012). *A systematic approach to software evolution*. Baden-Baden: Deutscher Wissenschafts-Verlag.

Dhillon, D. (2011). Developer-driven threat modeling: Lessons learned in the trenches. *IEEE Security and Privacy, 9*(4), 41–47. IEEE Computer Society.

Ekelhart, A., Fenz, S., & Neubauer, T. (2009). AURUM: A framework for information security risk management. In *Proceedings of the Hawaii International Conference on System Sciences (HICSS)* (pp. 1–10). IEEE Computer Society.

Fabian, B., Gürses, S., Heisel, M., Santen, T., & Schmidt, H. (2010). A comparison of security requirements engineering methods. *Requirements Engineering—Special Issue on Security Requirements Engineering, 15*(1), 7–40.

Haley, C. B., Laney, C. R., Moffett, D. J., & Nuseibeh, B. (2008). Security requirements engineering: A framework for representation and analysis. *IEEE Transactions on Software Engineering, 34*(1), 133–153.

Hatebur, D. (2012). *Pattern and component-based development of dependable systems*. Baden-Baden: Deutscher Wissenschafts-Verlag.

Howard, M., & Lipner, S. (2006). *The security development lifecycle: SDL: A process for developing demonstrably more secure software*. Redmond: Microsoft Press.

ISO/IEC. (2012). Common Criteria for Information Technology Security Evaluation (ISO/IEC 15408). Geneva, Switzerland: International Organization for Standardization (ISO) and International Electrotechnical Commission (IEC).

Jackson, M. (2001). *Problem frames. Analyzing and structuring software development problems*. New York: Addison-Wesley.

Lin, L., Nuseibeh, B., Ince, D. C., & Jackson, M. (2004). Using abuse frames to bound the scope of security problems. In *Proceedings of the Requirements Engineering Conference (RE)* (pp. 354–355). IEEE Computer Society.

Lund, M. S., Solhaug, B., & Stølen, K. (2010). *Model-driven risk analysis: The CORAS approach* (1st ed.). Berlin: Springer.

Massacci, F., Mylopoulos, J., & Zannone, N. (2010). Security requirements engineering: The SI* modeling language and the secure tropos methodology. *Advances in Intelligent Information Systems, 265*, 147–174.

Mayer, N. (2009). *Model-based management of information system security risk*. Unpublished doctoral dissertation, University of Namur.

Mayer, N., Heymans, P., & Matulevicius, R. (2007). Design of a modelling language for information system security risk management. In *Proceedings of the International Conference on Research Challenges in Information Science (RCIS)* (pp. 121–132). IEEE Computer Society.

Mellado, D., Fernandez-Medina, E., & Piattini, M. (2006a). A comparison of the common criteria with proposals of information systems security requirements. In *The First International Conference on Availability, Reliability and Security. ARES* (pp. 654–661). IEEE Computer Society.

Moffett, J. D., Haley, C. B., & Nuseibeh, B. (2004). Core security requirements artefacts. Technical Report No. 2004/23. Milton Keynes, United Kingdom: The Open University, UK.

Rottke, T., Hatebur, D., Heisel, M., & Heiner, M. (2002). A problem-oriented approach to common criteria certification. In *Proceedings of the 21st International Conference on Computer Safety, Reliability and Security* (pp. 334–346). Berlin: Springer.

Schmidt, H. (2010). *A pattern- and component-based method to develop secure software*. Baden-Baden: Deutscher Wissenschafts-Verlag.

Schneider, K., Knauss, E., Houmb, S., Islam, S., & Jürjens, J. (2012). Enhancing security requirements engineering by organizational learning. *Requirements Engineering, 17,* 35–56.

UML Revision Task Force. (2010a). OMG object constraint language: Reference.

UML Revision Task Force. (2010b, May). OMG unified modeling language: Superstructure.

van Lamsweerde, A. (2009). *Requirements engineering: From system goals to UML models to software specifications* (1st ed.). Chichester: Wiley.

Yin, L., & Qiu, F.-L. (2010). A novel method of security requirements development integrated common criteria. In *Proceedings of the International Conference on Computer Design and Applications (ICCDA)* (pp. 531–535). IEEE Computer Society.

Chapter 9
Supporting ISO 26262 Hazard Analysis with Problem Frames

Abstract Engineering safe software for the automotive domain is challenging due to ever more complex systems. The safety standard ISO 26262 provides car manu-factures in this effort with a guideline. In particular, the standard describes a hazard analysis and risk assessment for automotive systems to determine the necessary safety measures to be engineered for a specific feature. However, the standard contains ambiguous descriptions. We propose a structured method for hazard analysis and risk assessment based on a requirements engineering method that incorporates the problem frame approach to resolve these ambiguities. The concepts and terminology of the standard concepts and its terminology are represented by a UML profile that contains respective stereotypes. The UML model allows precise validation of several consistency constraints expressed in OCL. These artifacts are integrated in a structured method for safety engineering compliant with ISO 26262. We illustrate the application of our method with the example of an electronic steering column lock system.

9.1 Introduction

We introduced the concept of extending the problem frame notation and the ADIT process (Sect. 2.5.3) to support the Common Criteria standard (Sect. 2.2.4) in Chap. 8. This chapter proposes to transfer this idea to the knowledge area of safety. This shows the applicability of our concept to support the establishment of standards with ADIT and problem frames beyond the knowledge area of security. The difference between the areas is that in security we try to protect the machine from an attacker, in safety we try to protect the environment from the machine. In our example, we consider a standard for safety, as well. The released ISO 26262 standard (ISO 2011) requires a hazard analysis and risk assessment for automotive systems to determine the necessary safety measures to be implemented for a certain feature. This work benefited from the expertise of Thomas Frese and Denis Hatebur. Both are experts in the area of automotive safety for many years. They developed a structured method for hazard analysis and risk management in compliance to the ISO 26262 standard. In this work, we introduce UML models to the method, define OCL consistency checks for these models and apply our contribution to a real-life case study.

© Springer International Publishing Switzerland 2015
K. Beckers, *Pattern and Security Requirements*,
DOI 10.1007/978-3-319-16664-3_9

The hazard analysis and risk assessment are based on UML4PF, similar to the work presented in Chap. 8 for threat analysis. We created rigorous validation of several constraints expressed in OCL in this method, as well. We illustrate our method using an electronic steering column lock system.

This chapter is based on Beckers et al. (2013) and organized as follows. We introduce the background relevant for this chapter on the ISO 26262 standard in Sect. 2.3 and on UML4PF in Sect. 2.5.3. This chapter is organized as follows. We explain the difficulties of conducting an ISO 26262 Hazard Analysis in Sect. 9.2. Our method is presented in Sect. 9.3. This section also describes our UML profile, which is used to express the hazard analysis and risk assessment. Based on this profile, we define the validation conditions, as well. Tool support is outlined in Sect. 9.4. We introduce an example of an electronic steering column lock system as a case study in Sect. 9.5. Section 9.6 presents related work, while Sect. 9.7 summarizes our results.

9.2 Challenges in an ISO 26262 Hazard Analysis

The automotive standard for road vehicles ISO 26262 is an automotive industry standard for developing functional safety systems, and offers the ability to achieve a consistent functional safety process. The automotive standard for road vehicles has been released in November 2011. Its scope covers electronic and electric (E/E) systems for vehicles with a max gross weight up to 3500 kg. Since ISO 26262 is a risk-based functional safety standard addressing malfunctions, its process starts with a hazard analysis to determine the necessary risk reduction to achieve an acceptable level of risk. The necessary risk reduction is described by an automotive safety integrity level (ASIL).

Performing such a hazard analysis is a difficult task because

- The result of a hazard analysis needs to be safety goals with an appropriate ASIL that can be the starting point for further development.
- It should be possible to review the hazard analysis within a realistic time period.
- It should be comprehensible for different stakeholders, e.g., engineers, project leaders, and managers.
- Hazard analyses of different projects should be comparable.
- In a hazard analysis, all relevant faults or situations need to be considered.

We propose a structured method based on UML environment models supported by a tool. We start as required by ISO 26262 with the item definition. According to ISO 26262, the item is a set of functions realized by the system to be built. To support the item definition, we use a UML-based context diagram containing the item, its environment, and the system border. The requirements describe the functions of the item by referring to elements of the environment.

The main contributions of our structured method are:

- A focused analysis by eliminating faults or situations that are not relevant for a particular hazard analysis. The analysis is based on a set of *fault-type guide-words* and a hierarchically organized set of situations. We conduct our reasoning in the beginning of the hazard analysis, because it results in a limited set of relevant faults and situations. This helps safety experts using their time to focus only on this set instead of analyzing all possible faults and situations.
- A UML (UML Revision Task Force 2010c) Profile for expressing all elements of a hazard analysis in compliance with the ISO 26262 standard.
- Our UML profile provides the basis for object constraint language (OCL) (UML Revision Task Force 2010a) validation checks, as well. The OCL validation checks concern consistency and correctness of the hazard analysis model. Thus, we provide a computer-aided technique to discover errors in the hazard analysis caused by inconsistencies or errors in one or more UML diagrams.
- A mapping between the table-based representation of the hazard analysis and the UML model. The mapping allows the import of existing hazard analyses into the UML model, on which we use our validation conditions. We can also export the content of the UML model to a table-based representation for printing and reviewing.

9.3 A Hazard Analysis and Risk Assessment Method

We propose a method for performing a hazard analysis and risk assessment according to ISO 26262. The aim of the analysis is to identify and classify the potential hazards of the item and to formulate safety goals related to the prevention or mitigation of these hazards in order to achieve an acceptable residual risk. The method consists of the following steps, depicted in Fig. 9.1.

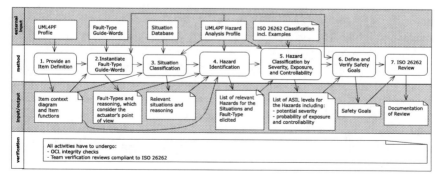

Fig. 9.1 A method for hazard analysis compliant to ISO 26262

1. Provide an Item Definition—ISO 26262 demands a definition of the item, its basic functionality, and its environment. We use the UML4PF profile to represent this description. The initial description of the item is provided in a context diagram that presents the item and the domains surrounding it, e.g., the driver, (see Fig. 9.6).

The functions of the item are defined by requirements referring to or constraining domains in its environment (Sect. 2.5.3). Instead of using the stereotype ≪machine≫, for the ISO 26262 item definition we use the stereotype ≪item≫ to define the domain to be developed.

It is also important to ensure that this step is performed on the right level of detail. It should be avoided to have a too detailed level with too many functions/subfunctions in order to make the hazard analysis assessable.

2. Instantiate Fault-Type Guide-Words—We propose a set of so-called *fault-type guide-words* inspired by the standard for Hazard and Operability Analysis (HAZOP) (IEC 2005). The guide-words help the developer to consider all relevant faults. Typical guide words are *no, unintended, early, late, more, less, inverted and intermittent*. Each guide-word has to be instantiated for the functions specified in the *item definition* in the previous step. In the context of a certain function, not all fault-types have to be considered. For example the fault-type "more" is not relevant for a function with a boolean output. For many time-critical functions, a fault-type "late" leads to the same hazard as the fault-type "no." Usually, it is helpful to start the fault-type consideration from the actuator's point of view and not from the sensor's, because the task of the fault-type consideration is not a verification of an existing design. This will be done with appropriate safety analyses, e.g., Failure Mode and Effects Analysis (FMEA) (DIN EN 2006) and Fault Tree Analysis (FTA) (U.S. Department of Defense 1998, Sect. 7.9), which take place after the ISO 26262 hazard analysis. For all combinations of function and fault, we describe how the system behaves in presence of the malfunction.

We support this step with a UML profile that can be used to express the different artifacts. Figure 9.2 shows the part of the profile that is used to express the faults of an actuator constrained by a functional requirement. A class with the stereotype ≪fault≫ is used to describe the faults. Each fault may be in one of the fault-types *no, unintended, early, late, more, less, inverted and intermittent*. For each requirement, all fault-types are checked if they have to be considered, are not possible, or are covered by other faults (this is checked by condition 2C01CF in Table 9.1[1]). A dependency with the stereotype ≪fails≫ is used to show the relation between requirements and faults (see Table 9.1, 2M02HF). Additionally, fault constraints can be described. For each considered fault, we describe the effect on the system level (see Table 9.1, 2C03EF) and not on component or vehicle level. The system level concerns effects based on relevant driving and operating situations from a birds perspective; the component level considers effects related to electronic control units (ECU); the vehicle level concerns the perspective of the driver of the vehicle.

---

[1]The first number refers to the step in our hazard analysis method, C is for consistency checks, M is for checks considering correct modeling, the next number is the number of the check within the step, and the last characters are an abbreviation of the description.

Fig. 9.2 Fault, hazard, assessment

On the system level, the elements of the item are visible, e.g., actuators. In vehicle level descriptions, only phenomena that can be observed or controlled by the driver or other persons are used. The component level contains descriptions of internal interfaces, e.g., CAN (ISO 2003) messages. For faults rated not to be considered, either a description why it is not relevant or a reference to at least one other fault using the attribute "covered by" has to be specified (see Table 9.1, 2C04DC).

3. Situation Classification—From project-experience, examples in the ISO 26262, national and international working groups, a hierarchically organized list of situations is created and stored in a database. Using this list, we rate if a situation is relevant for the described item with its requirements or not. If a more abstract situation is rated, it is not necessary to rate the special situations (see Table 9.1 3C01RE). If a situation is rated as not being relevant, either a reference to another situation that includes this situation is given, or a rationale is provided (see Table 9.1 3C02RR). The hierarchically organized list of situations is updated if new aspects are identified in projects in order to reduce the risk of forgetting hazardous situations.

In the UML profile, a number of situations are defined as stereotypes:

- Situations are, e.g., ≪standstillSituation≫, ≪maneuverSituation≫, and ≪drivingSituation≫.
- Special standstill situations are ≪standstillEngineOffSituation≫ (e.g., in parking lot) and ≪standstillEngineOnSituation≫.
- Special maneuver situations are for example ≪parkingManeuverSituation≫ and in addition the ≪drivingBackwardSituation≫.
- Special driving situations are on the one hand driving activities such as ≪brakingSituation≫, ≪accelerationSituation≫, ≪steeringSituation≫, and ≪rollingSituation≫.
- On the other hand, driving areas exist such as ≪citySituation≫, ≪countryRoadSituation≫, and also ≪highwaySituation≫.

4. Hazard Identification—For each fault/function combination, all situations that could lead to a potential hazard are identified in the list of situations being relevant. We describe the effect on the vehicle level, i.e., what behavior could occur in case of a potential item's malfunction. Based on the effect on the vehicle level, we describe the hazards and possible consequences. Hazards are defined in terms of the conditions or events that can be observed at the vehicle level (e.g, by the driver). A verbal description of consequences without ranking is given. In this step, also

Table 9.1 Validation conditions

Step	ID	Condition
2	2C01CF	All fault-types are considered for functional requirements in the item definition that constrain a domain
2	2M02HF	Dependencies with the stereotype ≪fails≫ only point from classes with the stereotype ≪requirement≫ to classes with the stereotype ≪fault≫
2	2C03EF	If a fault is rated as to be considered, the effect on system level is described (string not empty)
2	2C04DC	If a fault is rated as not to be considered, a reference to at least one other fault using the attribute covered by is provided or a description with a rationale is provided
3	3C01RE	All situation types are rated if they are relevant for the feature or not. If a more general situation is rated, it is not necessary to rate the detailed (derived) situations
3	3C02RR	If a situation is rated as not being relevant, a rationale is given (rationale string shall not be an empty string) or in the attribute covered by a reference to at least one other situation is provided
4	4M01CB	A dependency with the stereotype ≪when≫ only points from a class with the stereotype ≪Hazard≫ to a class with the stereotype ≪Situation≫
4	4M02WH	A dependency with the stereotype ≪causedBy≫ only points from a class with the stereotype ≪Hazard≫ to a class with the stereotype ≪Fault≫
4	4C03RW	Each situation being relevant is referenced by at least one hazard (using ≪when≫-dependency)
4	4C04CC	Each fault of a domain being marked as "to be considered" is referenced by at least one hazard (using ≪causedBy≫-dependency)
4	4C05WC	Each Hazard has at least one ≪when≫-dependency and at least one ≪causedBy≫-dependency
5	5C01SR	For a severity below S3 a rationale is provided
5	5C02SA	The same rationale for a severity does not lead to different rating in another assessment
5	5C03ER	For an exposure below E4 a rationale is provided
5	5C04EA	The same rationale for an exposure does not lead to different rating in another assessment
5	5C05CR	For a controllability below C3 a rationale is provided
5	5C06CA	The same rationale for a controllability does not lead to different rating in another assessment
6	6M01SH	A dependency with the stereotype ≪addresses≫ only points from a class with the stereotype ≪safetyGoal≫ to a class with the stereotype ≪Hazard≫
6	6C02QM	All ≪hazard≫s rated with QM are addressed by at least one ≪requirement≫
6	6C03HA	All ≪hazard≫s rated with ASIL A–D shall be addressed by at least one ≪safetyGoal≫
6	6C04SA	The ASIL of the ≪safetyGoal≫ is equal to the highest ASIL of the addressed ≪hazard≫s

assumptions (e.g., on driver actions to maintain controllability) should be considered and documented using our UML profile.

A hazard is ≪causedBy≫ a set of faults and refers to situations ≪when≫ it can occur (see Table 9.1, 4M01CB, 4M02WH). This is expressed by stereotyped dependencies. It is important that each relevant situation is referenced by at least one hazard (see Table 9.1, 4C03RW), and each of the faults of a domain that has to be considered is referenced by at least one hazard (see Table 9.1, 4C04CC). Additionally, each Hazard has at least one ≪when≫-dependency and at least one ≪causedBy≫-dependency (see Table 9.1, 4C05WC).

To describe the hazard, we use the stereotype ≪hazard≫ as depicted in Fig. 9.2 and start with the description of the effect on vehicle level. We may give details, examples, or remarks, and provide a description of the hazard that includes the situations and the fault effect. The hazard refers to the risk assessment to be performed as described in the next paragraph.

5. Hazard Classification by Severity, Exposure, and Controllability—The objective of the hazard classification is to assess the level of risk reduction required for the hazards. To classify the hazard, the following steps need to be performed according ISO 26262:

1. Estimate the potential severity and provide a rationale. ISO 26262 classifies the potential severity with the classes S0 (no injuries), S1 (light and moderate injuries), S2 (severe and life-threatening injuries, survival probable), and S3 (life-threatening injuries, fatal injuries).
2. Estimate the probability of exposure and provide a rationale. ISO 26262 classifies the exposure with the classes E0 (incredible, e.g., earthquake), E1 (very low probability, e.g., vehicle being towed), E2 (low probability, e.g., snow and ice on road), E3 (medium probability, e.g., heavy traffic with stop and go), E4 (high probability, e.g., highway).
3. Estimate the controllability and provide a rationale. ISO 26262 classifies the controllability with the classes C0 (controllable in general, e.g., maintain intended driving path in case of unexpected radio volume increase), C1 (simply controllable, e.g., brake to slow down/stop the vehicle in case of blocked steering column when starting the vehicle), C2 (normally controllable, e.g., maintain intended driving path in case of failure of ABS during emergency braking), C3 (difficult to control or uncontrollable, e.g., stay in lane in case of failure of ABS when braking on low friction road surface while executing a turn).

The description of and examples for the classes are taken from the standard (ISO 2011 Part 3, Appendix B). The risk assessment is documented in a class with the stereotype ≪RiskAssessment≫ as depicted in Fig. 9.2. This stereotype has the attributes S, E, and C, each of them typed with enumerations representing the classes. Additionally, it has the attributes sRationale, eRationale, and cRationale of type String that are used to provide a rationale for the selected class. For a severity below S3, an exposure below E4, and a controllability below C3, a rationale has to be provided (see Table 9.1, 5C01SR, 5C03ER, 5C05CR). The same rationale cannot lead to different rating in other assessments (see Table 9.1, 5C02SA, 5C04EA, 5C06CA). For example, the controllability rationale *no lateral control by steering is possible* cannot lead to C3 in one assessment and to C2 in another assessment.

Based on these estimations, the ASIL is determined automatically according to the corresponding ISO 26262 table. For example, a rating of S3, E4, and C3 leads to ASIL D. If one of the parameters is reduced to the lower class, ASIL C is derived. If the parameters are reduced more, ASIL B, ASIL A or QM is derived (Sect. 2.3 for a detailed explanation of the parameters). In case of S0, E0, or C0, no ASIL is assigned and n_a is inserted into the rating attribute ASIL.

6. Define and Verify Safety Goals—Safety requirements are special requirements with the attributes ASIL, safe state, and fault tolerance time (see Fig. 9.3). The ASIL is a measure of necessary risk reduction. The safe state is a state that shall be entered to avoid a hazard. The fault tolerance time is the time an actuator state can be unsafe before the situation becomes hazardous, e.g., an undue brake intervention may have a fault tolerance time of 100 ms in certain situations. Safety requirements are defined on different levels. A safety goal is a top-level safety requirement based on the hazards identified in this analysis. Functional safety requirements are derived from the safety goals. Technical safety requirements are derived from the functional safety requirements and specified for all components of the item considering the concrete architecture.

Safety goals have to be clear and precise, do not contain technical details, but have to be implementable by technical means (e.g. avoid referring to nonmeasurable data). Like functional requirements, safety goals refer to domains and constrain at least one domain.

ISO 26262 requires that at least one safety goal is assigned to each hazard rated as ASIL A, B, C, or D (see Table 9.1 6C03HA). It is not necessary to define safety goals for hazards rated as "QM" or "no assignment," but hazards rated with QM shall be addressed by at least one requirement of any kind (see Table 9.1 6C02QM).

One safety goal can address several hazards (see Table 9.1 6M01SH). A hazard can be addressed by more than one safety goal. ISO 26262 requires that if a safety goal can be achieved by transitioning to or by maintaining one or more safe states, then the corresponding safe states are specified.

In our tool, the ASIL of the safety goal is set automatically to the highest ASIL of the addressed hazards, as required by ISO 26262. This is also verified by a validation condition (see Table 9.1, 6C04SA). For each safety requirement, and therefore also for the safety goals, additionally the fault tolerance time has to be specified.

Fig. 9.3 Requirements

7. ISO 26262 Review—ISO 26262 requires that the results of the hazard analysis and risk assessment shall be reviewed by an independent person from a different department or organization, i.e., independent from the department responsible for the hazard analysis regarding management, resources, and release authority.

During the entire phase of the hazard analysis and risk assessment, correctness and consistency of intermediate results are checked by

- OCL integrity checks as described in Table 9.1.
- Team verification reviews compliant to ISO 26262.

It is possible to generate a document from the model that contains the models of the hazard analysis as tables. These tables show the relations between safety goals, faults, situations, and hazards.

9.4 Tool Support

We extended the UML4PF tool, a support tool for the requirements engineering process sketched in Sect. 2.5.3, to support our hazard analysis and risk assessment process described in Sect. 9.3, as well.

After the developer has drawn some diagram(s) using some EMF-based editor, for example Papyrus UML (Atos Origin 2011), UML4PF provides him or her with the following functionality: it checks if the developed model is valid and consistent by using our OCL constraints as described in Table 9.1, and it returns the location of invalid parts of the model.

Basis for our tool is the Eclipse platform (Eclipse Foundation 2011) together with its plug-ins EMF (Eclipse Foundation 2012) and OCL (UML Revision Task Force 2010a). Our UML profile is conceived as an Eclipse plug-in, extending the EMF meta-model. We store the data in the profile in XMI-format. We use UML with our own profile that extends UML with the ability to express requirements in a similar way as the SysML profile (UML Revision Task Force 2010b) extends UML. If a SysML model with blocks describing the context and requirements was given, we could use the same approach by extending SysML by the missing stereotypes and constraints.

We store all our OCL constraints in one file in XML-format. They are directly checked using the OCL executor, which is part of EMF.

For example, the OCL expression in Fig. 9.4 checks the condition that if a fault is rated as not to be considered, a reference to at least one other fault using the attribute covered by is provided or a description with a rationale is provided.

To perform the check, it first selects all classes with the stereotype ≪Fault≫ applied (using the EMF keyword getAppliedStereotypes in Line 1), and for each of these faults, it

- sets st to be the stereotype of the class (line 2).
- Using st, it retrieves the value of the boolean attribute toBeConsidered (line 4).

```
Class.allInstances()->select(getAppliedStereotypes().name      1
   ->includes('Fault'))->forAll(f|
  let st: Stereotype = f.getAppliedStereotypes()->select(name = 'Fault')   2
      ->asSequence()->first()
  in                                                            3
    not f.getValue(st,'toBeConsidered').oclAsType(Boolean) implies   4
   ( not f.getValue(st,'description').oclAsType(String).oclIsUndefined()   5
      and
    f.getValue(st,'description').oclAsType(String) <> '')  or    6
    f.getValue(st,'coveredBy')->size()>0)                       7
)                                                                8
```

Fig. 9.4 Validation condition 2C04DC

- If the attribute is not *true* (line 4),
- it is checked that a description (that should contain the rationale) is not undefined (line 5) and not empty (line 6), or
- a reference to at least one other fault is given (line 7).

The OCL expression in Fig. 9.5 checks if all fault-types are considered for each functional requirement, i.e., an object containing a rationale or reference exists for each fault-type. For all requirements (line 1), the set of fault-types the requirement refers to with a «fails» dependency are determined and named faults (lines 2 and 3). It is checked that more than 0 faults are in this set (line 4), and by using the fault stereotype st (lines 5 and 6) it is checked that all types of faults (no, unintended, ...) are in the set faults (lines 7–16). The type of the fault is an enumeration attribute of the stereotype «Fault» named type. The set of fault-type attributes ty is retrieved using the EMF keyword getValue in line 7. The other validation conditions given in Table 9.1 are implemented in a similar way.

```
Class.allInstances() ->select(getAppliedStereotypes().name ->     1
    includes('Requirement'))->forAll(req|
  let faults: Set(Class) = req.oclAsType(Class).clientDependency    2
      ->select(getAppliedStereotypes().name ->includes('fails')).target
      .oclAsType(Class)->asSet()
  in                                                                3
  faults ->size()>0 and                                            4
  (let st: Stereotype = faults.getAppliedStereotypes() ->select(name =   5
      'Fault') ->asSequence() ->first()
  in                                                               6
  let ty: Set(EnumerationLiteral) = faults .getValue(st,'type')     7
      .oclAsType(EnumerationLiteral) ->asSet()
  in of                                                            8
    ty.name -> includes('No') and                                 9
    ty.name -> includes('Unintended') and                         10
    ty.name -> includes('Early') and                              11
    ty.name -> includes('Late') and                               12
    ty.name -> includes('Less') and                               13
    ty.name -> includes('More') and                               14
    ty.name -> includes('Inverted') and                           15
    ty.name -> includes('Intermittent')))                         16
```

Fig. 9.5 Validation condition 2C01CF

9.5 Application

Our method was illustrated on an example of an electronic steering column lock (ESCL) system, which was presented at the "VDA Automotive SYS Conference 2012", June 18–20, 2012, in Berlin, Germany.

1. Provide an Item Definition—The main function of the ESCL is to provide lock and unlock commands to the lock actuator automatically to enhance theft protection for vehicles with a power button instead of a standard key. The context diagram in Fig. 9.6 shows the item, i.e., the ESCL and the elements in the environment, namely the driver, the lock actuator, and the vehicle. The item controls the lock and the unlock commands, and the lock actuator observes these phenomena. The driver presses the power button to crank or stop the engine. The vehicle moves at a certain speed and therefore controls the phenomenon Speed.

Table 9.2 shows the functional requirements of the ESCL. For both requirements R01 and R02, it shows which domains of the item's environment are constrained (see column ≪constrains≫) and which domain the requirements are referred to (see column ≪refersTo≫): The lock actuator is constrained. The driver and the vehicle are referred to since they together provide necessary information to deduce if the driver wants to drive or not.

2. Instantiate Fault-Type Guide-Words—To start with the hazard analysis, we look at the lock actuators constrained by the requirements R01 and R02, and investigate both functional requirements according to the guide-words.

Figure 9.7 shows some of the fault types assessed for R01. For the guide-word "no", the fault is that the ESCL does not lock in situations where it is expected. For the guide-word "unintended", the fault is that the ESCL locks in situations where

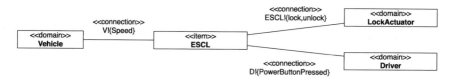

Fig. 9.6 ESCL context diagram

Table 9.2 ESCL requirements

No.	Text	≪constrains≫	≪refersTo≫
R01	The steering column shall be locked, when the driver wants to immobilize the vehicle	LockActuator	Driver, vehicle
R02	The steering column shall be unlocked, when the driver wants to drive	LockActuator	Driver, vehicle

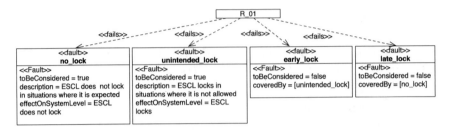

Fig. 9.7 ESCL faults

it is not allowed. For the guide-word "early", the fault is the same as described in fault unintended_lock. For the guide-word "late", the fault is either no problem, or in case of a long delay the same as described in fault no_lock. The faults related to all other guide-words are either mapped to "no", "unintended", or are not relevant since the locking is a binary decision and cannot be "less" or "more". The expression shown in Fig. 9.5 checks afterwards if we considered all fault-type guide words. The expression provides the information that the guide words "intermittent" and "inverted" were not considered. These are added afterwards and as stated before locking is a binary decision and it cannot be "intermittent". The "inverted" is not relevant, because it is already considered in the "unintended" fault.

Additionally, we have to describe the effect on system level. For example, for unintended locking, the effect on the system level is that the ESCL locks the steering column. The effect on the vehicle level is that the vehicle is not steerable. For R02, the procedure is the same.

3. Situation Classification—The situations are classified according to the item's functionality. In Fig. 9.8, the situation classification using our profile is depicted.

The situation "driving at speed"is classified as being relevant, because a hazard may occur if the vehicle is moving and the steering column is locked. The situation "maneuver" (including, e.g., parking, driving backwards) is marked as being not relevant, because for the ESCL system the maneuvering hazards are the same as for "driving at speed". The situation "standstill" is classified as being relevant, because a hazard may occur if the vehicle is "parked", and the steering column is not locked. The effort necessary for the situation classification is reduced, because it is not necessary to rate these detailed situations. The situations "being towed" and "rolling" are classified as being relevant because they consider the system state where the engine is off.

<<drivingSituation>> **DrivingAtSpeed**	<<standstillSituation>> **Parked**	<<manouverSituation>> **Maneuver**	<<beingTowedSituation>> **BeingTowed**	<<rollingSituation>> **Rolling**
<<DrivingSituation>> relevant = true	<<StandstillSituation>> relevant = true	<<ManouverSituation>> relevant = false includedIn = [DrivingAtSpeed]	<<beingTowedSituation>> relevant = true	<<rollingSituation>> relevant = true

Fig. 9.8 ESCL situations

Fig. 9.9 ESCL hazard example

4. Hazard Identification—In the next step, the hazards are identified. For this reason, all considered faults are combined with all situations where the fault leads to a problem. These are the ones having the attribute **toBeConsidered=true**. It is possible to have more than one fault that causes the hazard, and the hazard can be in place in different situations. Some combinations are not needed in the hazard analysis, e.g., the situation standstill does not need to be combined with the fault of unintended locking, because locking is intended in this situation.

Figure 9.9 shows the hazard that can occur when the vehicle is moving at speed. It may be caused by unintended steering column locking. To describe the hazard, first the effect on vehicle level is described. For the previously described fault, the effect on vehicle level is that the steering is locked and the vehicle is not steerable. The hazard is the "loss of steering control (locked steering) when driving at speed." The hazard refers to the risk assessment performed in the next step.

5. Hazard Classification by Severity, Exposure, and Controllability—The above-mentioned hazard is rated according to its severity, exposure, and controllability as depicted in Fig. 9.9. The highest severity level S3 is chosen, because a locked steering column lock at speed can lead to death or life-threatening injuries when the vehicle hits, e.g., obstacles near the road, pedestrians, oncoming traffic, or obstacles on the track. The exposure level is E4, because steering is necessary in the mentioned situations and in all situations with high speed, which are more than 10 % of the driving time. The highest controllability level C3 is chosen since no lateral control by steering is possible. The driver can only intervene by braking, but in case of high speed the driver cannot avoid the consequences of the hazard. Hence, ASIL D is automatically deduced from this classification.

6. Define and Verify Safety Goals—The described hazard can be addressed by a safety goal as depicted in Fig. 9.10. The same safety goal can also address other

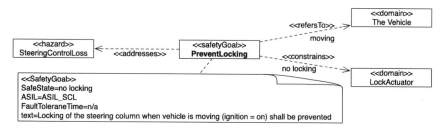

Fig. 9.10 ESCL safety goal

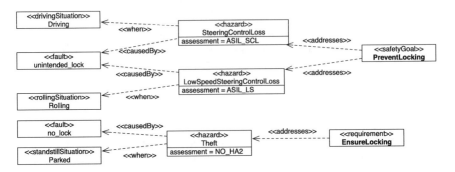

Fig. 9.11 Safety goal—hazard—fault relations

Table 9.3 Safety goal—hazard—fault relations

Fault	Situation	Hazard	Risk assessment	Safety goal/Requirement
unintended_lock	Driving	SteeringControlLoss	ASIL_SCL	PreventLocking
unintended_lock	Maneuver	LowSpeed SteeringControlLoss	ASIL_LS	PreventLocking
no_lock	Parked	Theft	NO_HA2	EnsureLocking

hazards (not shown here). To prevent the hazard, the safety goal is that "locking of the steering column when the vehicle is moving shall be prevented." This safety goal refers to the vehicle since it indicates if the vehicle is moving and constrains the steering column lock by preventing the lock. Since we cannot accept steering column locking even for a certain time, the fault tolerance time is not applicable. For other safety goals, e.g., prevention of braking intervention, the fault may be acceptable for a short time period. We model the relations between faults, hazards, and safety goals, depicted in Fig. 9.11. The information in this model can be converted into a table (see Table 9.3) for documentation purposes. The safety goal Prevent-Locking ≪addresses≫ the hazards SteeringControlLoss and LowSpeedSteeringControlLoss. The SteeringControlLoss occurs in the situation Driving and it is ≪causedBy≫ the fault unintended_lock. This fault also causes the hazard LowSpeedSteeringControlLoss, but it occurs in the situation Rolling. Since the hazard Theft (see Fig. 9.11) has no ASIL A, B, C, or D assigned, it is not addressed by a safety goal but by the functional requirement EnsureLocking. The hazard is ≪causedBy≫ the fault no_lock and occurs in the situation Parked.

7. ISO 26262 Review—To support the reviews, the validation conditions listed in Table 9.1 are executed on the complete case study. These validation conditions check the consistency and correctness of the model. That is, we check

- whether each relevant functional requirement in the item definition is considered,
- whether the hazard and risk assessment is aligned with the supplier's assessment, who delivers parts for building the item and
- whether the hazard and risk assessment is consistent with ISO 26262 description.

9.6 Related Work

We are not aware of any publications about a structured and model-based hazard analysis and risk assessment for automotive systems equipped with integrity checks.

Two hazard analysis methods are compared by Törner et al. (2006). The paper shows that the adapted functional failure analysis (FFA) is less time-consuming than the method of the European Space Agency (ESA method). Our method is based on the results of Törner et al. (2006).

Baumgart (2012) shows a safety lifecycle including hazard analysis and risk assessment for software product lines. The authors describe their lifecycle on a high level of abstraction and our hazard analysis method can be used to refine this part of their lifecycle.

The Safety Management System and Safety Culture Working Group provides guidance on hazard identification by different means, e.g., brainstorming, HAZOP, checklists, FMEA SMS WG (2009). Their results are considered in our method via the guide-words techniques.

Jesty et al. (2000) give a guideline for the safety analysis of vehicle-based systems, including system analysis, hazard identification, hazard analysis, identification of safety integrity levels, FMEA, and fault tree analysis. Their work also uses the HAZOP guide-words, but they focus on the safety integrity level as defined in the IEC 61508 and not on the ASIL from ISO 26262. Jesty et al. additionally address FMEA and fault tree analysis for analyzing existing systems, but do not consider a model or validation conditions.

In contrast to our work, which focuses on the determination of necessary risk reduction, following papers describe model-based approaches specific for later development phases, when the system is already designed and not the determination of necessary risk reduction:

Papadopoulos and Grante (2005) propose a process that addresses both cost and safety concerns and maximizes the potential for automation to address the problem of increasing technological complexity. It combines automated safety analysis with optimization techniques.

Li and Zhang (2011) present a comprehensive software hazard analysis method, which applies a number of hazard analysis techniques, and the proposed method is applied to a software development process of a control system. The described method for hazard analysis is similar but less detailed than ours.

Mehrpouyan (2011) proposes a model-based hazard analysis procedure (based on SysML models) for the early identification of potential safety issues caused by unexpected environmental factors and subsystem interactions within a complex safety-critical system. The proposed methodology additionally maps hazard and vulnerability modes to specific components in the designed system and analyzes the hazards.

Zhang et al. (2010) propose a comprehensive hazard analysis method based on functional models. It mainly addresses fault tree analysis and FMEA.

Giese et al. (2004) present an approach that supports the compositional hazard analysis of UML models described by restricted component and deployment diagrams. It also starts with environment models, but then focuses on the safety analysis of the design.

Hauge and Stølen (2012) introduce the SaCS method. The method provides guidance on how to select and use patterns for the development of safety control systems. The patterns are categorized into process and product patterns. For each of these categories patterns are further distinguished into process, solution, and safety case patterns. For example, a process requirement pattern is the hazard analysis, a process solution is HAZOP, and a process safety case pattern is safety management. This work differs from our own, because we focus specifically on early hazard analysis and provide detailed guidance.

9.7 Summary

Our method has been applied in several Ford of Europe projects. However, the formal validation conditions and tool support was not used in these projects and was developed as contribution for this work. We are confident that this contribution will ensure the same consistency and correctness of future hazard analyses with less effort than the manual approach currently used.

Our contribution has the following benefits:

- The guide-words approach avoids forgetting relevant faults and the selection of function/fault combinations beforehand helps to find the right level of detail.
- Selecting relevant situations from the hierarchically organized profile reduces the risk of forgetting a relevant situation and ensures to only consider situations that are relevant for the function in question.
- Structuring the analysis in different steps on different levels fosters an alignment between the analysis and the organizations (departments with experts regarding hardware/software, system level, vehicle/functional level) involved in the creation and review of the analysis.
- The rules for safety goal definitions help to define safety goals appropriately.
- Our UML profile contains all relevant elements for a hazard analysis in compliance with ISO 26262. The profile provides the basis for creating and validating hazard analysis models.
- The validation conditions support the review activities required by ISO 26262.
- The validation conditions expressed as formal OCL expressions give a precise definition of the necessary checks.
- The checking of the validation conditions can be performed automatically by using our support tool.

Hazard analysis in practice is currently table-based using spreadsheets such as Microsoft Excel. We are planning to work on an automatic import of these spreadsheets to our UML model. A text-to-model converter will convert entries of the spreadsheets to UML classes with the required stereotypes. The text-to-model converter will be based on this work. This makes it possible to apply our OCL validation conditions to the content of the spreadsheets. Hence, the spreadsheet-based hazard analysis can benefit from our method, as well.

References

Atos Origin. (2011). Papyrus UML Modelling Tool (http://www.papyrusuml.org/).
Baumgart, S. (2012). Investigations on hazard analysis techniques for safety critical product lines. In *Proceedings of the Workshop on Interesting Results in Computer Science and Engineering (IRCSE)*. ACM.
Beckers, K., Frese, T., Hatebur, D., & Heisel, M. (2013). A structured and model-based hazard analysis and risk assessment method for automotive systems. In *Proceedings of the 24th IEEE International Symposium on Software Reliability Engineering* (pp. 238–247). IEEE Computer Society.
DIN EN. (2006). *Analysetechniken für die Funktionsfähigkeit von Systemen-Verfahren für die Fehlzustandsart- und -auswirkungsanalyse (FMEA)* (DIN EN 60812). Berlin, Germany: Deutsches Institutür Normung (DIN EN).
Eclipse Foundation. (2011). Eclipse—An Open Development Platform (http://www.eclipse.org/).
Eclipse Foundation. (2012). Eclipse Modeling Framework Project (EMF). (http://www.eclipse.org/modeling/emf/).
Giese, H., Tichy, M., & Schilling, D. (2004). Compositional hazard analysis of UML component and deployment models. In *Proceedings of SAFECOMP* (pp. 166–179). Springer.
Hauge, A. A., & Stølen, K. (2012). A pattern-based method for safe control systems exemplified within nuclear power production. In *Proceedings of SAFECOMP* (pp. 13–24). Springer.
IEC. (2005). *Hazard and Operability Studies (HAZOP studies)* (ISO/IEC 62882). Geneva, Switzerland: International Electrotechnical Commission (IEC).
ISO. (2003). *ISO 11898 Road vehicles—Controller area network (CAN) Geneva*. International Organization for Standardization (ISO): Switzerland.
ISO. (2011). *ISO 26262—Road Vehicles— Functional Safety Geneva*. International Organization for Standardization (ISO): Switzerland.
Jesty, P. H., Hobley, K. M., Evans, R., & Kendal, I. (2000). Safety analysis of vehicle-based systems. In *Proceedings of the 8th Safety-Critical Systems Symposium* (pp. 90–110). Springer.
Li, W., & Zhang, H. (2011). A software hazard analysis method for automotive control system. In (pp. 744–748). IEEE Computer Society.
Mehrpouyan, H. (2011). Model-based hazard analysis of undesirable environmental and components interaction. Unpublished master's thesis, Linköpings Universitet.
Papadopoulos, Y., & Grante, C. (2005). Evolving car designs using model-based automated safety analysis and optimisation techniques. *Journal of Systems and Software-Special Issue: Computer Software and Applications, 76*(1), 77–89.
SMS WG. (2009). Guidance on hazard identification. Technical Report. San Francisco, U.S.: Safety Management System and Safety Culture Working Group (SMS WG).
Törner, F., Johannessen, P., & Öhman, P. (2006). *Evaluation of hazard identification methods in the automotive domain*. SAFECOMP 2006 (pp. 237–260). New York: Springer.

U. S. Department of Defense. (1998). Electronic Reliability Design Handbook (MIL-HDBK-338B). Washington, U.S.: U. S. Department of Defense. (http://www.everyspec.com/MIL-HDBK/MIL-HDBK-0300-0499/MIL-HDBK-338B15041/).

UML Revision Task Force. (2010a). OMG object constraint language: Reference.

UML Revision Task Force. (2010b). OMG systems modeling language (OMG SysML).

UML Revision Task Force. (2010c). OMG unified modeling language: Superstructure.

Zhang, H., Li, W., & Chen, W. (2010). Model-based hazard analysis method on automotive programmable electronic system. In *Proceedings of 3rd International Conference on Biomedical Engineering and Informatics (BMEI)* (pp. 2658–2661). IEEE Computer Society.

Chapter 10
A Catalog of Context-Patterns

Abstract In the beginning of every security analysis a Context Establishment aims at eliciting and understanding the system that shall be analyzed including its direct and indirect environment, the relevant stakeholders, other already established systems, and other entities that are directly or indirectly related to the system. For this purpose, we describe in this chapter a specific way of elicitation of the system context by introducing so-called context-patterns. The application of context-patterns helps to gather knowledge in a structured way about a specific domain such as cloud computing. These patterns contain graphical patterns and templates with elements that require consideration for a specific context. In addition, our context-pattern contains a method for eliciting domain knowledge using the graphical patterns and templates. In this chapter we present a catalog of context-pattern describing the following domains: Cloud Computing Systems, Peer-to-Peer Systems, Service-oriented Architectures, and Law. Furthermore, we distinguish our context-patterns from further existing patterns for system analysis.

10.1 Introduction

We present a set of so-called *context-patterns* in this chapter. Context-patterns are structural descriptions of kinds of systems including their environment, e.g., clouds (Sect. 12.2). These descriptions can be instantiated in order to describe a specific system of a certain kind. Our context-patterns are described based on existing context descriptions of these systems, as well as our experience and the experience of practitioners. This chapter is based on the following publications of which the author of this book is the main author (Beckers et al. 2011; Beckers and Faßbender 2012a, b). Stephan Faßbender is the main author of the work presented in the publications (Beckers et al. 2012a, b), which are also mentioned in this chapter. This chapter is organized as follows: In Sect. 10.2 we define *what* a context-pattern is and in Sect. 10.3 we distinguish the term *context-pattern* from other uses of the term *pattern* in software engineering. We present context-patterns for clouds in Sect. 10.4, for Peer-to-Peer systems in Sect. 10.5, for service-oriented architectures in Sect. 10.6, and for identifying laws in Sect. 10.7. In particular, we focus on the cloud system

© Springer International Publishing Switzerland 2015

K. Beckers, *Pattern and Security Requirements*,

DOI 10.1007/978-3-319-16664-3_10

analysis pattern, because the work in Chaps. 12 and 13 is based on it. This chapter provides an overview of these remaining context-patterns. They are illustrated in this chapter because our initial pattern language for context-patterns is based on them (Chap. 11).

10.2 Definition

Requirements define what properties and functionality a software should have. It is impossible to assess the quality of a software without requirements. Moreover, writing requirements is only possible if the domain knowledge of the system-to-be and its environment is known and considered thoroughly, otherwise severe problems can occur during software development, e.g., technical solutions to requirements might be impractical or costly. It is an open research question *how* to elicit domain knowledge correctly for effective requirements elicitation (Niknafs and Berry 2012). Moreover, several requirements engineering methods exist for security. Fabian et al. (2010) concluded in their survey about these methods that it is not yet state of the art to consider domain knowledge. We address these problems by describing common structures and stakeholders for several different domains in so-called *context-patterns* for structured domain knowledge elicitation. Depending on the kind of domain knowledge that we have to elicit for a software engineering process, we always have certain elements that require consideration. We base our approach on Jackson's work (Sect. 2.5.3) that considers requirements engineering from the point of view of a machine in its environment.

A context-pattern consists of the following parts:

Method A method contains a sequence of steps. Each step is described using well-defined activities, inputs, and outputs. Inputs are descriptions of the required artifacts to perform the activities of the method step. Activities are descriptions of the processing of all inputs into outputs. Outputs are the desired results of the activities of this step. A context-pattern has to contain a method that describes how to use the context-pattern.

Graphical pattern Our context-patterns do not enforce considering the machine meaning the thing we are going to build explicitly, but demand a description of its environment in graphical form. This environment contains domain knowledge. In particular, any given environment considers certain elements, e.g., stakeholders or technical elements. Moreover, we believe that every environment of a software engineering problem can be divided into parts that have direct physical contact with the machine and parts in the environment that have an effect on the machine without physical contact, e.g., laws. These relations between the environment and the machine have to be part of the graphic as well. A context-pattern has to contain at least one graphical pattern.

We use a UML-based notation for our graphical patterns that uses, e.g., stick figures for actors, but we also use notation elements that are not part of the UML such as rectangles that symbolize an environment, and all elements in the rectangle belong to this environment. We explain the concepts that we use in all our graphical patterns in Chap. 11 in detail. This is part of the effort to create a pattern language for context-patterns. Nevertheless, we show in Sect. 10.4.5 that a mapping from our graphical patterns to a model in strict UML notation is possible.

Templates Templates contain additional information about elements in the graphical pattern. For example, a graphical pattern contains a graphical figure of a stakeholder and a corresponding template for the stakeholder can contain, e.g., the motivation of the stakeholder for using the machine and the relations to other stakeholders. Templates provide the means to attach further information to the graphical pattern. The reason for adding this refinement in a template and not in the graphical pattern is not to overload the graphical pattern with too many elements. Templates are optional elements of context-patterns, because not all graphical patterns require a refinement.

In addition, it is possible to attach diagrams to our context-patterns, e.g., UML activity diagrams that refine interactions between different stakeholders of a context-pattern (Chap. 12 for an example). These diagrams can support different views and levels of granularity. All these diagrams are based on our patterns. Hence, we can apply consistency checks and traceability links with the elements in the attached diagrams and the created patterns.

Moreover, the domain knowledge elicited using our patterns is not limited to software engineering. We showed in previous work, e.g., Beckers et al. (2013a), that the knowledge can also support the establishment of security standards of the ISO 27000 family of standards (ISO/IEC 2009).

10.3 Related Work

Alexander (1977) developed patterns for building houses and even towns. The author proposed to use patterns to support architects in the process of building houses and towns. Alexander used textual templates for describing proven design practices for a specific domain in a pattern. Architects can use his patterns and look for specific design problems when building houses and find solutions in his patterns. We build on Alexander's idea of making knowledge reusable by describing it in patterns.

Jackson (2001) describes Problem Frames as follows: "A problem frame is a kind of pattern. It defines an intuitively identifiable problem class in terms of its context and the characteristics of its domains, interfaces and requirement". Hence, Jackson added the notion that a pattern has to be defined in relation to a context. Our work is inspired by this idea and provides the means to describe patterns of software engineering contexts.

We were also inspired by Fowler (1996), who developed patterns for the analysis phase of a given software engineering process. Fowler defined so-called *analysis patterns*, which describe organizational structures and processes, e.g., accounting, planning, and trading. The patterns are derived from experience in projects. Fowler did not rely on a fixed structure for his analysis patterns. He states that each pattern should have a name, but he defines no further structure. Nevertheless, Fowler stated in a later work (Fowler 2002) that a *pattern structure* is essential for expressing a pattern. For Fowler a pattern structure should consist of a *sketch* (Fowler 2002), which is a structural and graphical description, and a textual description of behavior and relations. The author also uses a graphical notation he defines in the book that contains, e.g., class and interaction diagrams. He uses this graphical notation to create sketches for his patterns. A similar understanding of how to describe a pattern can be found in Hafiz et al. (2012) and Fernandez and Pan (2001). We agree with the works presented here and base our structure for context-patterns (Sect. 10.2) on these works. Our graphical pattern is similar to Fowler's sketch and our templates are similar to Fowler's textual descriptions. In addition, our context-patterns require a method that describes how to use the pattern.

Inspired by Alexander's work (1977), Gamma et al. (1994) propose simple textual patterns for the design phase of software engineering. The authors argue that "Graphical notations, while important and useful, are not sufficient. They simply capture the end product of the design process as relationships between classes and objects. To reuse the design, we must also record the decisions, alternatives, and tradeoffs that lead to it" (Gamma et al. 1994, p. 6). The authors describe design problems they encountered frequently during software engineering projects in templates. These textual patterns (or templates) contain a structure that is similar for each pattern. For example, every pattern has a name, a motivation for its usage, and consequences. A complete description can be found in Gamma et al. (1994). These patterns support software engineers during the design and implementation phases of a given software engineering process. The authors decided not to rely only on graphical models, because these cannot capture the decisions, alternatives, and tradeoffs that led to successful design decisions. Nevertheless, graphical representations of the structure of design solutions in classes are based on the object modeling technique (OMG). The behavior of diagrams is shown in interaction diagrams. Both kinds of diagrams are explained in detail in Gamma et al. (1994). In addition, Schumacher et al. (2006) propose a similar work on patterns specifically for security. The authors defined simple solutions to security problems during the software engineering design and implementation phases. In contrast, our *context-patterns* for the analysis phase differ from patterns concerning solutions for the design phase of software engineering like the Gang of Four patterns (Gamma et al. 1994) or the security patterns by Schumacher et al. (2006). The reason is that we provide a means for a structured elicitation of domain knowledge for systems. We do not provide solutions for the implementation phase of software engineering.

Schmidt (2010) proposes the *Security Engineering Process using Patterns (SEPP)*, which refines security issues into isolated concerns, e.g., a specific threat toward a specific part of a system. Each security concern is analyzed in relation to existing functional requirements and expressed as a security requirement. These security concerns are presented in a so-called *security problem frame*. This is a kind of pattern that can be reused for different situations. However, security problem frames capture detailed concerns in isolation and do not try to create patterns for entire kinds of systems such as clouds.

Withall (2007) provides guidelines and examples for formulating software requirements based on project experience. Withall also explains the need for documentation of requirements including assumptions, glossary, document history, and references. Withall's work aims at writing textual requirements, which also consider domain knowledge in the form of assumptions to these requirements. Our work differs from Withall's, because we do not provide patterns for requirements. Our work focuses on the elicitation of domain knowledge.

Fernandez et al. (2007) design several UML models of some aspects of Voice-overIP (VoIP) infrastructure, including architectures and basic use cases. The authors also present security patterns that describe countermeasures to VoIP attacks. In addition, Hafiz described four privacy design patterns for the network level of software systems. These patterns solely focus on anonymity and unlinkability of senders and receivers of network messages from protocols, e.g., HTTP (Hafiz 2006). The patterns are specified with several categories. Among them are intent, motivation, context, problem, and solution, as well as forces, design issues, and consequences. Forces are relevant factors for the applicability of the pattern, e.g., number of users or performance. Design issues describe how the forces have to be considered during software design. For example, the number of stakeholders has to have a relevant size for the pattern to work. Consequences are the benefits and liabilities the pattern provides. For example, an anonymity pattern can disguise the origin of a message, but the pattern will cause a significant performance decrease (Hafiz 2006). This work focuses on privacy issues on the network layer. The works of Fernandez et al. and Hafiz focus exclusively on specific kinds of systems: Voice-over-IP and network-based software systems. Our *context-patterns* concern different kinds of systems.

Hatebur and Heisel proposed patterns for expressing and analyzing dependability requirements (Hatebur and Heisel 2009). The approach defines important elements of these requirements and checks for consistency in these requirements. In addition, the authors support reasoning for these requirements based on Jackson's problem frame method (Jackson 2001). Beckers and Heisel (2012) proposed a similar approach for privacy. Both methods can complement our own in the sense that we can use our context-patterns as input for these requirements reasoning methods. Hence, we can create methods for expressing and analyzing privacy and dependability requirements for specific contexts, e.g., clouds.

10.4 Cloud System Analysis Pattern

We present our cloud system analysis pattern in Sect. 10.4.1 that helps to systematically describe cloud computing scenarios and identify assets in these scenarios (Sects. 10.4.1–10.4.3). We illustrate the instantiation of the pattern using an online

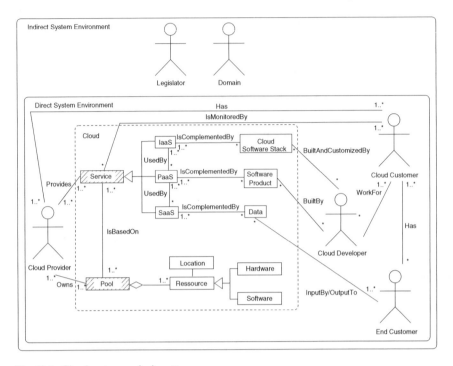

Fig. 10.1 Cloud system analysis pattern

banking service example (Sect. 10.4.4) and report on tool support (Sect. 10.4.5). This context-pattern is based on the background presented in Sect. 12.2.

10.4.1 Graphical Pattern

The *cloud system analysis pattern* shown in Fig. 10.1 provides a conceptual view on cloud computing systems and serves to systematically analyze stakeholders and technical cloud elements. The notation used to specify the pattern is based on UML[1] notation, i.e., the stick figures represent roles, the boxes represent concepts or entities of the real world, the named lines represent relations (associations) equipped with cardinalities, the unfilled diamond represents a *part-of* relation, and the unfilled triangles represent generalization.

A *Cloud* is embedded into an environment consisting of two parts, namely the *Direct System Environment* and the *Indirect System Environment*. The *Direct System Environment* contains stakeholders and other systems that directly interact with the *Cloud*, i.e., they are connected by associations. Moreover, associations between

---

[1] Unified Modeling Language: http://www.omg.org/spec/UML/2.3/.

stakeholders in the *Direct* and *Indirect System Environment* exist, but not between stakeholders in the *Indirect System Environment* and the cloud. Typically, the *Indirect System Environment* is a significant source for compliance and privacy requirements.

The *Cloud Provider* owns a *Pool* consisting of *Resources*, which are divided into *Hardware* and *Software* resources. The provider offers its resources as *Services*, i.e., *IaaS*, *PaaS*, or *SaaS* (Sect. 12.2). The boxes *Pool* and *Service* in Fig. 10.1 are hatched, because it is not necessary to instantiate them. Instead, the specialized cloud services such as *IaaS*, *PaaS*, and *SaaS* and specialized *Resources* are instantiated. The *Cloud Developer* represents a software developer assigned by the *Cloud Customer*. The developer prepares and maintains an *IaaS* or *PaaS* offer. The *IaaS* offer is a virtualized hardware, in some cases equipped with a basic operating system. The *Cloud Developer* deploys a set of software named *Cloud Software Stack* (e.g., web servers, applications, databases) into the *IaaS* in order to offer the functionality required to build a *PaaS*. In our pattern *PaaS* consists of an *IaaS*, a *Cloud Software Stack* and a *cloud programming interface (CPI)*, which we subsume as *Software Product*. The *Cloud Customer* hires a *Cloud Developer* to prepare and create *SaaS* offers based on the CPI, finally used by the *End Customers*. *SaaS* processes and stores *Data* input and output from the *End Customers*. The *Cloud Provider*, *Cloud Customer*, *Cloud Developer*, and *End Customer* are part of the *Direct System Environment*. Hence, we categorize them as *direct stakeholders*. The *Legislator* and the *Domain* (and possibly other stakeholders) are part of the *Indirect System Environment*. Therefore, we categorize them as *indirect stakeholders*.

10.4.2 Templates

We accompany this cloud system analysis pattern by templates to systematically gather domain knowledge about the direct and indirect system environments based on the stakeholders' relations to the cloud and other stakeholders. The first template serves to describe stakeholders contained in the direct system environment:

Name State the identifier of the stakeholder or group of stakeholders, e.g., company name or group of end customers.

Description Describe the stakeholder informally, e.g., whether the stakeholder is a natural or a legal person.

Relations to the cloud Describe the inputs and outputs represented as a relation (line from this stakeholder to the cloud) between the stakeholder and the cloud, e.g., the kind of data or software.

Motivation State the motivation of the stakeholder for using the cloud based on the previously elicited relations to the cloud, e.g., business goals such as profit and cost reduction.

Relations to other direct stakeholders For each relation (line from this stakeholder to another direct stakeholder), name the kind of dependency between the stakeholders, e.g., controlled by contract, served by, indirectly influenced by customer-demand.

Assets   Document the already known assets of this stakeholder. Note that these assets are not represented in the graphical cloud pattern but are just added to the list of assets created in Step 2 of our method.

Compliance and Privacy   Document the already known compliance and privacy laws as well as regulations.

The second template serves to describe stakeholders contained in the indirect system environment:

Name   (see direct stakeholder template).

Description   (see direct stakeholder template).

Relations to other stakeholders   For each relation from this stakeholder to another direct or indirect stakeholder (no line explicitly shown), name the kind of dependency between the stakeholders, e.g., protected by, controlled by law, implement laws.

Motivation   State the motivation of the stakeholder for having any kind of indirect relation to stakeholders of the direct or indirect environment or technical cloud elements, e.g., protect privacy of citizens or enforce concrete laws of an economic community.

Compliance and Privacy   Identify relevant compliance and privacy laws and regulations for the cloud scenario, e.g., the German Federal Data Protection Act (BDSG) and the German Law on Monitoring and Transparency in Businesses (KonTraG).

10.4.3 Method

We illustrate our method in Fig. 10.2, which consists of two steps: The context establishment of a cloud scenario and the asset identification for this scenario. The steps are explained in the following.

Step 1: Context Establishment -

Input:	Unstructured scenario description, Cloud system analysis pattern and templates
Output:	Instantiated cloud system analysis pattern and templates

We now generally explain the process of instantiating our cloud system analysis pattern. The instantiation of the direct cloud environment is done first and the indirect environment afterwards. The reason for this sequence is that we need to know the stakeholders using the cloud and software type running in the cloud to determine the relevant rules and regulations relevant for instantiating the indirect cloud environment.

Fig. 10.2 A method for
cloud context elicitation

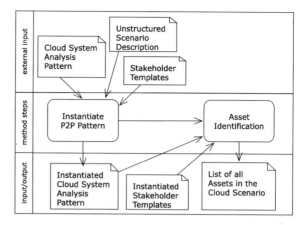

1. Instantiate the direct system environment

   a. State the instantiations of the cloud stakeholders of the direct system environment, e.g., which company is the cloud provider.
   b. Moreover, the stakeholders defined by the cloud system analysis pattern can be complemented by further stakeholders.
   c. For each direct stakeholder instantiate the direct stakeholder template.

2. Instantiate the cloud

   a. Provide a functional description of the software offered by the cloud. Define the service layer, e.g., SaaS the software is located in, the input and output of the service(s), and the connections to the stakeholders of the direct system environment.
   b. The *Data* might be analyzed more precisely using, e.g., class diagrams. This helps to apply asset identification and analysis techniques in the second step of this method.
   c. Provide the geographical location(s) of the cloud.
   d. State the deployment scenario of the cloud (private, public, hybrid).
   e. State the technical implementation behind the system, e.g., the required applications in the cloud software stack to provide an SaaS offer.

3. Instantiate the indirect system environment

   a. Determine the relevant domains (e.g., finance, medical, insurance) for the cloud scenario by considering the outsourced processes of the cloud customer and select the relevant legislators (e.g., Germany, US). Such legislators or jurisdictions are relevant where the resources are located, i.e., where the data is physically stored and processed, and where users, providers, and cloud customers are based.
   b. In addition to the indirect stakeholders covered by the cloud system analysis pattern, further indirect stakeholders can be added.
   c. For each indirect stakeholder instantiate the indirect stakeholder template.

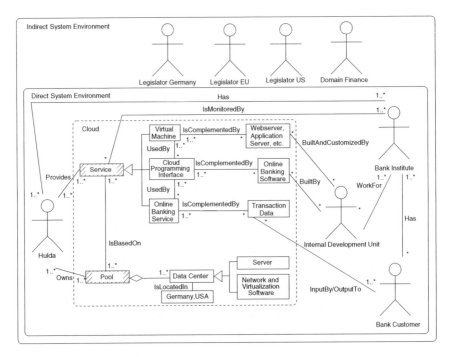

Fig. 10.3 Concrete cloud computing system for an online banking service

In our example in Fig. 10.3, the domain is the *Domain Finance*, and the relevant legislators are *Germany*, *EU*, and *US* (since the data centers of *Hulda* are located in Germany as well as in the US).

The cardinalities contained in the pattern can be instantiated in a restricting way only. When instantiating the cloud system analysis pattern, one also fills in the corresponding templates.

Step 2: Identify Assets -

Input:	Instantiated cloud system analysis pattern and templates
Output:	List of assets

We explain how to identify assets in the scope of the organization at an appropriate level of detail based on the ISO 27005 standard (ISO/IEC 2008). We distinguish between *primary assets* and *supporting assets*. Primary assets are *business processes*, *activities* and *information*. Supporting assets are *hardware*, *software*, *network*, *staff*, *site* or *organization's structure*. The input for the identification of assets are the scope and boundaries identified in the context establishment activity.

The cloud system analysis pattern helps in identifying the primary and supporting assets by considering the instantiated boxes and the associations between the direct stakeholders and the cloud. The associations indicate the flow of information into and out of the cloud and therefore help to analyze the information assets processed and stored in the cloud.

Primary asset information comprises *vital information*, *personal information*, *strategic information* and *high-cost information*. Vital information is relevant for running the organization's business, personal information comprises personal data or privacy relevant data, strategic information is required for achieving business goals, and high-cost information is information whose gathering or processing requires a long time or high acquisition cost.

10.4.4 Example

Step 1: Context Establishment

We illustrate our pattern-based support for security standards, using the example of a bank offering an online-banking service for their customers. This bank plans to outsource the affected IT processes to reduce costs and scale-up their system for a broader amount of customers. Customer data such as account number, amount, and transaction log history are stored in the cloud, and transactions like credit transfer are processed in the cloud. The bank authorizes the software department to design and build the cloud-specific software according to the interface and platform specification of the cloud provider. Figure 10.3 shows an instance of our cloud system analysis pattern regarding our online banking service example.

1. Instantiate the direct system environment
   In our example in Fig. 10.3, we assume that the cloud provider is *Hulda*, the cloud customer is the *Bank Institute*, the end customers are *Bank Customers*, and the cloud developer is the *Internal Development Unit* of the *Bank Institute*.
2. Instantiate the cloud
   In our example in Fig. 10.3, we consider the *Online Banking Service* as an SaaS offer, and the data is instantiated with *Transaction Data*. The cloud of the provider *Hulda* is public, and the data centers are located in Germany as well as in the US.
3. Instantiate the indirect system environment
   In our example in Fig. 10.3, the domain is the *Domain Finance*, and the relevant legislators are *Germany*, *EU*, and *US* (since the data centres of *Hulda* are located in Germany as well as in the US).

We present an example of an indirect stakeholder template instance:

Name   *Legislator Germany.*

Description   The *Legislator Germany* represents all German laws relevant for this cloud scenario.

Motivation   The German laws try to control the risks of companies (*Hulda* and *Bank Institute*) and to protect the privacy of the *Bank Customers* by regulating disclosure of personal data.

Relations to other stakeholders   Controlled by law: The laws have to be obeyed by all stakeholders of the *Direct System Environment*.

Compliance and Privacy   The following regulations might be considered:

- Privacy protection: e.g., BDSG
- Risk management: e.g., German Stock Corporation Act (AktG).

We present an example of a direct stakeholder template instance, as well:

Name   *Bank Customer.*

Description   The *Bank Customer* uses the online banking service of the *Bank Institute*.

Motivation   The *Bank Customer* wants cheap and secure financial transactions via the bank's online banking service.

Relations to the cloud   Financial data is *InputBy* the *Bank Customer*. Financial data is personal information that the *Bank Institute* acquires from the *Bank Customer*. The information is required by the *Bank Institute* for billing purposes.

Relations to other stakeholders   *Has*: *Bank Institute* as SaaS provider.

Assets   Financial data, data related to the person.

Compliance and Privacy   The following regulations might be considered:

- Privacy protection: BDSG Sections. 3, 4, 9 and 11
- Risk management: AktG Sections. 91 and 93.

Step 2: Asset Identification

The cloud system analysis pattern instance in Fig. 10.3 helps identifying the primary and supporting assets. For instance, Transaction Data is related to the Bank Customer in Fig. 10.3. Thus, Transaction Data that can be refined to more concrete assets such as account number or account balance can be identified as a primary information asset. The owner of that asset is the Bank Customer. We categorize the assets in our example as follows: Transaction Data as personal information due to the relation to the Bank Customer. Here, privacy related data, e.g., account balance, are exchanged with the cloud. Further, billing data can be regarded as vital information for the Cloud Provider to run its business. In addition, Transaction Data can also be classified as vital information, because it is essential for running the organization's business.

The cloud in Fig. 10.3, surrounded by the dashed line, structures the cloud system and the instantiated boxes represent supporting assets. Different refinements of the supporting asset software can be found: Virtualization Software as well as the

Online Banking Software. The latter is deployed into the cloud by the Internal Development Unit, which could be wrongly determined as the asset owner. Using the associations in our pattern, it can be analyzed that the Internal Development Unit WorkFor the Bank Institute, which is therefore a more appropriate owner of the Online Banking Software, because it has contracted the software.

The personnel type consists of all the groups of people involved in the information system, for example *users*, *developers* or *decision makers*. The stakeholders in the Direct System Environment as well as the Indirect System Environment support the refinement of this supporting asset in the standard. We consider the location of elements in the scope, as well. We have to identify the location of the Data Centers. This means for our example that personal data processed or stored in data centers of the US do not comply to German privacy laws, and therefore this fact has to be documented here.

The organization's structure contains for example *subcontractors* of the organization. Here, the Direct Stakeholder Environment could help identifying them. The Cloud Provider as well as the Internal Development Unit can be documented here, as well.

10.4.5 Tool Support

We present our support tool for the *cloud system analysis pattern* in the following. The pattern (see Fig. 10.1) provides a conceptual view on cloud computing systems and serves to systematically analyze stakeholders and requirements as presented in the previous sections. This section is based on Beckers et al. (2013b, 2014) (Figs. 10.4 and 10.5).

We provide modeling support as part of our support tool that allows users to extend the cloud system analysis pattern with additional direct or indirect stakeholders, as well as further cloud elements, and the relations between them. Figure 10.6 shows the additional stakeholder *Cloud Phone Support*.

We also added a validation functionality in the tool that checks, e.g., if newly added elements to the pattern have been given a name. In addition, the tool checks if the instantiation process has worked correctly. The tool also has a validation functionality for the instantiation of the pattern. For example, Fig. 10.7 shows several warnings, which state that the attribute compliance in several stakeholder templates is not instantiated, meaning entered, by the user.

Basis for our tool is the Eclipse platform (Eclipse 2011a) together with its plug-ins Eclipse Modeling Framework (EMF) (Eclipse 2012a) and the Graphical Editing Framework (GEF) (Eclipse 2012b). We further use the Graphical Modeling Framework (GMF) (Eclipse 2011b), which provides a set of generative components for developing a graphical editor based on EMF and GMF.

We present our architecture in Fig. 10.4. The Cloud Pattern Analysis Plugin uses the GMF framework for creating the graphical user interface for the Cloud Pattern Analysis Plugin. We created an EMF model of the Cloud Pattern that we will explain in the following. The model itself is stored in the CAP module. CAP stands for

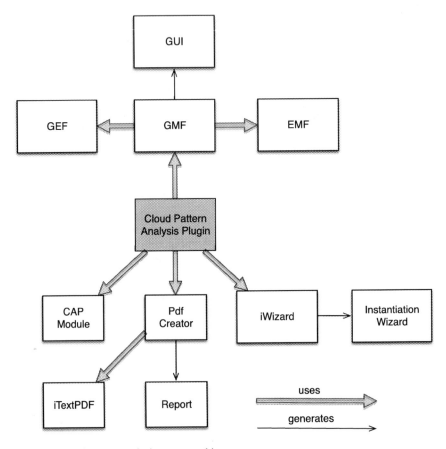

Fig. 10.4 Cloud system analysis pattern architecture

Cloud Analysis Pattern and stores the diagram parts of the tool. This part also stores models that are created using the graphical user interface (GUI). This model is used by the GMF framework via the Cloud Pattern Analysis Plugin, which uses in turn the combination of the EMF and the GEF framework to generate the source code for the cloud pattern analysis GUI. The Cloud Pattern Analysis Plugin provides the functionalities to create cloud patterns and to instantiate existing cloud patterns. The Cloud Pattern Analysis Plugin also uses an *Instantiation Wizard*, which uses the eclipse interface *IWizard* in turn. The *Instantiation Wizard* supports the instantiation of an existing cloud pattern. The wizard provides a graphical interface that asks the user for the information necessary to instantiate a cloud element or a stakeholder, e.g., the name of a stakeholder. In addition, the wizard supports the instantiation of several instances of a cloud element or a stakeholder. For example, the wizard can instantiate five cloud customers in sequence. Furthermore, the wizard can execute validation conditions of an instantiated cloud model. At the moment, this validation checks if all template fields of stakeholders have entries.

The tool can also export instantiated models to a pdf file. This file is called *Report* and contains the graphical model of the instantiated pattern, as well as the text in the stakeholder templates. The Cloud Pattern Analysis Plugin uses the *PdfCreator* to generate the Report. The PdfCreator uses the *iTextPDF* interface of Eclipse for the PDF creation.

We present a UML class model (UML Revision Task Force 2010) of our Cloud Analysis pattern in Fig. 10.5 that is created using EMF. The internal representation of the cloud pattern in the tool is based on this UML model. The *DirectEnvironment* is part of the *IndirectEnvironment*. The classes *DirectStakeholder* and *IndirectStakeholder* inherit from the class *Stakeholder*. *IndirectStakeholder*s are a part of the *IndirectEnvironment* and the *DirectStakeholder*s are a part of the *DirectEnvironment*. The *Cloud* is a part of the *DirectEnvironment*, as well. *CloudElement*s are a part of the *DirectEnvironment*. *Asset*s are a separate class. Our classes have certain attributes. *Stakeholders* have the attributes *name, description, motivation,* and *complianceAndPrivacy*. All of these attributes are of type String, because we do not want to restrict the values of these attributes. This makes it easy to accommodate later changes in our methodology. The *IndirectStakeholder* has an *instanceType*, which can be set with the values of the *IndirectStakeholderType* ≪enumeration≫: *legislator, domain,* and *contract*. The enumeration restricts the possible types of indirect stakeholders that can be selected. This prevents the selection of an indirect stake-

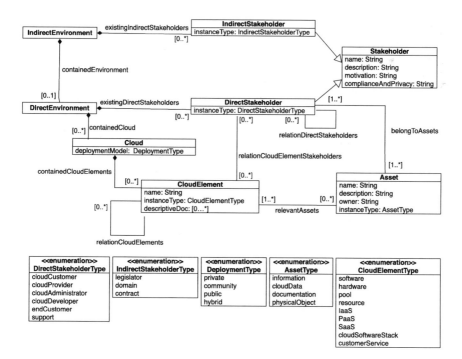

Fig. 10.5 A UML meta-model of the cloud system analysis pattern

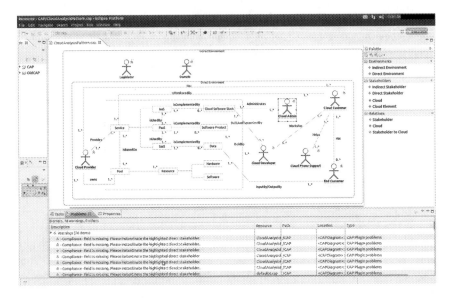

Fig. 10.6 Cloud system analysis pattern with phone support

holder type that does not exist. In addition, this enumeration can be extended with a new indirect stakeholder type if this is necessary. Similar, the *DirectStakeholder* has an *instanceType* which can be set with the values of the *DirectStakeholderType* ≪enumeration≫: *cloudCustomer, cloudProvider, cloudAdministrator, cloudDeveloper, endCustomer,* and *support.* The *Cloud* has the attribute *deploymentModel* and its ≪enumeration≫ *DeploymentType* has the values: *private, community, public,* and *hybrid.* The *CloudElement* has the attributes *name* and the type is String. The instanceType can be set to one of the values of the ≪enumeration≫ *CloudElementType,* which are: *software, hardware, pool, resource, IaaS, PaaS, SaaS, cloudSoftwareStack,* and *customerService.* An *Asset* has the attributes *description, owner,* and *name* of type String. The attribute *instanceType* can be set to one of the values of the ≪enumeration≫ *AssetType,* which are: *information, cloudData, documentation,* and *physicalObject.*

Moreover, this UML model shows that is possible to represent context-patterns completely in UML and it is a proof of concept that a transformation from our context-pattern description to UML is possible.

10.5 Peer-to-Peer System Analysis Pattern

Aligning software systems to meet requirements is hard, which is even more difficult when a Peer- to-Peer (P2P)-based software system shall be developed. For example, the effects of churn, the random leaving or joining of peers in the system, can cause

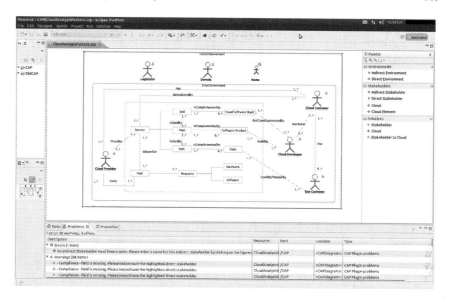

Fig. 10.7 Cloud system analysis pattern validation functionality

data loss. If a requirement exists stating that no data loss shall occur in the system, then churn presents a challenge that has to be considered for this requirement. Software engineers can design a countermeasure if they are aware of the challenge. This, however, is difficult, because of numerous challenges that are caused by attributes of the P2P protocol or even the network layer.

We present a pattern-based method to identify existing challenges in a P2P-based software system. An initial pattern considers all layers of a P2P architecture and offers more detailed patterns for, e.g., P2P protocols. The instantiation of these patterns enables an analysis of the system's challenges and reveals the information in which layer each challenge originates. An extensive description of the context-pattern can be found in Beckers and Faßbender (2012a).

10.5.1 Graphical Pattern

Our *Peer-to-Peer (P2P)* pattern (see top of Fig. 10.8) is based on the P2P architecture from Lua et al. (2005), which is derived from a survey of existing *P2P* systems. This survey describes *P2P* systems as layered architectures that contain at least the following layers.

The *Application Layer* concerns applications that are implemented using the underlying *P2P* overlay, for example, a Voice-over-IP (VoIP) application. The *Service Layer* adds application-specific functionality to the *P2P* infrastructure. For example, for parallel and computing-intensive tasks or for content and file management.

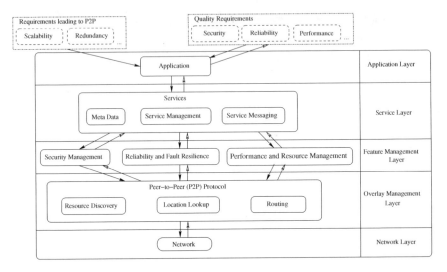

Fig. 10.8 P2P pattern

Meta-data describe what the service offers, for instance, content storage using *P2P* technology. Service messaging describes the way services communicate. The *Feature Management Layer* contains elements that deal with security, reliability, and fault resiliency, as well as performance and resource management of a *P2P* system. All these aspects are important for maintaining the robustness of a *P2P* system. The *Overlay Management Layer* is concerned with peer and resource discovery and routing algorithms. The *Network Layer* describes the ability of the peers to connect in an ad hoc manner over the Internet or small wireless or sensor-based networks.

10.5.2 Templates

We created P2P templates for the layers of a P2P system, e.g., the network layer. For details we refer to Becker and Faßbender (2012a).

10.5.3 Method

Step 1: Instantiate P2P-Pattern -
   We illustrate our method in Fig. 10.9, which consists of instantiating the P2P pattern and assess the feasibility of the quality requirements for the P2P system. The steps are described in the following.

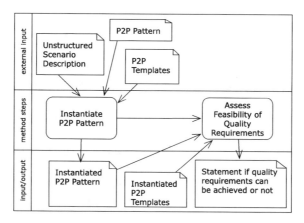

Fig. 10.9 Method for P2P pattern usage

Input:	Unstructured scenario description
	P2P pattern and templates
Output:	Instantiated P2P pattern and templates

Peer-to-Peer (P2P) architectures design distributed systems in which identical software works on every peer. These systems are distributed without any centralized control or hierarchical organization that form a self-organizing overlay network on top of the Internet Protocol (IP). An overlay is a network which is built on top of one or more existing networks. This adds additional properties to the underlying network, e.g., more efficient search of data or adding locality information to peers. It provides a communication infrastructure for all peers in the P2P architecture. P2P systems can be clients and servers at the same time. They provide access to their resources by other systems and support resource sharing on an internet scale. This requires fault tolerance, self-organization, and significant scalability properties. One of the obstacles a P2P system has to overcome is churn, the joining and leaving of peers in a P2P architecture without prior notification. Two fundamentally different types of P2P systems exist. Structured P2P systems organize peers via an algorithm, which leads to an overlay with specific properties. They are often based on a distributed hash table. Unstructured P2P systems organize their peers in a random graph in a flat or hierarchical manner (e.g., Supernodes exist that outrank normal nodes). They are based on techniques such as flooding, random walks, etc. Hence, P2P architectures are fundamentally different from stand-alone or client server architectures. The pattern contains the layers of a P2P architecture, shown in Fig. 10.8. The instantiation of the pattern begins with one or more requirement(s) that lead to the decision to use a P2P protocol. For example, scalability or redundancy requirements can result in this decision. The pattern is derived of a survey of existing P2P systems. The requirements and the layers of a P2P system are connected with arrows. Arrows with a black arrowhead present the collection of information through the P2P layers, getting more detailed on every level. The white arrowheads present challenges that are derived from all the information collected before.

The Application Layer concerns applications that are implemented using the underlying P2P overlay, for example, a Voice-over-IP (VoIP) application. The Service Layer adds application specific functionality to the P2P infrastructure. For example, for parallel and computing intensive tasks or for content and file management. Meta-data describes what the service offers, for example, content storage using P2P technology. Service messaging describes the messages and protocols services use to communicate. The Feature Management Layer contains elements that deal with security, reliability and fault resiliency, as well as performance and resource management of a P2P system. All these aspects are for maintaining the robustness of a P2P system. We renamed the resource management from the original architecture from Lua et al. (2005) into performance and resource management, because the quality of the access and distribution of resources is the main performance property of P2P architectures. The Overlay Management Layer is concerned with peer and resource discovery and routing algorithms. We aim to explain these in terms of software engineering and not network engineering. The Network Layer describes the ability of the peers to connect in an ad hoc manner over the Internet or small wireless or sensor-based networks.

Step 2: Assess Feasibility of Quality Requirements -

Input:	Instantiated P2P pattern and templates
Output:	Statement that clearly describes if and why the quality requirements can or cannot be achieved

We consider the impact of each challenge on each quality requirement. The challenges are investigated considering the chosen technologies for each of the layers of the P2P system. The analysis is based on the instantiated P2P-Pattern. We use this information to determine if the quality requirements are achievable (see Beckers and Faßbender 2012a for details).

10.6 Service-Oriented Architecture Pattern

Our Service-oriented Architectures (SOA) pattern concerns eliciting domain knowledge for SoA. A detailed description of the context-pattern can be found in Beckers et al. (2012a).

10.6.1 Graphical Patterns

An SOA spans different layers (Beckers et al. 2012a), which form a pattern on an SOA with technological focus, as depicted in Fig. 10.10 on the top. The first and top

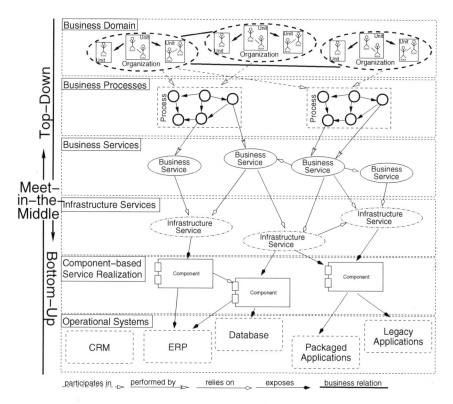

Fig. 10.10 SOA layer pattern

layer is the *Business Domain* layer, which represents the real world. It consists of *Organizations*, their structure and actors, and their *business relations* to each other. The second layer is the *Business Process* layer. To run the business, certain *Processes* are executed. Organizations *participate in* these processes. These processes are supported by *Business Services*, which form the *Business Service* layer. A business service encapsulates a business function, which *performs* a process activity within a business process. All business services rely on *Infrastructure Services*, which form the fourth layer. The infrastructure services offer the technical functions needed for the business services. These technical functions are either implemented especially for the SOA, or they expose interfaces from the *Operational Systems* used in an organization. These operational systems, such as databases or legacy systems, are part of the last SOA layer at the bottom of the SOA stack. These layers form a generic pattern, the SOA layer pattern, to describe the essence of a SOA.

In Fig. 10.10 at the bottom, we adapted problem-based methods, such as problem frames by Jackson (2001), to enrich the SOA layer pattern with its environmental context. The white area in the bottom of Fig. 10.10 spans the SOA layers that form the machine. The business processes describe the behavior of the machine. The business

services, infrastructure services, components, and operational systems describe the
structure of the machine. Note that the business processes are not part of the machine
altogether, as the processes also include actors, which are not part of the machine.
Thus, the processes are the bridge between the SOA machine and its environment.
The environment is depicted by the gray parts of the bottom of Fig. 10.10. The light
gray part spans the *Direct Environment* and includes all entities, which participate
in the business processes or provide a part, like a component, of the machine. An
entity, for example, is something that exits in the environment independently of
the machine or other entities. The dark gray part in the bottom of Fig. 10.10 spans
the indirect environment. It comprises all entities not related to the machine but
to the direct environment. The Business Domain layer is one bridge between the
direct and indirect environment. Some entities of the Direct Environment are part
of organizations. Some entities of the Indirect Environment influence one or more
organizations. The machine and the Direct Environment form the *inner system*, while
the *outer system* also includes the Indirect Environment (Fig. 10.11).

The entities we focused on for the stakeholder SOA pattern are stakeholders,
because all requirements to be elicited stem from them. There are two general kinds
of stakeholders. The *direct stakeholders* are part of the direct environment, while the
indirect stakeholders are part of the indirect environment. We derived more specific
stakeholders from the direct and indirect stakeholders, because these two classes are

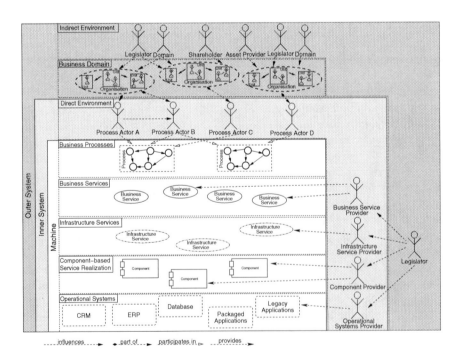

Fig. 10.11 SOA layer stakeholder pattern

very generic. Process actors and different kinds of providers are part of the direct environment. Legislators, domains, shareholders, and asset providers are part of the indirect environment. At the bottom of Fig. 10.10, the resulting stakeholder classes are depicted as stick figures. For a detailed description of these stakeholders we refer to our previous work (Beckers et al. 2012a).

10.6.2 Templates

We created templates for direct and indirect stakeholders similar to the ones described in Sect. 10.4.2. For details we refer to Beckers et al. (2012a).

10.6.3 Method

Step 1: Information Structuring -
We illustrate our method for SOA knowledge elicitation in Fig. 10.12 and explain its steps in the following.

Input:	Unstructured scenario description
	SOA layer pattern
Output:	Instantiated SOA layer pattern

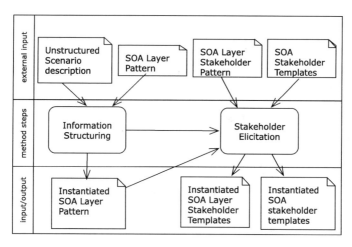

Fig. 10.12 SOA knowledge elicitation method

For structuring the information necessary to design an SOA according to the SOA layer pattern, we suggest a meet-in-the-middle procedure. The reason is that the business services and infrastructure services layers are intertwined with the business elements. Therefore, we need information both about the technology-related and the business-related layers. Furthermore, our method provides validity checks for the relations between the different layers. The external input for all steps of this phase is the unstructured scenario description. The SOA Layer Pattern is an additional external input for the first step Describe Organizations. It is instantiated layer by layer in separate steps, based on the Unstructured Scenario Description. In the step Describe Organizations, we have to collect all relevant organizations. For each such organization, we have to collect statements about this organization, describing it further. Next, we have to analyze these statements if they describe business relations between organizations. Lastly, we have to check for inconsistencies. For example, we have to ensure that no organization is isolated. Being isolated means that there are no business relations to other organizations. In case we find an isolated organization, this organization is either not of relevance for our scenario, or we are missing important information.

In the next step, Describe Choreography and Coarse Grained Processes, we have to structure the interaction between organizations. Additionally, we structure the available information about internal processes of organizations. The input for this step is the partly instantiated SOA Layer Pattern. We start with the organizations and their choreography. The choreography describes the interaction between the organizations. We recommend to document the choreography using an appropriate notation. For example, UML and SoaML collaboration diagrams can be used. UML and SoaML activity diagrams are of use for more detailed interaction and internal process descriptions. The described processes do not have to be complete, but processes and process steps already mentioned or explained in the scenario description should be captured by such diagrams. Next, we have to ensure the coherence of the SOA Layer Pattern instance for finishing the step Describe Choreography and Coarse Grained Processes. The interactions described by the choreography should reflect the business relations found for the organizations. Moreover, the detailed process descriptions must be coherent at the points of transitions between organizations. For the step Describe Operational Systems, we look for statements mentioning IT systems already used within one of the organizations. The operational systems found have to be analyzed for those that are to be replaced, and those that should be wrapped in the new SOA. Only the latter kind of operational system should be added to the SOA Layer Pattern instance. At least one component should be described in the step Describe Components for each operational system. Whenever for an operational system there is no statement about a component, which should be reused, the information is missing in the Unstructured Scenario Description, or the operational system is unnecessary. Note that components can exist, which are not part of an operational system, but are already mentioned in the scenario description. But to be sure not to miss any relevant operational system, for each component mentioned in the Unstructured Scenario Description, we have to check if it exposes such a system. Up to this point, we have structured the business and the technical parts of the description. Next, we close the gap between those parts. We search for statements, which describe business services for the step Describe Business Services. We add those services to the business service layer. For each business service, we have to check if there is a corresponding process step in the processes described in the step Describe Orchestration and Coarse Grained Processes. If such an activity is missing, it should be added. Additionally, we have to check if the business service at hand directly exposes a component, which is already part of the SOA Layer Pattern instance. Last, we have to check whether the business service at hand is atomic or a composition of other business services. If it is a composition, the business services used to compose it have to be added to the business service layer, too. The procedure for Describe Infrastructure Services is almost the same. One difference is that infrastructure services are mapped to business services instead of activities within the process, and that they can be orphans. It is not necessary that an infrastructure service is used by an already known business service. The reason is that infrastructure services may provide a more general functionality.

Step 2: Stakeholder Elicitation -

Input:	Instantiated SOA layer pattern
	SOA layer stakeholder pattern and templates
Output:	Instantiated SOA layer stakeholder pattern and templates

In this phase, we elicit the stakeholders of our SOA. Therefore, we inspect each element of the SOA Layer Pattern instantiated in the previous phase. First, we instantiate the direct system environment. We start with the organizations given in the SOA Layer Pattern Instance. For each process related to an organization, we identify the process actors, which act on behalf of the organization in this particular process. There has to be at least one process actor for each organization-process-relation. For all process actors, we have to instantiate the corresponding Direct Stakeholder Templates. Finally, we have to establish the relations between associated process actors. Next, we inspect each business service, infrastructure service, component, and operational system, whether there are already known provider(s) or not. When the providers are already known, we instantiate Direct Stakeholder Templates for them and add them and the corresponding relations to the SOA Stakeholder Pattern instance. Further, we instantiate the indirect system environment. We also start with the organizations. We analyze for each organization at hand, if there are relevant legislators, domains, shareholders, and asset providers. For each identified indirect stakeholder, we instantiate the corresponding Indirect Stakeholder Template, and we add the indirect stakeholder and their relations to the SOA Stakeholder Pattern instance. We repeat this procedure for all providers we find in the direct system environment.

10.7 Law Pattern

Commonly, laws are not adequately considered during requirements engineering. Therefore, they are not covered in the subsequent system development phases. One fundamental reason for this is that the involved engineers are typically not crossdisciplinary experts in law and software and systems engineering. Hence, we present in this section a context-pattern for identifying laws and regulations including a method to systematically consider laws in the requirements engineering process. For our method we chose the German law as the binding law.

10.7.1 Graphical Patterns

Based on the structure of laws (details can be found in Beckers et al. (2012b)) we define a *law pattern* shown in Fig. 10.13. The pattern consists of three parts: the dark gray part represents the *Law Structure*, the light gray part depicts the *Classification* to consider the specialization of the elements contained in the *Law Structure* in related laws or sections, and the white part considers the *Context*. The *Context* part of the law pattern contains the *Legislator(s)* defining the jurisdiction, and the *Domain(s)* clarifying for which domain the law was established.

As it is necessary to know in which context and relation a law is used, we introduce *Regulation(s)*, which are *Related To* the section at hand. *Regulation(s)*, *Legislator(s)*, and *Domain(s)* can be ordered in hierarchies, similar to classifiers. For instance, Germany is part of the EU and consists of several states.

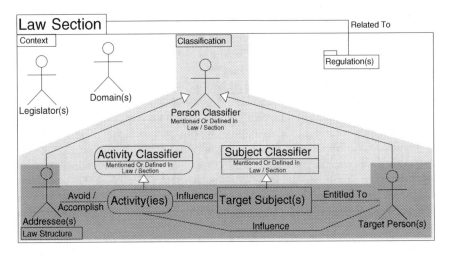

Fig. 10.13 Law pattern

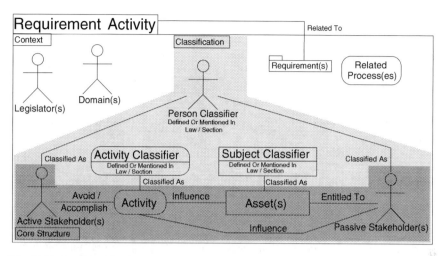

Fig. 10.14 Law identification pattern

Figure 10.14 shows our law identification pattern. The structure is similar to the law pattern in Fig. 10.13 to allow a matching of instances of both patterns. In contrast to the legal vocabulary used in the *Law Structure* of our law pattern, the wording for the elements in the dark gray colored *Core Structure* of our law identification pattern is based on terms known from requirements engineering. For example, the element *Asset(s)* in our law identification pattern represents the element *Target Subject(s)* in our law pattern.

Our law identification pattern takes into account that requirements are often inter-dependent (*Requirement(s)* in the *Context* part). Given a law relevant to a requirement, the same law might be relevant to the dependent requirements, too. Furthermore, the pattern helps to document similar dependencies for a given *Activity* using the *Related Process(es)* in the *Context* part.

10.7.2 Templates

We did not define templates for this context-pattern. Note that templates are optional in context-patterns (Sect. 10.2).

10.7.3 Method

Our general process for identifying relevant laws consists of five steps, which is depicted in Fig. 10.15. For this process law experts and software engineers have to

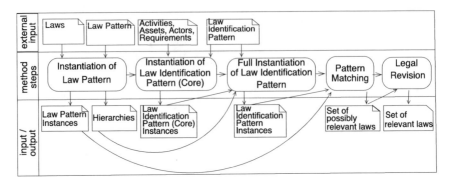

Fig. 10.15 Law identification method

work together for the necessary knowledge transfer. Step one can be done alone by legal experts and for step two only software engineers are needed. But in step three and four both groups are needed to bridge the gap between the legal and technical world. The last step can be accomplished alone by legal experts.

Step 1: Law Pattern -

Input:	Laws
Output:	Instantiated law patterns

In the first step we instantiate law patterns with the relevant laws for our scenario. We now describe the instantiation process for our law pattern using Sect. 4b BDSG as an example. We explained the importance of this particular law in the beginning of this section. Section 4b BDSG regulates the abroad transfer of data. The resulting instance is shown in Fig. 10.16. Our method starts based on the first sections of the law to be analyzed. These sections are self-contained, i.e., they define all necessary elements of our *Law Structure*. Additionally, the *Legislator(s)* and *Domain(s)* can be instantiated according to the considered law (e.g., *Germany* and *General Public* in the *Context* part). Given a section of a law not yet captured by our law pattern, we identify and document the related laws and sections referred to by the given section (e.g., *BDSG Section. 1* in the *Context* part). Then we search for the *Law Structure* directly defined in this section. In Section. 4b BDSG, we find *Abroad Transfer*, and we use it to instantiate *Activity(ies)*. *Address(es)*, *Target Subject(s)*, and *Target Person(s)* are not defined in Section. 4b BDSG. Therefore, related sections defining these terms have to be discovered. In our example, we find *Private Bodies* for the *Address(es)*, *Personal Data* for the *Target Subject(s)*, and *Individual* for the *Target Person(s)* in Sect. 1 BDSG (according to *BDSG Section. 1* in the *Context* part). We arrange these specialization in the appropriate parts of the hierarchies in Fig. 10.17. The classifier is instantiated with the parent node of the corresponding hierarchy, which is for instance *Transfer* for *Abroad Transfer*.

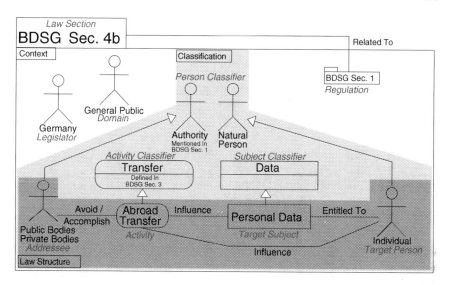

Fig. 10.16 Law pattern instance

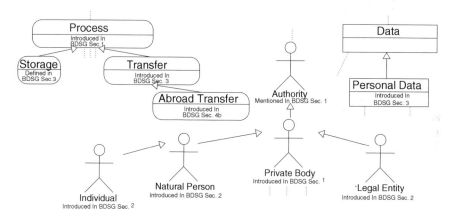

Fig. 10.17 Examples for person (*left*), subject (*middle*), and activity (*right*) legal hierarchies

Step 2: Law Identification Pattern -

Input:	Scenario description and requirements
Output:	Partially instantiated law identification patterns

Identifying relevant laws based on functional requirements is difficult because functional requirements are usually too imprecise, they contain important information only implicitly and use a different wording than in laws. To bridge between

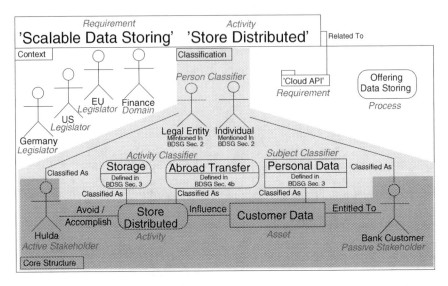

Fig. 10.18 Law identification pattern instance

the gap of wording and to facilitate the discussion between requirements engineers and legal experts, we define a *law identification pattern* to support identifying relevant laws.

As our example in Fig. 10.18 shows, we select *Hulda* as the cloud provider, then we choose the functional requirement *Scalable Data Storing*. One of the activities associated with this requirement is the activity *Store Distributed*, which refers to the asset *Customer Data* of the *Bank Customer*. Moreover, we instantiate the elements *Legislator(s)* and *Domain(s)*. In our example in Fig. 10.18, we include the legislators *Germany*, *US*, *EU*, and the domain *Finance*. In addition, we discover the related requirement *Cloud API* and the process *Offering Data Storing*, and document them in the instance of our law identification pattern.

Step 3: Establishing the Relation between Laws and Requirements -

Input:	Partially instantiated law identification Patterns
Output:	Completely instantiated law identification patterns

To instantiate the *Classification* part, legal expertise is necessary. According to the *Core Structure* of the instance of our law identification pattern and the hierarchies built when instantiating our law pattern, legal experts classify the elements of the *Core Structure*. For example, the activity *Store Distributed* is classified as *Abroad Transfer* based on a discussion between the legal experts and software engineers.

Step 4: Deriving relevant Laws -

Input:	Instantiated law identification patterns, instantiated law patterns
Output:	Set of possibly relevant Laws

The identification of relevant laws is based on matching the classification part of the law identification pattern instance (light gray part) with the law structure and classification part of the law pattern instance (light and dark gray parts), and thereby considering the previously documented hierarchies. If all elements match, the law is identified as relevant. For example, we find direct matches in the law pattern instance in Fig. 10.16 for the elements *Abroad Transfer*, *Personal Data*, and *Individual* contained in the law identification pattern instance shown in Fig. 10.18. *Hulda* is classified as *Legal Entity* and the only element that does not directly match with *Private Bodies* in the law structure of Section. 4b BDSG. In this case, the hierarchy in Fig. 10.17 helps to identify that *Legal Entity* is a specialization of *Private Bodies*, and thus, we identify Section. 4b BDSG as relevant. Finally, we check for all laws identified to be relevant if *Legislator(s)* and *Domain(s)* are mutually exclusive. In our example, the legislator *Germany* contained in *Context* of the law pattern instance depicted in Fig. 10.16 can be found in *Context* of the law identification pattern instance shown in Fig. 10.18. The domain *General Public* in the law pattern instance can be considered as a generalization of the domain *Finance* in the law identification pattern instance. The resulting set of laws relevant for the given development problem serves as an input for the next step.

Step 5: Legal Revision -

Input:	Set of possibly relevant Laws
Output:	Set of relevant Laws

This last step covers the identification and specification of requirements based on laws identified to be relevant by our method, e.g., using existing methods such as the one from Breaux and Antón (2008).

10.8 Summary

The long known credo of requirements engineering states that it is impossible to build the right system, if you do not know what is right. Requirements engineering methods have to consider domain knowledge, otherwise severe problems can occur during software development, e.g., technical solutions to requirements might be impractical or costly. It is an open research question of *how* to elicit domain knowledge correctly for effective requirements elicitation (Niknafs and Berry 2012). Several requirements engineering methods exist, e.g., for security. Fabian et al. 2010 concluded in their survey about these methods that it is not yet state of the art to consider domain knowledge.

We built a catalog of context-patterns for a structured domain knowledge elicitation to address this problem. We described common structures and stakeholders and technical elements of these systems for several different domains in our context-patterns. Our main contributions of this sections are:

- Domain-specific context establishment based on patterns.
- Systematic pattern-based identification of relevant stakeholders and technical components of a system.
- The patterns can be accompanied by further diagrams to support different views and detail levels, for example, using UML sequence diagrams (Chap. 12).
- The patterns can be integrated into existing software development processes in order to improve context elicitation activities.
- The patterns are useful beyond software engineering. For example, they can be applied to create the documentation for implementing security standards, e.g., ISO 27001 (ISO/IEC 2005) (Chap. 12).

Our catalog can be extended with the identification of further context-patterns by using our initial pattern language in Chap. 11.

References

Alexander, C. (1977). *A pattern language: Towns, buildings, construction.* Oxford: Oxford University Press.

Beckers, K., & Faßbender, S. (2012a). Peer-to-peer driven software engineering considering security, reliability, and performance. In *Proceedings of the International Conference on Availability, Reliability and Security (ARES)—2nd International Workshop on Resilience and IT-Risk in Social Infrastructures (RISI 2012)* (pp. 485–494). IEEE Computer Society.

Beckers, K., & Faßbender, S. (2012b). Supporting the context establishment according to ISO 27005 using patterns. In: *Software Engineering 2012—Workshopband* (pp. 141–146). Berlin, Germany: GI. (Workshop Zertifizierung und modellgetriebene Entwicklung sicherer Software, Software Engineering 2012 (ZeMoSS'12)).

Beckers, K., & Heisel, M. (2012). A foundation for requirements analysis of privacy preserving software. In *Proceedings of the International Cross Domain Conference and Workshop (CD-ARES 2012)* (pp. 93–107). Springer.

Beckers, K., Küster, J.-C., Faßbender, S., & Schmidt, H. (2011). Pattern-based support for context establishment and asset identification of the ISO 27000 in the field of cloud computing. In: *Proceedings of the International Conference on Availability, Reliability and Security (ARES)* (pp. 327–333). IEEE Computer Society.

Beckers, K., Faßbender, S., Heisel, M., & Meis, R. (2012a). Pattern-based context establishment for service-oriented architectures. *Software service and application engineering* (pp. 81–101). Berlin: Springer.

Beckers, K., Faßbender, S., Küster, J.-C., & Schmidt, H. (2012b). A pattern-based method for identifying and analyzing laws. In *Proceedings of the International Working Conference on Requirements Engineering: Foundation for Software Quality (REFSQ)* (pp. 256–262). Springer.

Beckers, K., Faßbender, S., & Schmidt, H. (2012c). An integrated method for pattern-based elicitation of legal requirements applied to a cloud computing example. In *Proceedings of the International Conference on Availability, Reliability and Security (ARES)—2nd International Workshop on Resilience and IT-Risk in Social Infrastructures (RISI 2012)* (pp. 463–472). IEEE Computer Society.

Beckers, K., Côté, I., Faßbender, S., Heisel, M., & Hofbauer, S. (2013a). A pattern based method for establishing a cloud-specific information security management system. *Requirements Engineering, 18*(4), 1–53.

Beckers, K., Côté, I., Goeke, L., Güler, S., & Heisel, M. (2013b). Structured pattern-based security requirements elicitation for clouds. In *Proceedings of the International Conference on Availability, Reliability and Security (ARES)—7th International Workshop on Secure Software Engineering (SecSE 2013)* (pp. 465–474). IEEE Computer Society.

Beckers, K., Côté, I., Goeke, L., Güler, S., & Heisel, M. (2014). A structured method for security requirements elicitation concerning the cloud computing domain. *International Journal of Secure Software Engineering (IJSSE), 5*(2), 20–43.

Breaux, T. D., & Antón, A. I. (2008). Analyzing regulatory rules for privacy and security requirements. *IEEE Transactions on Software Engineering, 34*(1), 5–20.

Eclipse Foundation. (2011a). Eclipse—An open development platform (http://www.eclipse.org/).

Eclipse Foundation. (2011b). Eclipse graphical modeling framework (GMF). (http://www.eclipse.org/modeling/gmf/).

Eclipse Foundation. (2012a). Eclipse modeling framework project (EMF). (http://www.eclipse.org/modeling/emf/).

Eclipse Foundation. (2012b). Graphical editing framework project (GEF). (http://www.eclipse.org/gef/).

Fabian, B., Gürses, S., Heisel, M., Santen, T., & Schmidt, H. (2010). A comparison of security requirements engineering methods. *Requirements Engineering—Special Issue on Security Requirements Engineering, 15*(1), 7–40.

Fernandez, E. B., & Pan, R. (2001). A pattern language for security models. In *8th Conference of Pattern Languages of Programs (PLoP)*.

Fernandez, E. B., Pelaez, J. C., & Larrondo-Petrie, M. M. (2007). Security patterns for voice over ip networks. In *Proceedings of the International Multiconference on Computing in the Global Information Technology* (pp. 19–29). IEEE Computer Society.

Fowler, M. (1996). *Analysis patterns: Reusable object models*. Boston: Addison-Wesley.

Fowler, M. (2002). *Patterns of enterprise application architecture*. Boston: Addison-Wesley Longman Publishing Co., Inc.

Gamma, E., Helm, R., Johnson, R., & Vlissides, J. (1994). *Design patterns: Elements of reusable object-oriented software*. Boston: Addison-Wesley.

Hafiz, M. (2006). A collection of privacy design patterns. In *Proceedings of the 2006 Conference on Pattern Languages of Programs* (pp. 1–13). ACM.

Hafiz, M., Adamczyk, P., & Johnson, R. E. (2012). Growing a pattern language (for security). In *Proceedings of the ACM International Symposium on New Ideas, New Paradigms, and Reflections on Programming and Software* (pp. 139–158). ACM.

Hatebur, D., & Heisel, M. (2009). A foundation for requirements analysis of dependable software. In *Proceedings of the International Conference on Computer Safety, Reliability and Security (SAFECOMP)* (p. 311–325). Springer.

ISO/IEC. (2005). Information technology—Security techniques—Information security management systems—Requirements (ISO/IEC 27001). Geneva, Switzerland: International Organization for Standardization (ISO) and International Electrotechnical Commission (IEC).

ISO/IEC. (2008). Information technology—Security techniques—Information security risk management (ISO/IEC 27005). Geneva, Switzerland: International Organization for Standardization (ISO) and International Electrotechnical Commission (IEC).

ISO/IEC. (2009). Information technology—Security techniques—Information security management systems—Overview and Vocabulary (ISO/IEC 27000). Geneva, Switzerland: International Organization for Standardization (ISO) and International Electrotechnical Commission (IEC).

Jackson, M. (2001). *Problem frames. Analyzing and structuring software development problems*. Boston: Addison-Wesley.

Lua, E. K., Crowcroft, J., Pias, M., Sharma, R., & Lim, S. (2005). A survey and comparison of peer-to-peer overlay network schemes. *IEEE Communications Surveys and Tutorials, 7*, 72–93.

Niknafs, A., & Berry, D. M. (2012). The impact of domain knowledge on the effectiveness of requirements idea generation during requirements elicitation. In *Proceedings of the 20th IEEE International Requirements Engineering Conference (RE)*, (pp. 181–190). IEEE Computer Society.

Schmidt, H. (2010). *A pattern—and component-based method to develop secure software*. Deutscher Wissenschafts-Verlag (DWV) Baden-Baden.

Schumacher, M., Fernandez-Buglioni, E., Hybertson, D., Buschmann, F., & Sommerlad, P. (2006). *Security patterns: Integrating security and systems engineering*. New York: Wiley.

UML Revision Task Force. (2010, May). OMG unified modeling language: Superstructure [Computer software manual].

Withall, S. (2007). *Software requirement patterns*. Redmond: Microsoft Press.

Chapter 11
Initiating a Pattern Language
for Context-Patterns

Abstract A context-pattern describes common elements, structures, and stakeholders for a specific domain such as cloud computing. These commonalities for a
context were obtained from observations about the domain from sources such as
standards, domain specific-publications, domain experts, and guidelines. Existing
context-patterns can be used for a structured elicitation and analysis of domain
knowledge by instantiating the respective context-pattern. In the previous chapter,
we showed a catalog of context-patterns. In this chapter, we aim at broadening the
context-pattern approach by initiating a pattern language for context-patterns, which
will be continuously improved. The aim of this language is to gain an understanding of common elements in context-patterns and support engineers in applying this
knowledge for describing their own context-patterns. For this purpose, we analyzed
the common concepts in our context-patterns and created a meta-model to show the
common elements and their relations. Moreover, we defined a pattern language syntax for context-patterns based on the insights we gathered from analyzing possible
relations and relation types between context-patterns. Furthermore, we compared
our pattern language with different existing pattern languages. For this comparison,
we contribute a structured template for describing pattern languages and instantiate
the template for well-known pattern languages.

11.1 Motivation

In the previous chapter, we introduced a catalog of context-patterns (see Chap. 10).
We described common structures and stakeholders for several different domains
in our context-patterns. The common elements of the context were obtained from
observations about the domain in terms of standards, domain specific-publications,
and implementations. Whenever a system-to-be is already described by a context-
pattern, one can use this context-pattern to elicit domain knowledge via instantiation
of the context-pattern.

Our context-patterns support the structured elicitation of domain knowledge and
we showed a number of these in the previous chapter. However, these patterns are
isolated from each other in the pattern catalog and a common language is missing

© Springer International Publishing Switzerland 2015
K. Beckers, *Pattern and Security Requirements*,
DOI 10.1007/978-3-319-16664-3_11

that describes how a problem that requires multiple patterns can be solved by their combination. Our intention is that the pattern catalog should grow, and providing an easy way for engineers to describe further context-patterns is a step toward this aim. Hence, our pattern language has the intention to support engineers in a better understanding of context-patterns, describing further context-patterns, and finding a useful way to combine context-patterns to solve a problem. Furthermore, this chapter is based on the publications (Beckers et al. 2013, 2014a). These publications are the result of an active discourse with the pattern community at the EuroPlop conference and researchers in the field.

In this chapter, we contribute a template for describing pattern languages in Sect. 11.2 and instantiate the template with our pattern language for context-patterns. The elements of our pattern language include a meta-model, shown in Sect. 11.3, and relations between context-patterns in Sect. 11.4 and an overview of all possible sequences of context-patterns.

11.2 A Template for Pattern Languages

We propose a template to describe and compare pattern languages. This template is based on the idea that a pattern language consists of the same elements the natural language does: Vocabulary, syntax, and grammar. We looked into the works of Alexander (1977) and Buschmann et al. (2007) and their views on pattern languages (see Sect. 11.2.1). We discuss our template based on inspirations by their fundamental work on the area of pattern languages. In addition, we instantiated our template with several influential existing software engineering pattern languages (see Sect. 11.2.3) and discuss the results. Finally, we describe in Sect. 11.2.4 how we show an instantiation of our template for our pattern language for context-patterns.

11.2.1 Viewpoints of Pattern Languages

Alexander (1977) described the term *pattern language*, which is a structured method for describing common design practices for a knowledge area. Alexander described a pattern language for creating towns and buildings in Alexander (1977) and he wanted to empower ordinary people to successfully solve very large, complex design problems. "This language is extremely practical. It is a language that we have distilled from our own building and planning efforts over the last 8 years. You can use it to work with your neighbours, to improve your town and neighbourhood. You can use it to design a house yourself, with your family; or to work with other people to design an office or a workshop or a public building like a school. And you can use it to guide you in the actual process of construction." (Alexander 1977, p. x).

Inspired by the work of Alexander we looked into the essential elements of a pattern language and state that these elements are vocabulary, syntax, and grammar. Note that Alexander did not explicitly state in his work that these are elements of a pattern language, but we argue in the following that these elements are referenced in his work. Beforehand, we define these terms for human language based on the Oxford English Dictionary (OED). The term language in the OED[1] is defined as follows: "The system of spoken or written communication used by a particular country, people, community, etc., typically consisting of words used within a regular grammatical and syntactic structure". In addition, the OED[2] defines the vocabulary of a language as: "the body of words used in a particular language". Moreover, the OED[3] defines the term semantic as: "relating to meaning in language or logic". Alexander states in regard to a pattern language that "the elements of this language are entities called patterns." (Alexander 1977, p. x). Hence, patterns are the vocabulary of a pattern language. In addition, Alexander states that "A pattern language has the structure of a network. [...] However, when we use the network of a language, we always use it as a sequence, going through the patterns, moving always from the larger patterns to the smaller, always from the ones which create structures, to the ones which then embellish those structures, and then to those which embellish the embellishments. ... Since the language is in truth a network, there is no one sequence which perfectly captures it. But the sequence which follows, captures the broad sweep of the full network; in doing so, it follows a line, dips down, dips up again, and follows an irregular course, a little like a needle following a tapestry." (Alexander 1977, p. xviii).

Furthermore, Alexander reasons about the use of his language in comparison to the use of the English language. "This language, like English, can be a medium for prose, or a medium for poetry. The difference between prose and poetry is not that different languages are used, but that the same language is used, differently. In an ordinary English sentence, each word has one meaning, and the sentence too, has one simple meaning. In a poem, the meaning is far more dense. Each word carries several meanings; and the sentence as a whole carries an enormous density of interlocking meanings, which together illuminate the whole." (Alexander 1977, p. xli).

Buschmann et al. (2007) formulate a hypothesis in their work that a pattern language is built up from over several stages. First, pattern stories describe specific examples of the application of patterns in combination. Second, the experiences from the stories are abstracted into pattern sequences. Third, numerous sequences of patterns form a pattern language. These show that the patterns can be combined in a way that helps engineers to solve problems with different solutions.

---

[1]The definition of the term language in the Oxford English Dictionary: http://www.oed.com.

[2]The term vocabulary defined in the Oxford dictionaries http://www.oxforddictionaries.com/definition/english/vocabulary?q=vocabulary.

[3]The definition of the term semantic in the Oxford dictionaries http://www.oxforddictionaries.com/definition/english/semantic.

11.2.2 A Template for Describing a Pattern Language

Note that the difference between a language and a pattern language is that a language focuses on communication. In addition, a pattern language focuses on complex engineering activities. Complex engineering problems are often split up into subproblems, which are addressed separately. Different patterns contain problems and solutions for the different granularity levels of a problem (its subproblems). Hence, the solution to a design problem often requires the combination of different patterns applied in sequence. Moreover, in a pattern language there often exist multiple solutions to a problem, which means multiple pattern sequences. A pattern language contains all these sequences of patterns.

We propose the following template for describing a pattern language:

Patterns (Vocabulary)   "The elements of this language are entities called patterns. Each pattern describes a problem which occurs over and over again in our environment, and then describes the core of the solution to that problem, in such a way that you can use this solution a million times over, without ever doing it the same way twice." (Alexander 1977, p. x). Moreover, patterns have to be presented in a consistent format "For convenience and clarity " (Alexander 1977, p. x). The same format also makes them easier to understand and browse.

Patterns have to describe how a solution solves a problem. This solution has to be described in a way that makes it possible to decide if this solution creates an added value (or a benefit) for the user of the pattern. Hence, the engineer can decide if the solution does create the added value he/she is looking for or if the solution should not be implemented to save time and resources. It is vital "to present the problem and solution of each pattern in such a way that you can judge it for yourself, and modify it, without losing the essence that is central to it." (Alexander 1977, p. x). Note that Alexander does not state in one precise sentence that problem–solution pairs are an element of a pattern language. However, Alexander states (Alexander 1977, p. x) that a pattern language consists of patterns and in turn that a pattern contains the essence of a problem–solution pair. Moreover, Alexander states that all patterns of a pattern language should be described using the same format. In the following page (Alexander 1977, p. xi), Alexander states that: "There are two essential purposes behind this format. First, to present each pattern connected to other patterns, so that you grasp the collection of all 253 patterns as a whole, as a language, within which you can create an infinite variety of combinations. Second, to present the problem and solution of each pattern in such a way that you can judge it for yourself, and modify it, without losing the essence that is central to it." (Alexander 1977, p. xi). Hence, it is our understanding that patterns in a pattern language should have the same format and the problem–solution pair is essential to a pattern. We conclude that in turn the problem–solution pair is essential to a pattern form used in a particular pattern language, as well.

Connections Between Patterns (Syntax)   Each solution includes syntax, a description that shows where the solution fits in a larger, more comprehensive design and which other solutions can refine this design. This relates the solution into a network of other needed solutions. For example, a larger solution might be a house for a place people want to live in. The rooms are part of the house and require ways to get light. One way to get light into a room is an electronic lamp. Another way is a candle. "In short, no pattern is an isolated entity. Each pattern can exist in the world, only to the extent that is supported by other patterns; the larger patterns in which it is embedded, the patterns of the same size that surround it, and the smaller pattern which are embedded in it." (Alexander 1977, p. xiii).

Pattern Sequences (Grammar) The grammar provides the meaning of sequences of patterns. Meaning with regard to patterns is a solution to a problem, which is derived by applying a sequence of patterns. Note that a pattern language allows that different sequences of patterns exist, which all solve the same problem. This is in line with Buschmann et al. (2007), who state that numerous pattern sequences form a pattern language. This can be compared to a language in which different sentences can have the same meaning, while being syntactically different. For example, the following sentences are syntactically different but have the same meaning; (1) I have not seen the sun in a long time, and (2) It has been ages since I saw the sun. In short, a grammar explains in which places of what sequences a pattern is useful (Eloranta et al. 2014).

In several books regarding pattern languages, all possible sequences of patterns are shown in a diagram, such as Eloranta et al. (2014), Buschmann et al. (1996), Gamma et al. (1994), Schumacher et al. (2006). In some cases such as Eloranta et al. (2014) the number of possible sequences of patterns lead to a large diagram. To address this problem, the authors show only a partial view of the diagram in the book and reference the entire diagram on a corresponding homepage. Thus, the scalability issue of the diagram size can be solved.

11.2.3 Software Engineering Definitions of a Pattern Language

We describe the vocabulary, syntax, and semantics of a pattern language for the related work on patterns for software engineering. We focus in particular on the works of Fowler (1996), Gamma et al. (1994), Schumacher et al. (2006). We consider the works of Fowler, because he presented analysis patterns for capturing domain knowledge of enterprise systems. His pattern language for analysis pattern has some impact, to be precise 223 citations are listed in the ACM digital library.[4] Fowler's analysis patterns refer to the analysis phase of software engineering and support in particular the structured reuse of elicited domain knowledge. His work has the closest similarity to our work concerning the reuse of elicited domain knowledge using context-pattern. To the best of our knowledge, no work about patterns for reusing elicited domain knowledge for software engineering with more citations

---

[4]ACM citation count of analysis patterns: reusable objects models source: http://dl.acm.org/citation. cfm?id=265172.

exists; this is why we consider Fowler's work. In addition, we decided to consider patterns for the design phase of software engineering with significant impact. The work on design patterns of Gamma et al. (1994) has a citation count of 4524 in the ACM digital library[5] and we are not aware of a work regarding patterns in software design with a higher citation count. Similar works, for example, the work of Buschmann et al. (1996) regarding pattern-oriented software architecture, have a lower citation count. The work of Buschmann et al. has 837 citations.[6] We selected the work of Schumacher et al. (2006) as a representative pattern-based work concerning a specific knowledge area in software design, in this case security. Note that recently Fernandez (2013) released an updated version of Schumacher et al. (2006). However, while numerous new patterns are presented in this work, the structure of patterns and views toward pattern language stay mostly the same. Hence, we base our work on the original publication of Schumacher et al. (2006), due to the fact that more engineers are familiar with it than the new version of Fernandez (2013).

In the future, we are planning to include further work regarding pattern languages, e.g., the previously mentioned work of Buschmann et al. in an extended comparison of pattern languages using the structure following parts of a pattern language.

Patterns (Vocabulary)   Fowler agrees with Alexander that a pattern language requires to have a common way to describe patterns (Fowler 1996, 2002). His common way for describing analysis pattern contains a unique name, structural and graphical description, and a textual description of behavior and relations to other patterns. A very similar understanding of how to describe a pattern can be found in Hafiz et al. (2012), Fernandez and Pan (2001).

Even though Fowler does not follow it strictly in his analysis patterns, he identified a meta-structure of design patterns: "It is commonly said that a pattern, however it is written, has four essential parts: A statement of the context where the pattern is useful, the problem that the pattern addresses, the forces that play in forming a solution, and the solution that resolves those forces." (Fowler 1996, p. 6).

Gamma et al. (1994) also state that patterns need a consistent format in agreement with Alexander. The format of the author's design patterns is defined by a template, which is structured into different sections. For example, every pattern has among others a section for its name, intent, motivation, solution, forces, consequences, and known uses.

Schumacher et al. (2006) use a similar template as Gamma et al. (1994) for their security design pattern. It is interesting to note that the patterns of the authors have no security specific sections in their template such as security goals, e.g., confidentiality. This allows the assumption that the template could also be used in a more general sense for nonsecurity-related design patterns.

To sum up, Gamma et al. (1994), Schumacher et al. (2006) follow a well-defined set of sections in their template that describes each pattern. In contrast, (Fowler

---

[5]ACM citation count of Design patterns: Elements of reusable object-oriented software source: http://dl.acm.org/citation.cfm?id=186897.

[6]ACM citation count of Pattern-oriented software architecture: A system of patterns source: http://dl.acm.org/citation.cfm?id=249013.

1996, 2002) defines the structure of his analysis patterns more abstract. His structure just requires a name and some form of structural and behavioural description. Fowler (1996) refrains from restricting his *analysis pattern* to a fixed form of a single problem–solution relationship in contrast to design patterns. "A fixed form carries its own disadvantages, however. In this book, for instance, I do not find that a problem–solution pair always makes a good unit for a pattern. Several patterns in this book show how a single problem can be solved in more than one way, depending on various tradeoffs. Although this could always be expressed as separate patterns for each solution, the notion of discussing several solutions together strikes me as no less elegant than pattern practice. Of course, the contents of the pattern forms make a lot of sense-any technical writing usually includes context, problem, forces, and solution. Whether this makes every piece of technical writing a pattern is another matter for discussion." (Fowler 1996, pp. 6–7). However, Fowler states in his meta-structure for *design patterns* of other authors (introduced above in the syntax part) that: "This form appears with and without specific headings but underlies many published patterns. It is an important form because it supports the definition of a pattern as 'a solution to a problem in context', a definition that fixes the bounds of the pattern to a single problem-solution pair." (Fowler 1996, p. 6).

The pattern template of Gamma et al. (1994) describes the problem in several separate sections. The section *Intent* describes the general design problem. The section *Motivation* states a scenario in which the design problem occurs and the section *Applicability* refers to specific situations in which the design pattern is useful. The solutions are described in several sections, as well. The section *Structure* describes the graphical representation of the pattern. *Participants* illustrates the elements in the graphical representation. *Collaborations* states the elements' collaborations and responsibilities. *Implementations* and *Sample Code* illustrate how to represent the pattern in source code.

Schumacher et al. (2006) consider the sections *Problem* and *Solution* explicitly in their template. The sections are also paired in the sense that the Solution section follows the Problem sections without any section in between. However, several sections refine the solution such as descriptions of *Structure*, *Dynamics*, and *Implementation*.

Overall, design patterns such as the ones from Gamma et al. (1994), Schumacher et al. (2006) seem to follow the guideline of describing one problem and one solution in a pattern. Nevertheless, Fowler (1996) still embeds a problem solution relationship in his patterns, but not as strict. In some cases, he refers to multiple solutions or problems.

Connections Between Patterns (Syntax)   Fowler (1996, 2002) states that the relations between patterns are important. Note that Alexander uses the term connection instead of relation. We argue that connection is a synonym for relationship[7] and use the term relationship for the remainder of this chapter. The reason is

---

[7] Connection and relationship are synonyms according to dictonary.com: http://dictionary.reference. com/browse/relationship.

that software engineering works such as (Fowler 1996, 2002), Schumacher et al. (2006) use the term relationship, as well.

Moreover, according to (Fowler 1996, 2002) a pattern language is indeed about the relations between patterns. Fowler dedicates an entire part of his book (Fowler 1996) to *Support Patterns*, which define the relationships between organizational patterns for, e.g., accounting to software architecture patterns such as the *Layered Architecture* pattern.

In addition, Hafiz et al. (2012) agree and elaborate that an enumeration of patterns without defined relations among them is just a pattern catalog. Both, Hafiz et al. and Fowler, basically adopt the view of (Alexander 1977) toward connections between patterns being an essential part of a pattern language.

Gamma et al. (1994) state also that the relationship between patterns is important. They even create a figure to illustrating the relations between their patterns. Moreover, a section in their template defines the relations to other patterns. The relations between design patterns have many different names such as *single instance*, *adding operations*, or *defining algorithm's steps*.

Schumacher et al. (2006) dedicate two sections in their template to documenting the relations between patterns. The *Variants* section contains descriptions of variants and specialisations of a pattern. In addition, the *See Also* section in the template references patterns that solve similar problems and patterns that refine the pattern.

In summary, Fowler (1996), Gamma et al. (1994), Schumacher et al. (2006) agree that relations between patterns have to be defined. However, each pattern language uses different kinds of relations and different ways to document these relations.

Pattern Sequences (Grammar)   Fowler uses two different types of patterns: Analysis patterns that refer to a particular business domain and supporting patterns that describe how to apply the analysis patterns. The pattern sequences in Fowler's work can relate different analysis patterns or analysis patterns and supporting patterns. However, Fowler's books do not contain a diagram that shows all pattern sequences, instead the sequences are written in texts of the individual patterns (Fowler 1996, 2002).

Gamma et al. (1994) show diagrams in their books, which contain all possible sequences of their patterns. In contrast, Schumacher et al. (2006) use a taxonomy for security and map their patterns to the respective parts of the taxonomy.

11.2.4 A Pattern Language for Context-Patterns

We describe in the following how our work done so far on context-patterns fits into the required elements of a pattern language and discuss how close we are to having a pattern language for context-patterns.

Patterns (Vocabulary)   We analyze in Beckers et al. (2013) which elements and concepts we used in the context-patterns presented in different works of ours (Beckers et al. 2012a, b, c; Beckers and Faßbender 2012). Further, we described the relations between the identified elements and concepts in a meta-model in Beckers et al. (2013). A short summary is given in Sect. 11.3 and we showed *how* to describe a context-pattern using the meta-model. We claim our meta-model and the pattern catalog contain the vocabulary for our pattern language and published the claim in Beckers et al. (2013).

Our context-patterns each address a particular problem and have a method that states how to solve this problem. The pattern form which reflects this information is introduced in Beckers et al. (2014b). We explicitly state the problem and forces for the problem. The grammar is contained in the method description which is part of each of our patterns. Describing the solution in a method provides engineers with descriptions of how to apply the solutions.

Connections Between Patterns (Syntax)   We analyzed the relations between our context-patterns and present the results in Sect. 11.4. Context-patterns *can refine* each other. Moreover, the domain knowledge elicited and stored in one context-pattern can be used by another pattern as *input*. This requires that both patterns are combined in a new method and that their elements have to be mapped. We showed how this can work in Beckers et al. (2012c) and combined the cloud system analysis pattern with the law pattern adding a new method. How to derive relations between context-pattern and the relations found for the existing context-pattern are the contribution of this work.

Pattern Sequences (Grammar)   We are showing all possible sequences between our context-patterns in Fig. 11.2, similar to the related works (Eloranta et al. 2014; Buschmann et al. 1996; Gamma et al. 1994; Schumacher et al. 2006). The figure shows sequences of context-patterns of the types technical, organizational and technical, and organizational. The sequences contain context-patterns that are used jointly, refine or are input for other context-patterns.

As a result, we claim to have defined the vocabulary of a pattern-language via our meta-model and pattern catalog in Sect. 11.3 and Beckers et al. (2013), and the syntax via our defined relations between the context-patterns in Sect. 11.4. We rely on the methods in our context-patterns as grammar for the pattern language (Beckers et al. 2014b). This is quite different from the design patterns by Gamma et al. (1994), Schumacher et al. (2006), which use more explicit sections of their templates to describe problems and solutions. However, the analysis patterns by Fowler (1996) also contain problems and solutions, but in a less strict format. Our context-patterns also focus on the analysis phase of software engineering and we claim to be in alignment with Fowler. Nevertheless, our view of integrating a method as solution is novel and we have to discuss further with the pattern community if this satisfies as a grammar for our pattern language.

In contrast, not all works follow Alexander's definition of a pattern language. Jackson's work on problem frames (Jackson 2001) provides a different view. He considers his problem frames as a *kind of pattern* and also presents a pattern language. This language for expressing problem frames contains domain-types and interfaces between problem frames. However, Jackson avoids the term *pattern language* in his definition and it is an open debate if his language qualifies as a pattern language, because it differs from Fowler (1996), Gamma et al. (1994), Schumacher et al. (2006). This example shows that not all patterns require a pattern language in Alexander's sense. However, we will focus on defining an accepted pattern language for context-patterns via publishing papers in the community and discussing it with the experts of the pattern community.

11.3 A Meta-Model for Context-Pattern

In this section, which is a summary of a previous work (Beckers et al. 2013), we present a meta-model for building context-patterns that consider domain knowledge during the analysis phase of software engineering. We consider different kinds of domain knowledge, e.g., technical domain knowledge. Therefore, we used a bottom-up approach, starting with a set of previously and independently developed context-patterns.

We identified the common concepts in our existing context-patterns (Beckers et al. 2012a, b, c; Beckers and Faßbender 2012), and aggregate this knowledge into a meta-model of elements one has to talk and think about when describing a new context-pattern (Beckers et al. 2013).

This is quite similar to what Jackson (2001) proposed for requirements. He defined a meta-model of reoccurring domains, like causal, biddable, and lexical domains. These domains are used to define basic requirements patterns, so-called *Problem Frames* (Jackson 2001).

In this section, we show a similar meta-model for context-patterns. We show how we derived it from already existing context-patterns and how it can be used to describe the structural part of a new context-pattern. This section summarizes the results from one previous works of ours (Beckers et al. 2013).

This meta-model has several benefits. First, it forms a uniform basis for our context-patterns, making them comparable. Second, findings and results for one pattern can be transferred to the other patterns via a generalization. Third, the meta-model contains the important conceptual elements for context-patterns. Fourth, it enables us to form a pattern language for context-patterns. However, in this work we focus on the aspects of the meta model which create the basis of a pattern language for context-patterns.

Using this meta-model, we empower requirements and software engineers to describe their own context-patterns, which capture the most important parts for understanding the context of a system-to-be. The meta-model was derived in a bottom-up way from the different patterns we described independently for different domains. For the process of deriving the general elements, which then form the meta-model, we started to analyze each context-pattern in isolation. For each element in a context-pattern we discussed what the general concept behind this element is or if it is a general concept in itself. In a next phase we harmonized the conceptual elements by comparing the found elements, merging them if needed and setting up their

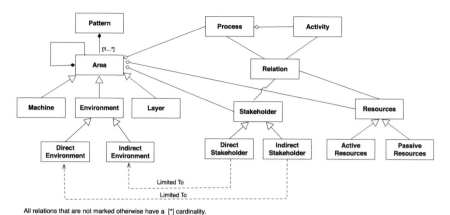

All relations that are not marked otherwise have a [*] cardinality.

Fig. 11.1 Context-pattern meta-model

relations. This way we got a coherent set of conceptual elements over all patterns. In the last phase we had to choose which conceptual elements should be part of the meta-model. Finally, we formed the meta-model as depicted in Fig. 11.1 out of the selected conceptual elements. The meta-model was modeled using the UML notation.

The root element is the `Pattern` itself. Each pattern contains at least one `Area`. In general, an area contains elements of either a technical or organizational view. An area can contain other areas, which do not need to be the same view. An area can concern either a `Machine`, i.e. the thing to be developed, or an `Environment`, which in turn contains elements that have some kind of relation to the machine, or a `Layer`, which encapsulates a set of elements.

The environment can be further refined. There are elements which directly interact with the machine, captured in the `Direct Environment`. And there are elements which have an influence on the system via elements of the direct environment, captured by the `Indirect Environment`.

An element which is part of an `Area` can be a `Process`, a `Stakeholder`, or a `Resource`. A process describes some kind of workflow or sequence of activities. Therefore, it can contain `Activities`. A stakeholder describes a person, a group of persons, or organizational units, which have some kind of influence on the machine. A stakeholder can be refined to a `Direct Stakeholder` who interacts directly with the machine, and an `Indirect Stakeholder` who only interacts with direct stakeholders but has some interest in or influence on the machine. A `Resource` describes some physical or non physical (e.g., information) element which is needed to run the machine or which is processed by the machine and which is not a stakeholder. A resource can be an `Active Resource` with some behavior or a `Passive Resource` without any behavior.

This meta-model has several benefits. First, it forms a uniform basis for our context-patterns, making them comparable. If a method already makes use of one of the patterns, it is now easy to generalize the usage to the elements of the meta-model.

This enables one to replace a given used pattern by another one easily. Second, findings and results for one pattern can be transferred to the other pattern via a generalization to the meta-model elements. Third, the meta-model contains the important conceptual elements for context-patterns. Thus, it is helpful to know these elements and search for them in a specific domain when setting up a new context-pattern for a domain. Fourth, it enables to form a pattern language for the context-pattern. The common meta-model eases relating the patterns to each other.

11.4 Relations Between Existing Context-Patterns

For forming a context-pattern language, we investigated the commonalities and differences between the patterns as enumerated in Chap. 10. We also identified how one context-pattern can be used in combination with another context-pattern.

We defined relations between our patterns and identified the following kinds of relations based on the insights gathered in the work showed in the previous sections:

Refines   The refines relation describes that one pattern refines another pattern or parts thereof. For example, one pattern can refer to services, while another describes how these services are composed.

Input   The information contained in the instantiation of one pattern can be the input for another pattern. For example, our law pattern can use the information of other patterns to identify relevant laws. We show how the information in the cloud pattern can be used as scenario description for the law pattern in Beckers et al. (2012c).

We describe each relation using a template (see Table 11.1 for an example) that states first the *Direction* of the relation, second the *Relation Type*, and third the *Reasoning* why the relationship holds. Relations between patterns can consist of several relations.

SOA Pattern ↔ Cloud Pattern—The SOA pattern refines the Cloud pattern (see Table 11.1).

P2P Pattern ↔ Cloud Pattern—The P2P Pattern refines the Cloud pattern (see Table 11.2).

Table 11.1 Pattern relation SOA to cloud pattern

Direction	SOA to cloud pattern
Relation type	Refines
Reasoning	The services deployed in a cloud can be created or composed in a SOA. Hence, the information in the SOA patterns can be seen as a refinement of the services in the cloud pattern. In addition, the stakeholders involved in the creation and maintenance of the service can be cloud stakeholders

Table 11.2 Pattern relation P2P to cloud pattern

Direction	P2P to cloud pattern
Relation type	Refines
Reasoning	Specific technologies that form the core of the cloud, e.g., the cloud database or the hypervisor, are likely based on P2P-architectures. In addition, the services deployed in the cloud can also be based on P2P-architectures. In both cases the P2P pattern can refine the description of these services or cloud components

SOA Pattern ↔ P2P Pattern—The SOA patterns can be refined by a P2P Pattern (see Table 11.3).

Cloud Pattern ↔ Law Pattern—Our cloud pattern can be input for the law pattern. We show an example of how to use this relation in Beckers et al. (2012c) (Table 11.4).

P2P Pattern ↔ Law Pattern—Our P2P pattern can be input for the law pattern to identify relevant laws for a cloud computing scenario (Table 11.5).

SOA Pattern ↔ Law Pattern—Our SOA pattern can be input for law patterns to identify relevant laws for SOA scenarios. A relation towards the other direction is not possible for the same reason provided in the previous relations (Table 11.6).

We present an overview of all defined pattern relations in Table 11.7. The table shows relations from a pattern in the horizontal to a pattern in the vertical axis. The fields of the table refer to the relation types defined in the beginning of this section.

Table 11.3 Pattern relation SOA to P2P pattern

Direction	P2P to SOA pattern
Relation type	Refines
Reasoning	The services described in the SOA pattern can rely on P2P-architectures. The P2P pattern can be used to create a refined description of these services and also reason if these services can fulfil certain quality requirements, e.g., security. The isolated analysis of services in a SOA can be helpful when services shall be evaluated for the question if they can fulfil certain requirements at all. Hence, the P2P pattern can help excluding certain services from the SOA patterns

Table 11.4 Pattern relation cloud pattern to law pattern

Direction	Cloud to law pattern
Relation type	Input
Reasoning	The cloud pattern can be used as the input for identifying relevant laws using the law patterns. Moreover, the creation of legal hierarchies can be based on the cloud pattern. The hierarchies are essential for the mapping of law patterns to law identification patterns

Table 11.5 Pattern relation P2P pattern to law pattern

Direction	P2P pattern to law pattern
Relation type	Input
Reasoning	The P2P pattern can be used as the input for identifying relevant laws and the creation of legal hierarchies. This relation is similar to the relation between the cloud pattern and the law patterns

Table 11.6 Pattern relation SOA pattern to law pattern

Direction	SOA Pattern to law pattern
Relation type	Input
Reasoning	The SOA patterns can be used as the input for identifying relevant laws and the creation of legal hierarchies. This relation is similar to the relation between the cloud pattern and the law patterns

We marked all fields that do not contain a relation with a "–" symbol. Further relations may exist, but at this point we have not identified them.

Noble (1998) classified relations between object-oriented design patterns and identified the following main relationships:

Uses   A pattern uses another pattern
Refines   A more focused pattern refines a general pattern
Conflicts   Two patterns address the same problem.

Our relations for context-patterns map to Nobel's design pattern relations as follows. The *used jointly* and *input for* relations both map to the *uses* relations. For context-patterns it is important to distinguish if two patterns are used together or if they are used in sequence. In a sequence one is used first and the output of the pattern instantiation serve as the input of the other pattern, which is used afterwards. We also have identified the refines relations with the same meaning as Noble's relation. In contrast to Noble's work we do not have a *conflicts* relation, because each context requires a separate context-patterns. For instance, a cloud computing scenario requires the cloud system analysis pattern and no other context-pattern such as the SOA Layer pattern can be applied to this particular context.

The resulting relations between context-patterns are shown in Fig. 11.2. In general, we have three groups of context-patterns. Context-patterns which only focus on the technical context, context-patterns that only focus on organisational aspects, and context-patterns which combine those two views. The groups and the information to which group a pattern belongs are one result of our previous work (Beckers et al. 2013). In this work we observed that some patterns only describe the indirect environment of the system-to-be (organisational), some only describe the direct environment and the system-to-be itself (technical), and some mix both views.

Table 11.7 Relations between context-patterns

From/to	Type	Technical	Technical	Technical, organizational	Technical, organizational	Organizational	Organizational
Type	Pattern/pattern	P2P pattern	SOA layer pattern	Stakeholder SOA pattern	Cloud pattern	Law identification pattern	Law pattern
Technical	P2P pattern	–	Refines	Refines	Refines	Input	Law pattern
Technical	SOA layer pattern	–	–	–	Refines	Input	–
Technical, organizational	Stakeholder SOA pattern	–	–	–	Refines	Input	–
Technical, organizational	Cloud pattern	–	–	–	–	Input	–
Organizational	Law identification pattern	–	–	–	–	–	–
Organizational	Law pattern	–	–	–	–	–	–

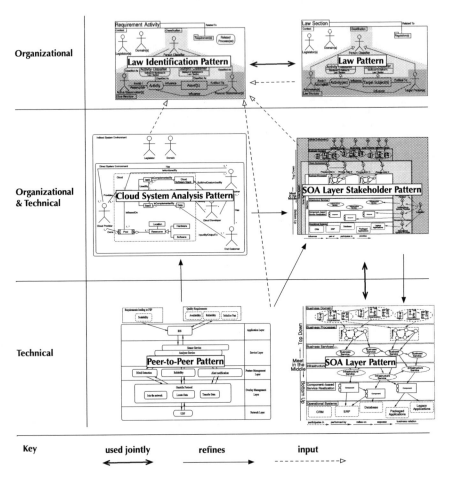

Fig. 11.2 Relations between context-patterns

The relations are shown using directed arrows. Arrows indicate *refine* relations and *input* relations. Some of the patterns are used jointly which means that those patterns are usually used together and closely related. Figure 11.2 shows that the SOA stakeholder layer pattern and SOA layer pattern, as well as the law pattern and law identification pattern, are used jointly. Moreover, the Cloud System Analysis Pattern, Peer-to-Peer Pattern and the SOA Layer Stakeholder Pattern are input for the Law Identification Pattern. In addition, the Peer-to-Peer Pattern refines the Cloud System Analysis Pattern and the SOA Layer Stakeholder Pattern.

11.5 Summary

We have presented a pattern language for context-patterns, which provides the basis for describing and analysing context-patterns. The pattern language consists of a catalog of context-patterns, a meta-model for context-patterns, and the explicit description of relations between context-patterns. Finally, all possible sequences of context-patterns are documented in this chapter. In the future, we will update this pattern language considering further context-patterns.

We illustrated our approach by showing context patterns, e.g., patterns that consider specific technologies such as Peer-to-Peer networks, specific types of architectures like cloud computing, and specific domains, e.g., the legal domain. All of these patterns relate to our meta-model.

We can use instantiated patterns as a basis for writing requirements, deriving architectures or structured discussions about a specific domain. In addition, our patterns can be used outside the domain of software engineering, for example for scope descriptions, asset identification, and threat analysis, when building an ISO 27001 (ISO/IEC 2005) compliant Information Security Management System (Chap. 12).

We showed the following in this chapter:

- A meta-model for describing common elements in context-patterns for various kinds of domain knowledge.
- We defined relation types and analysed all relations between our existing context-patterns.
- The relations between context-patterns, in combinations with the meta-model and an overview of all possible sequences for context-patterns are the foundation of our pattern language for context-patterns.
- We contribute a template for describing pattern languages and apply the template to our pattern language for context-patterns, as well as several other well-known pattern languages.

In summary, we presented in this chapter a pattern language for context-patterns, which supports software and security engineers in describing their own context-patterns and identifying useful sequences of context-patterns that can be applied to describe a complex system context such as a legal compliance analysis for cloud computing systems.

References

Alexander, C. (1977). *A pattern language: Towns, buildings, construction*. Oxford University Press.
Beckers, K., & Faßbender, S. (2012). Peer-to-peer driven software engineering considering security, reliability, and performance. In *Proceedings of the International Conference on Availability, Reliability And Security (ARES)—2nd International Workshop on Resilience and It-Risk in Social Infrastructures* (RISI 2012) (pp. 485–494). IEEE Computer Society.

Beckers, K., Faßbender, S., & Heisel, M. (2013). A meta-model approach to the fundamentals for a pattern language for context elicitation. In *Proceedings of the 18th European Conference on Pattern Languages of Programs (Europlop)*.

Beckers, K., Faßbender, S., & Heisel, M. (2014a). Deriving a pattern language syntax for context-patterns. In *Proceedings of the 19th European Conference on Pattern Languages of Programs (Europlop)*.

Beckers, K., Faßbender, S., & Heisel, M. (2014b). A meta-pattern and pattern form for context-patterns. In *Proceedings of the 19th European Conference on Pattern Languages of Programs (Europlop)*.

Beckers, K., Faßbender, S., Heisel, M., & Meis, R. (2012a). Pattern-based context establishment for service-oriented architectures. In *Software service and application engineering* (pp. 81–101). Springer.

Beckers, K., Faßbender, S., Küster, J.-C., & Schmidt, H. (2012b). A pattern-based method for identifying and analyzing laws. In *Proceedings of the International Working Conference on Requirements Engineering: Foundation for Software Quality (REFSQ)* (pp. 256–262). Springer.

Beckers, K., Faßbender, S., & Schmidt, H. (2012c). An integrated method for pattern-based elicitation of legal requirements applied to a cloud computing example. In *Proceedings of the International Conference on Availability, Reliability and Security (ARES)—2nd International Workshop on Resilience and It-Risk in Social Infrastructures* (RISI 2012) (pp. 463–472). IEEE Computer Society.

Buschmann, F., Meunier, R., Rohnert, H., Sommerlad, P., & Stal, M. (1996). *Pattern-oriented software architecture volume 1: A system of patterns*. New York: Wiley.

Buschmann, F., Henney, K., & Schmidt, D. C. (2007). *Pattern-oriented software architecture volume 5: On patterns and pattern languages*. New York: Wiley.

Eloranta, V.-P., Koskinen, J., Leppänen, M., & Reijonen, V. (2014). *Designing distributed control systems: A pattern language approach*. New York: Wiley.

Fernandez, E. B. (2013). *Security patterns in practice: Designing secure architectures using software patterns*. Boston: Addison-Wesley.

Fernandez, E. B., & Pan, R. (2001). A Pattern Language for Security Models. In *8th conference of pattern languages of programs (plop)*.

Fowler, M. (1996). *Analysis patterns: Reusable object models*. Reading: Addison-Wesley.

Fowler, M. (2002). *Patterns of enterprise application architecture*. Boston: Addison-Wesley.

Gamma, E., Helm, R., Johnson, R., & Vlissides, J. (1994). *Design patterns: Elements of reusable object-oriented software*. Reading: Addison-Wesley.

Hafiz, M., Adamczyk, P., & Johnson, R. E. (2012). Growing a pattern language (for security). In *Proceedings of the ACM International Symposium on New Ideas, New Paradigms, and Reflections on Programming and Software* (pp. 139–158). ACM.

ISO/IEC. (2005). Information technology—Security techniques—Information security management systems–Requirements (ISO/IEC 27001). Geneva, Switzerland: International Organization for Standardization (ISO) and International Electrotechnical Commission (IEC).

Jackson, M. (2001). *Problem Frames. Analyzing and structuring software development problems*. Reading: Addison-Wesley.

Noble, J. (1998). Classifying relationships between object-oriented design patterns. In *Proceedings of the Australian Software Engineering Conference* (pp. 98–107). IEEE Computer Society.

Schumacher, M., Fernandez-Buglioni, E., Hybertson, D., Buschmann, F., & Sommerlad, P. (2006). *Security patterns: Integrating security and systems engineering*. New York: Wiley.

Chapter 12
Supporting the Establishment of a Cloud-Specific ISMS According to ISO 27001 Using the Cloud System Analysis Pattern

Abstract Our context-patterns describe common elements, structures, and stakeholders for a specific domain such as cloud computing. In the previous chapters, we introduced our catalog of context-patterns, and our pattern language for context-patterns. The pattern language helps engineers to describe their own context-patterns and to understand the relations between our existing context-patterns. In this chapter, we show how to conduct a cloud-specific security analysis based on a specific context-pattern, namely our cloud system analysis pattern. Initially, we discuss the larger issue of governance, risk management, and compliance for cloud computing and argue why the ISO 27001 security standard certification is one possible choice to establish security management for clouds. Furthermore, we analyzed the entire ISO 27001 standard and show how each demanded activity can be conducted using our context-pattern and how information can be reused between different security analyses. As a result of our analysis, we created the PACTS methodology, which also dedicates individual activities concerning legal compliance and privacy management. Finally, we illustrate how all of the ISO 27001 documentation demands can be met by using our method.

12.1 Introduction

We provide an example in this chapter of how context-patterns can support the establishment of an ISO 27001 compliant Information Security Management System (ISMS). We are concerned with a cloud-specific ISMS, which also considers privacy and legal compliance. This work is based on the publication (Beckers et al. 2013a). The author of this book is the main author of this publication. Isabelle Côté provided valuable feedback for the overall method and helped with the argumentation for the introduction, Stefan Hofbauer provided valuable insights for the industrial example, Stephan Faßbender helped with the legal compliance part of PACTS and Maritta Heisel provided also feedback on the overall method. This work is based on the background of the ISO 27001 standard (Sect. 2.2.2), clouds (Sect. 12.2), and our context-pattern catalog (Chap. 10).

© Springer International Publishing Switzerland 2015
K. Beckers, *Pattern and Security Requirements*,
DOI 10.1007/978-3-319-16664-3_12

12.2 Governance, Risk, and Compliance for Clouds

The term *cloud computing* describes a technology as well as a business model (Armbrust et al. 2009). According to the *National Institute of Standards and Technology (NIST)*, cloud computing systems can be defined by the following properties (Mell and Grance 2009): the cloud customer can acquire resources of the cloud provider over *broad network access* and *on-demand*, and pays only for the used capabilities. Resources, i.e., storage, processing, memory, network bandwidth, and virtual machines, are combined into a so-called *pool*. Thus, the resources can be virtually and dynamically assigned and reassigned to adjust the customers' variable load and to optimize the resource utilization for the provider.

The virtualization causes a location independence: the customers generally have no control or knowledge of the exact location of the provided resources. Another benefit is that the resources can be quickly scaled up and down for customers and appear to be unlimited, which is called *rapid elasticity*. The pay-per-use model includes guarantees such as availability or security for resources via customized *Service Level Agreements (SLA)* (Vaquero et al. 2008).

The architecture of a cloud computing system consists of different service layers and allows different business models: on the layer closest to the physical resources, the *Infrastructure as a Service (IaaS)* provides pure resources, for example virtual machines, where customers can deploy arbitrary software including an operating system. Data storage interfaces provide the ability to access distributed databases on remote locations in the cloud. On the *Platform as a Service (PaaS)* layer, customers use an API to deploy their own applications using programming languages and tools supported by the provider. On the *Software as a Service (SaaS)* layer, customers use applications offered by the cloud provider that are running on the cloud infrastructure. Furthermore, cloud providers require a layer that monitors their customers' resource usage, e.g., for billing purposes and service assurances. Buyya et al. (2009) introduce this layer as a middleware in their cloud model. Cloud computing offers different *deployment scenarios*: *private clouds* are operated solely for an organization, *public clouds* are made available to the general public or a large industry group and are owned by a third party selling cloud services. In between these scenarios are *hybrid clouds* where internal IT resources upon demand are complemented with resources from an external vendor (Armbrust et al. 2009).

Cloud computing offers elasticity, services can be added or deleted on demand by a customer. Thus, any part of the cloud computing system has to support this scalability. In Fig. 12.1 we present several areas of Governance, Risk, and Compliance identified challenging for cloud computing, considering the previously raised trust issues in the cloud computing domain. This work is based on the publication (Beckers and Jürjens 2010).

In *Governance* policies have to be written that accommodate the permanent change of a system. Today's policies are written for comparable stable systems and are tightly focused on them. This will not suffice for a cloud computing system, because of the elasticity. Furthermore, a detailed classification of data is required.

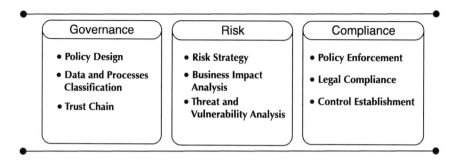

Fig. 12.1 Governance, risk, and compliance management in clouds

Data in a closed company network needs to be protected and the responsible staff in a company is in full control of the infrastructure. In a cloud computing scenario data will be given to a cloud provider and the customer has to trust in the security capabilities of the vendor. Thus, for instance highly sensible data or processes that are vital for a company should not enter the cloud. Moreover, multiple customers or even cloud vendors might be involved in a business process that involves cloud usage. In these cases the trust of each partner has to be taken into account of security considerations. The interested reader is referred to our work eliciting and calculating trust-based evaluation of cloud providers (Moyano et al. 2014).

Risk management requires a strategy that includes every possibility of security failures with a cloud computing integration. This becomes increasingly difficult due to the almost infinitive number of possible scenarios, because of the change of the company's structure when cloud computing is used in a part of it. In addition, possible impacts of cloud computing on the effectiveness and efficiency of business processes of a company have to be evaluated on a similar scale. Numerous new threats and vulnerabilities of cloud computing have to be analyzed and the risk analysis itself for each cloud scenario has to be verified continuously.

Compliance in a cloud computing scenario is difficult, due to the fact that companies rely on cloud vendors for security policy enforcement, the adherence to regulations, e.g., Sarbanes-Oxley Act (SOX).[1] Furthermore, companies have to find ways to gain control of specific scenarios. For instance, in case highly classified data made it into the cloud by accident an emergency procedure should exist that erases the data completely from the cloud's systems.

Evaluating business benefits against privacy, security, and compliance concerns of clouds is difficult, because implementation and operational details are often not transparent to cloud customers or end customers. These stakeholders entrust their data to a cloud provider, which leads to concerns regarding data integrity, recovery and location, as well as legal issues (Gartner 2008).

Cloud Computing Governance, Risk, and Compliance Management (GRC) can be supported by standards. These provide processes or guidelines for achieving

---

[1]Sarbanes-Oxley Act: https://www.sec.gov/about/laws/soa2002.pdf.

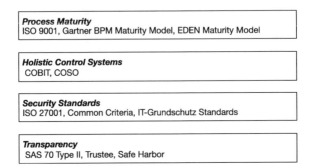

Fig. 12.2 Standards in support of cloud governance, risk, and compliance management

objectives such as process maturity. A certification according to one of these standards is of particular relevance for a cloud computing scenario, due to the limited knowledge of the internal organization of cloud providers and their systems for cloud customers. We show relevant standards for cloud GRC in Fig. 12.2. Note that this list is by no means exhaustive and should simply show an overview of relevant standards.

Process Maturity   On an economic level the integration of cloud and company has to result in effective and efficient business processes. This can be verified, for instance with the ISO 9001 (ISO 2008), the Gartner BPM Maturity Model,[2] and the EDEN Maturity Model for Business Processes.[3]

Holistic Control Systems   Holistic control systems provide a way to react to security and similar incidents within a system. Cobit (I. G 2007) and Coso (Moeller 2007) offer models for documenting, analyzing, and creating internal control systems.

Security Standards   We described a meta-structure of security standards and several standards such as ISO 27001 (ISO/IEC 2005b) and Common Criteria (ISO/IEC 2012) (cf. Chap. 4).

Transparency   A topic specifically relevant for cloud computing is transparency. In many cases customers of cloud computing offers have to trust the vendor that the GRC demands are met. Several standards evaluate the capabilities of the cloud system. Statement on Auditing Standards (SAS) Number 70 Type II confirms a cloud vendor from an external party that monitoring activities for IT technologies and processes are present and documented. One result of an SAS 70 II certification is a security report, which is filed under the individual standards of the evaluated organization. It is not a certificate with a predefined set of terms (Streitberger and Ruppel 2009). A further type of compliance in the cloud area are so-called Trust Audit Frameworks, e.g., SysTrust or WebTrust. These frameworks focus on internal controls in a company based upon financial systems. These systems were

---

[2] Gartner BPM Maturity Model: https://www.gartner.com/doc/497289/bpm-maturity-model-identifies-phases.

[3] The EDEN Maturity Model for BPM: http://www.bpm-maturitymodel.com/eden/opencms/en/What_is_eden/.

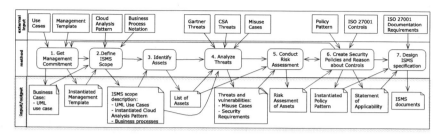

Fig. 12.3 The steps of our PACTS method concerning security

developed for eCommerce applications (Streitberger and Ruppel 2009). Further agreements, for instance safe harbor, ensure that certain privacy demands are met.[4]

ISO 27001 is of particular relevance for clouds, because it creates an information security management system (ISMS), which is a process for security for an organization that can be certified. The ISO 27001 standards mentioned in the process maturity category can serve as input for an ISMS. Furthermore, the holistic control systems can help to refine the controls defined in ISO 27001. An example for a mapping between the control systems and ISO 27001 is provided in Sect. 12.10.3. The standards in the transparency section offer an evaluation process without a detailed evaluation criteria.

The Common Criteria and the IT-Grundschutz standards are other relevant security standards. However, the IT-Grundschutz standards focus on German organizations and we argue why ISO 27001 fits a cloud scenario better than Common Criteria in the next section.

We address these concerns by proposing our PACTS method (see Fig. 12.3) for creating a cloud-specific ISMS compliant to the ISO 27001 standard with a particular focus on legal compliance and privacy. PACTS considers either the *cloud provider* or the *cloud customer* as possible stakeholders, who build an ISMS. The reason is that these are organizations that should earn the trust of their customers via certifying an ISMS.

12.3 Motivation for a Cloud-Specific ISMS Establishment Method

The possibility of quickly acquiring or disposing of resources such as storage and memory provides a great attraction for a variety of customers and constitutes the means for this kind of acquisition or deposition. However, potential customers are still reserved when it comes to using cloud resources. In 2009, a study was conducted

---

[4]Safe Harbor: http://www.export.gov/safeharbor/.

by the International Data Corporation[5] about this issue. It pointed out that security
is a significant barrier for the acceptance of clouds in companies. The lack of trust
in cloud security lies within the nature of clouds: storing and managing critical data
and executing sensitive IT-processes is performed beyond the company's/customer's
control. To gain the customer's trust and to illustrate that security is taken seriously,
cloud providers have to certify their services with respect to security. One way of
doing so is to turn to standards that put security at the center of interest. Examples for
such standards are the ISO 27000 (ISO/IEC 2009) standards family and the Common
Criteria (ISO/IEC 2012) (CC). The Common Criteria is a document-driven standard.
It is necessary to specify a *target of evaluation* (ToE), which can be a security
system or a security product. The ToE must be described completely. Whenever a
part of the ToE changes, it is necessary to recertify the system or product. These
two points constitute a problem when dealing with clouds. A cloud consists of a
significant amount of hard- and software parts. Describing the ToE may therefore
be a challenging task. Furthermore, due to the fact that resources of clouds can
be dynamically scaled, the ToE changes with every scaling. Thus, a recertification
would be triggered each time the customer initiates a change in the resource usage.
We therefore conclude that performing a CC certification for a complete cloud system
is rather not practical. We are currently not aware of any company that completed
a CC evaluation for an entire cloud computing system. However, it can be used to
certify specific parts within a cloud. For example, the hypervisor is a prominent
candidate for a CC evaluation.

The ISO 27001 standard—in contrast to the CC—is process-driven. This applies
well to the service concept of a cloud. Several well-known companies have adopted
this approach, such as Microsoft,[6,7] Amazon,[8] Google,[9,10] and Salesforce.[11] The
aim of the ISO 27001 standard is to establish an Information Security Management
System (ISMS). To use this standard for cloud computing systems is in accordance
with the German Federal Office for Information Security (BSI).[12] The current version
of the standard does not take cloud-specific security issues into consideration. The
BSI recommends to consider cloud-specific threats when dealing with cloud systems.
The Cloud Security Alliance (CSA) (2010) and Gartner (2008) have identified several
of these threats. We take their findings and use them in our work. Assembling an
ISMS according to the ISO 27001 standard is a nontrivial task. This is due to the fact

---

[5]https://www-304.ibm.com/isv/library/pdfs/cloud_idc.pdf.

[6] http://blogs.msdn.com/b/windowsazure/archive/2011/12/19/windows-azure-achieves-is0-2700
1-certification-from-the-british-standards-institute.aspx.

[7]http://www.windowsazure.com/en-us/support/trust-center/compliance/.

[8]http://aws.amazon.com/security/.

[9]http://googleenterprise.blogspot.com.br/2012/05/google-apps-receives-iso-27001.html.

[10] http://www.computerweekly.com/news/2240150882/Google-Apps-for-Business-wins-ISO-27
001-certification.

[11]http://www.salesforce.com/platform/cloud-infrastructure/security.jsp.

[12] https://www.bsi.bund.de/SharedDocs/Downloads/EN/BSI/Publications/Minimum_information/
SecurityRecommendationsCloudComputingProviders.pdf.

that descriptions for system development and documentation are rather sparse. For example, the required input for the *scope and boundaries* description is to consider "characteristics of the business, the organization, its location, assets and technology" (ISO/IEC 2005b, p. 4). No further information beyond that is given.

We present our *PAttern-based method for establishing a Cloud-specific informa-Tion Security management system (PACTS)*. We analyzed the activities demanded by the standard to build an ISMS and present patterns for these, incorporating existing security requirements approaches where applicable. We also provide a structured method that shows how the different elements described above have to be applied in order to create the required ISMS documentation admissible for certification. We use existing research on context descriptions for clouds in our method in order to provide a domain-specific approach. The patterns define stakeholders and techno-logical artifacts that are used in the context description and all subsequent patterns and models, e.g., security policies. Furthermore, we provide relations from these patterns to cloud-specific lists of threats proposed by the Cloud Security Alliance (CSA) (2010) and Gartner (2008). In addition, our approach provides a structured refinement of the cloud system's and stakeholder's information to assess the threats for a particular instantiation of our cloud pattern. Our method uses this informa-tion for risk assessment and security control selection according to the ISO 27001 standard. Moreover, the ISO 27001 standard demands consideration of privacy and legal compliance. We integrated existing pattern-based research on compliance and privacy requirements into our approach in order to satisfy these demands.

The main contributions of our PACTS method are:

- A structured method to build an ISMS considering security, compliance, and pri-vacy
- Detailed sub-methods for each step of PACTS
- Patterns and templates to support the documentation of management commitment, scope descriptions, asset documentation, and defining security policies
- Reuse of these patterns and templates for different projects via instantiation
- Integration of our patterns and templates into existing methods for the identification of relevant laws, eliciting and verifying privacy requirements, and risk management
- Supporting the ISO 27001 documentation demands.

Running Example We illustrate our approach by the example of a bank providing an online banking service to their customers. The bank uses a cloud for providing the service. We consider transaction services for two particular kinds of customers, namely a bank customer and a VIP bank customer who have specific service level agreements with the bank regarding the availability of the service. These customers' data such as account number, balance, and transaction log history are stored in the cloud. The bank authorizes its software department to design and build the cloud-specific software according to the interface and platform specification of the envi-sioned cloud provider. The cloud provider orders its data center administration to configure the cloud accordingly, and the support team of the cloud provider helps the bank with problems.

The remainder of the work is organized as follows: We explain our method in Sect. 12.4. The method consists of steps for management commitment (Sect. 12.5), ISMS scope definition (Sect. 12.6), asset identification (Sect. 12.7), threat analysis (Sect. 12.8), risk management (Sect. 12.9), control reasoning (Sect. 12.10), and creating ISMS design specifications (Sect. 12.11). We propose an addition to our method (Sect. 12.4) concerning legal compliance (Sect. 12.12) and privacy (Sect. 12.13). Section 12.14 presents related work, and Sect. 12.15 summarizes.

12.4 Overview of Our PACTS Method

Evaluating business benefits against privacy, security, and compliance concerns of clouds is difficult because implementation and operational details are often not transparent to cloud customers or end customers. These stakeholders entrust their data to a cloud provider, which leads to concerns regarding data integrity, recovery and location, as well as legal issues (Gartner 2008).

We address these concerns by proposing our PACTS method for creating a cloud-specific ISMS compliant to the ISO 27001 standard with a particular focus on legal compliance and privacy. PACTS considers either the *cloud provider* or the *cloud customer* as possible stakeholders, who build an ISMS. The reason is that these are organizations that should earn the trust of their customers via certifying an ISMS.

Our *cloud system analysis pattern* (Sect. 10.4) provides a basic structure of a cloud computing architecture, which considers the relations between stakeholders and the cloud. The pattern can be instantiated for any given cloud scenario and if required extended with little effort. The pattern provides a basis for cloud-specific asset identification, threat analysis, risk management, and control selection. For example, several threats are already mapped to the cloud pattern and can be analyzed based upon the pattern's instantiation. The instantiated pattern is also the input for our identification of relevant laws and analysis of privacy requirements.

Our cloud pattern reduces the effort for creating a description of a cloud. We can simply instantiate the pattern in order to obtain a description. The benefit of basing our method on the cloud pattern is also that knowledge collected using the pattern can be reused for different instantiations of the pattern. For example, assets identified using the pattern can be instantiated for different projects, e.g., the *Data* in the cloud pattern has been identified as an asset. Hence, all instantiations of *Data* are assets as well. In addition, experiences from using our method can also lead to an improved pattern, e.g., the pattern can be extended with further stakeholders.

We present an overview of our method for establishing a cloud-specific ISMS in this section. In the remainder of the section we provide detailed descriptions of each step of our PACTS method. We begin by describing the steps concerning security, depicted in Fig. 12.3.

Step 1: Get Management Commitment—The precondition for building an ISMS is that the management commits to it. Hence, we dedicate the first step of our method to get management commitment for the ISMS and the provision of adequate resources to establish it. We describe the characteristics of the business via UML use case diagrams (UML Revision Task Force 2010). The use cases are accompanied by our management templates, which have to be instantiated with relevant information for building the ISMS, e.g., high level security goals, cloud-specific management concerns, and resource management.

Step 2: Define ISMS scope—The scope for building the ISMS shall be described using the initial use cases. These are refined using our cloud system analysis pattern for a structural description of the cloud scenario and a business process notation for behavioral description. In our examples, we choose UML activity diagrams (UML Revision Task Force 2010) as business process notations.

Step 3: Identify Assets—The entire ISMS scope description is the input for the asset identification. We identify all items of value to the cloud stakeholders by iterating over the relations from cloud stakeholders to cloud elements in the cloud system analysis pattern and activity diagrams. This results in a list of assets and the stakeholders that own them.

Step 4: Analyze Threats—We conduct a threat analysis using the list of threats released by the Cloud Security Alliance (CSA) (2010), an industrial consortium that investigated practical security issues with clouds, and the threats that Gartner (2008) considers. We propose to identify threats to the previously identified assets using our cloud system analysis pattern. This activity includes an investigation of vulnerabilities of cloud components. We use the identified threats as an input for misuse cases. The results of the misuse cases are specific threats and security requirements.

Step 5: Conduct Risk Assessment—The assets, threats, vulnerabilities, and security requirements serve as an input for risk assessment. We conduct an asset-based method that uses the previously elicited knowledge to derive likelihood and consequences scales, as well as acceptable risk levels. This information is used to determine, which cloud threats cause unacceptable risks.

Step 6: Create Security Policies and Reason about Controls—Controls in the ISO 270001 standard reduce risks to assets. The reasoning about controls considers the risks to each asset and supports the decision if a control is needed or not. For each asset, we propose to compile a list that states why a control in the normative Annex A of the ISO 27001 should or should not be applied to that asset. We instantiate our policy patterns to ease this activity. The policy patterns help to define precisely which elements of the cloud pattern the control refers to and the security goal the control shall achieve. If the decision is made that a control has to be introduced, we go back to the previous step of our method in order to adjust the risk assessment for that particular asset. This information is in turn used to check if the control already results in an acceptable risk level or if it has to be modified or another control should be introduced. The resulting information is used to compile the so-called *Statement of Applicability*, which is a mandatory document for reasoning about the ISO 27001 controls.

We also have to check carefully if our use cases defined in Step 1 can lead to acceptable risk levels using reasonable and affordable controls. Hence, we also have to consider changing the use cases in case acceptable risk levels cannot be achieved with reasonable effort.

Step 7: Design ISMS specification—The final step of our method concerns the ISO 27001 specification, an implementable description of the ISMS. We consider the ISO 27001 documentation demands and use the information elicited and documented in the previous steps of our method. This information is mapped to the required document types. These documents are also the basis for a certification of an ISO 27001 compliant ISMS.

The ISO 27001 standard has demands for quality requirements beyond security, namely compliance and privacy. We refer to Sects. 12.12 and 12.13 for definitions and detailed discussion of these terms. We provide support for eliciting and analyzing these requirements as part of our approach. We show how compliance and privacy concerns are addressed in Fig. 12.4. The consideration of compliance issues for cloud computing systems is also a key recommendation of Gartner's (Gartner 2008) analysis.

In our PACTS method compliance identifies relevant laws and regulations and defines corresponding requirements. The relation between compliance and privacy is that the compliance part identifies relevant laws that concern privacy. The privacy part of our method uses these laws as an input.

Step 8: Identify Relevant Laws and Regulations—We use the information from the asset identification. That is, we identify relevant laws and regulations with this information. This activity also has to identify assets in terms of laws and regulations, which can be related to assets in terms of security. We discuss the differences in Sect. 12.12. Moreover, laws and regulations can regulate privacy concerns. This information is used during the *instantiate privacy patterns* step of our method.

Step 9: Define Compliance Controls—Once laws and regulations are identified, they have to be translated into ISO 27001 compliance controls. This translation is difficult, because in some cases laws or regulations demand reasoning about a specific concern, or they demand a specific functionality. We discuss this issue also in Sect. 12.12. In addition, compliance controls can have relations to other ISO 27001 controls. For example, a law could demand a specific control in a certain situation, while the risk assessment results would not.

The ISO 27001 standard also demands the consideration of privacy in the informative ANNEX B. We propose the following steps to address this concern.

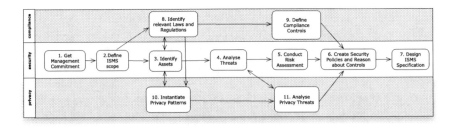

Fig. 12.4 The steps of PACTS concerning compliance and privacy

Table 12.1 Template for relation to the ISO 27001 standard

Significance for the ISO 27001 standard	Define the general relation of the section to the ISO 27001 standard
Related section(s) of the standard	State the related sections of the standard

Step 10: Instantiate Privacy Patterns—We use textual privacy patterns based upon the ISMS scope definition and relevant laws and regulations. These patterns can be instantiated, and they give rise to initial privacy requirements. In addition, the identified assets for security can be considered, because if these contain personal information, they can also support instantiating further privacy patterns.

Step 11: Analyze Privacy Threats—We use a privacy threat analysis based on information flow between requirements. We analyze the flow of personal information based on the previously instantiated privacy patterns and functional requirements of the cloud scenario. We also refine the initial privacy requirements. The information flow among the requirements shows, which stakeholders potentially have access to which personal information. Afterwards, software engineers have to check if the requirements have to be modified in order to be privacy preserving.

In the following, we begin each section with an instantiation of the template presented in Table 12.1. The rows of the template state the relations of the topic of the section to the standard, e.g., threat analysis, and the standard's relevant section for that topic. We assume that users of our method may look for support for establishing a specific ISO 27001 section instead of establishing an entire ISMS. Hence, the information in the last column can be used to identify the part of our method that supports a specific ISO 27001 section.

We use several artifacts from previous research as part of this work. We explicitly state the background and contributions in the beginning of these sections.

Mapping to the Conceptual Framework for Security Standards—We present a mapping in Table 12.2 from this method to the conceptual framework for security standards, which is the foundation for our PEERESS framework (Chap. 3). The table lists the activities of the conceptual framework for security standards on the left column and the steps of the PACTS method that concern these activities in the right column.

12.5 PACTS Step 1: Get Management Commitment

The ISO 27001 standard dedicates its entire Section 5 to the importance of management commitment for implementing an ISMS, shown in Table 12.3. ISO 27001 Section 5 contains subsections for management commitment proofs and provisioning of sufficient resources.

Table 12.2 A Mapping between the standard activities in the conceptual framework for security standards and the steps of the PACTS method

Activities in the conceptual framework for security standards	Steps in the PACTS method
Environment description	Step 1: Get management commitment
	Step 2: Define ISMS scope
Stakeholder description	Step 1: Get management commitment
	Step 2: Define ISMS scope
Asset identification	Step 3: Identify assets
Risk level description	Step 5: Conduct risk assessment
Security property description	Step 4: Analyze threats
Control assessment	Step 4: Analyze threats
Vulnerability and threat analysis	Step 4: Analyze threats
Risk determination	Step 5: Conduct risk assessment
Security assessment	Step 6: Create security policies and reason about controls
Security measures	Step 6: Create security policies and reason about controls
Risk acceptance	Step 5: Conduct risk assessment
Documentation	Step 7: Design ISMS specification

Table 12.3 Management commitment demands within the ISO 27001 standard

Significance for the ISO 27001 standard	The ISO 27001 standard demands documentation of management commitment for the establishment of an ISMS
Related section(s) of the standard	*Section 5.1 Management commitment* concerns proof the management shall provide for establishing an ISMS objectives, plans, responsibilities, and accepting risks. *Section 5.2 Resource management* concerns the provision of resources for establishing the ISMS and the training of the members of the organization for security awareness and competence

---

Changes for ISO 27001:2013 Compliance

The management commitment is described in ISO27001:2013 in Section 5.1 and called *Leadership and Commitment*. The use case descriptions and the template for Management Approval of the ISMS is still appropriate for this part of the standard.

The management commitment for implementing an ISMS according to the ISO 27001 standard is of utmost importance, because without the commitment of sufficient staff and resources the ISMS implementation is doomed to fail. In addition,

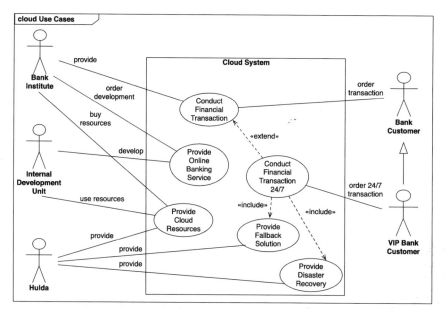

Fig. 12.5 Use case diagram for a cloud-based online banking system

the publicly available examples of ISMS documentations, e.g., the so-called *ISMS toolkit*[13] defines this also as the first step of establishing an ISO 27001 compliant ISMS. The management commitment is based upon business cases concerning a cloud scenario. We defined a set of UML use cases (UML Revision Task Force 2010) to illustrate business cases for our running example, depicted in Fig. 12.5. The *Bank Institute* offers a service to *Conduct Financial Transactions*. The institute orders the development of the *Online Banking Service* by the *Internal Development Unit*. The cloud provider *Hulda* provides the cloud resources to implement the services. The *Bank Customer* conducts *Financial Transactions*. The *VIP Bank Customer* conducts a specific kind of financial transaction, a so-called *24/7 Financial Transaction* that has a guaranteed availability on all days of the week of 99.9999 %. This is guaranteed by an SLA (Sect. 12.2). This service uses a specific fallback solution and disaster recovery.

We provide a template for management approval of the ISMS, presented in Table 12.4. The template structure is inspired by Section 5 of the ISO 27001 standard. The first column in the template lists the management commitment and the second the natural person who is responsible for this concern or the tasks or resources required to address the concern. The template consists of three parts: *Management Commitment* states the responsible persons for the overall establishment of the ISMS and vital concerns toward its success, e.g., criteria for risk acceptance. *Cloud-specific Management Commitment* defines responsible persons for cloud-specific concerns

---

[13]http://www.iso27001security.com/html/iso27k_toolkit.html.

Table 12.4 Template for management approval of the ISMS

Management commitment	
ISMS security goal	State the referenced security goal from the policy pattern
Establish responsibilities	State which person is responsible for the overall ISMS establishment
Communicate importance of security	Define the actions taken to communicate the importance of security
Criteria for risk acceptance	Define worst case scenarios
Conduct ISMS audits	Define audit responsibilities
ISMS management reviews	Define ISMS management audit responsibilities
Cloud-specific management commitment	
Decide for a cloud deployment scenario	Define responsibilities for deciding to use a public, private or hybrid cloud deployment scenario
Check security assurances of cloud provider	Define responsibilities for analyzing the security assurances of the cloud provider
Conduct on site reviews of cloud provider	Define responsibilities for on site auditing of the cloud provider's data center(s)
Conduct neutral security assessment of cloud provider	Define who has to contract an external security team to validate the security audit of the cloud provider. If no external security team is contracted, this management decision has to be justified
Resource management	
Provided resources for the ISMS	List the provided resources for establishing, implementing, operating, monitoring, reviewing, maintaining, and improving an ISMS and reason why the listed resources are sufficient
Security supports business needs	List the resources that allow security support without interfering business needs
Competent personal	List the provided resources for establishing, implementing, operating, monitoring, reviewing, maintaining, and improving an ISMS
Provide training	List the training programs initiated to ensure acceptable security levels
Effectiveness evaluation	List the measures taken to check the effectiveness of the measures
Records of education, training, skills, experience, and qualification	Define responsibilities for documentation of education, training, skills, experience, and qualification of staff with regard to security

such as deployment scenarios and audit management. *Resource Management* states the required resources for establishing an ISMS. We use our running example with the template to show an integrity and a confidentiality goal in Table 12.5. Cloud computing relies on the evaluation of security controls at the site of the cloud provider (Gartner 2008). We need to assign responsibilities for checking the security assurances of possible cloud providers and for on-site evaluations of these providers. Moreover, a decision has to be made if a neutral third party performs the security assessment or if this is done with internal staff.

12.6 PACTS Step 2: Define ISMS Scope

The relevance of the scope definition of the ISO 27001 standard is already mentioned on page 1 of the standard. The ISMS establishment description in Section 4 of the ISO 27001 standard contains numerous references to the scope description and thus highlights its importance. We list all appearances in Table 12.6.

We introduced our *cloud system analysis pattern* in Sect. 10.4. The pattern supports the scope definition for the ISO 27000 series of standards (cf. Beckers et al. 2011). The contribution in this section is an updated cloud system analysis pattern and templates. We also devise a technique to refine the pattern and add further details. For example, one can add behavioral descriptions using UML activity diagrams to refine the pattern. We recently developed tool support[14] for our pattern, which was not part of previous publications. We also updated the cloud pattern with further elements and included behavioral descriptions in it.

12.6.1 The Extended Cloud Pattern

We show the updated version of our *Cloud System Analysis Pattern* in Fig. 12.6. We included a *Location* element for the *Resources* to define in which countries the cloud hardware and software is located. This supports the ISO 27001 demands for including locations in the context description found in ISO 27001 Section 4.2.1a. In order to identify relevant laws, we need the location information, as well.

Moreover, we added two cloud-specific software components to the pattern (see Fig. 12.6). The first component is a *Cloud database*. This provides scalability in terms of data storage. Cloud databases differ from traditional SQL databases. The main differences are that cloud databases allow inconsistencies in the data storage for a short period of time and a decreased control over the data in the cloud due to data distribution. The latter manifests itself in the fact that cloud providers are often not able to provide detailed information on the location of their customers' data (Jansen

---

[14]http://www.uml4pf.org/cloudtool/cloudSystemAnalysisTool.zip.

Table 12.5 Instantiated template for management approval of the ISMS

Management commitment

ISMS security goal	The transaction data of bank customers shall be kept confidential
	The integrity of the transaction data shall be preserved
	The online banking shall be available 24/7
Establish responsibilities	The responsible person for the fulfillment of all security goals is Mr. Jones from the bank institute
Communicate importance of security	The employees of the bank institute receive an education about the consequences to the bank caused by a loss of integrity
Criteria for risk acceptance	The bank wants to avoid insolvency
Conduct ISMS audits	Mr. Jones is responsible for building the ISMS, hence he should not be responsible for hiring or conducting the audits. Mr. Smith is responsible for conducting internal and external audits
ISMS management reviews	Neither Mr. Smith nor Mr. Jones should be responsible for the management reviews, because they are part of it. Instead this tasks is assigned to Mr. Shell

Cloud-specific management commitment

Decide a cloud deployment scenario	Mr. Schneider is responsible for deciding to use a public, private or hybrid cloud scenario
Check security assurances of cloud provider	Mr. Schneider is responsible for this task, because he is the resident security expert
Conduct on site reviews of cloud provider	Ms. Schneider is also responsible for this task
Conduct neutral security assessment of cloud provider	Ms. Smith is responsible for the selection of an external security evaluator for the selected cloud provider

Resource management

Provided resources for the ISMS	The ISMS requires external parties to conduct the checking of the file integrity of transaction information using, e.g., Hmac a keyed-hashing for message authentication (IETF 1997) by the cloud provider. The resources for these integrity checks have to be provided
Security supports business needs	The integrity checking of the files should not make the transactions impossible or decrease the transaction time significantly
Competent personal	List all resources necessary for conducting integrity checks. These are financial resources for hiring security experts to conduct integrity checks

(continued)

Table 12.5 (continued)

Resource management

Provide training	The training program in this case is for auditing the cloud provider and the external party that conducts the integrity checks. The bank institute requires skilled parties to conduct these audits
Effectiveness evaluation	Have an audit that checks all taken measures. In this case, audit training programs and personal. A specific audit for that case has to be taken
Records of education, training, skills, experience and qualification	Mr. Jones is responsible for fulfilling documentation demands, e.g., which external party was hired and the reasons for hiring this particular party

Table 12.6 Relevance of the scope definition within the ISO 27001 standard

Significance for the ISO 27001 standard	The ISMS scope definition of the ISO 27001 standard is a vital step for its successful implementation, because all subsequent steps use it as an input
Related section(s) of the standard	Section 4 describes the Information security management system and in particular in *Section 4.2—Establishing and managing the ISMS* states the scope definition. Section 4.2.1a demands to "Define the scope and boundaries of the ISMS in terms of the characteristics of the business, the organization, its location, assets and technology, and including details of and justification for any exclusions from the scope" (ISO/IEC 2005b, p. 4). Section 4.2.1d concerns risk identification and the section recommends to consider the scope definition for identifying assets. Section 4.2.3 demands management reviews of the ISMS that also includes to check for possible changes in the scope of the ISMS. Section 4.3 lists the documentation demands of the standard, and Section 4.3.1d requires a documentation of the scope of the ISMS

2011). An example for a cloud database is Google's Bigtable (Fay Chang and Jeffrey Dean 2006). An open question is how a cloud provider can prove that data has been deleted (Chow et al. 2009). Cloud databases are particularly relevant for the threat *Data Loss or Leakage* in Sect. 12.8.1.

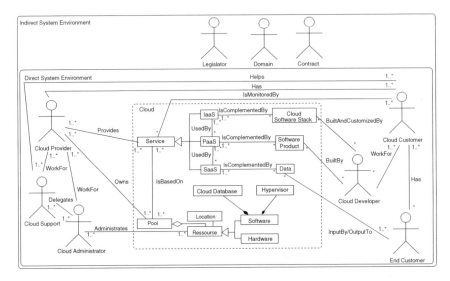

Fig. 12.6 Extended cloud system analysis pattern

Changes for ISO 27001:2013 Compliance

In ISO27001:2013 the scope definition is still essential. Section 4 of the standard describes the demands for the scope definition. ISO27001:2013 demands an internal and external context of the organization in Section 4, which shall be the basis for determining the boundaries of the scope. The scope is still the input for subsequent steps. This is explicitly demanded in Section 6.1.2c as part of the risk identification. The following use of the cloud pattern is still appropriate and supports the identification of an internal and external context, as well as the scope.

The second component we add to our cloud pattern is the *Hypervisor*. According to Scarfone et al. (2011), a hypervisor "*controls the flow of instructions between the guest OS and the physical hardware, such as CPU, disk storage, memory, and network interface cards. The hypervisor can partition the systems' resources and isolate the guest OS so that each has access to only its own resources, as well as possible access to shared resources such as files on the host OS. Also, each guest OS can be completely encapsulated, making it portable. Some hypervisors run on top of another OS, which is known as the host operating system*". Therefore, the hypervisor is of particular interest considering the threat *Shared Technology Issues* (Sect. 12.8.1).

Furthermore, we added two stakeholders to the *Direct System Environment*. The *Cloud Support Helps* the *Cloud Customer* when using the cloud and *Works-For* the *CloudProvider*. The stakeholder is relevant, e.g., for the threat *Malicious Insiders* (Sect. 12.8.1). We also introduce the stakeholder *Cloud Administrator* who

Administrates the cloud's *Resources* and *WorksFor* the *CloudProvider*. This stakeholder is relevant, e.g., for the threat *Account or Service Hijacking* mentioned in Sect. 12.8.1.

Cloud Stakeholder Templates—We supplement the cloud system analysis pattern by templates to systematically gather domain knowledge about the direct and indirect system environments based upon the stakeholders' relations to the cloud and other stakeholders. We updated the templates with location, cloud deployment scenarios, and privacy concerns with respect to a previous version in Chap. 10 and the publication (Beckers et al. 2011).

The first template serves to describe stakeholders contained in the direct system environment, shown in Table 12.7, which is an updated version from Beckers et al. (2011). The second template describes the stakeholders contained in the indirect system environment (see Table 12.8), which is also an updated version from Beckers et al. (2011). In addition to our method, we use a hierarchical structure of models, which lets us analyze the cloud system at different decomposition levels or views. The use case diagrams in Sect. 12.5 are the initial model, and the *Cloud System Analysis Pattern* is the first refinement level. Beyond the *Cloud System Analysis Pattern*, there

Table 12.7 Direct stakeholder template—updated version

Name	State the identifier of the stakeholder or group of stakeholders, e.g., company name or group of end customers
Description	Describe the stakeholder informally, e.g., if the stakeholder is a natural or a legal person
Relations to the cloud	Describe the input and output represented as relation (line from this stakeholder to the cloud) between the stakeholder and the cloud, e.g., the kind of data or software
Cloud deployment scenarios	State the deployment scenarios the cloud stakeholder demands: public, private, or hybrid. Also state the reason for the particular deployment scenario
Location	State the country the stakeholder works in
Motivation	State the motivation of the stakeholder for using the cloud based on the previous considered relations to the cloud, e.g., business goals such as profit increase
Relations to other direct stakeholders	For each relation (line from this stakeholder to another direct stakeholder), name the kind of relation between the stakeholders, e.g., indirectly influenced by customer-demand
Assets	State the assets of the stakeholder that are already known
Compliance	State relevant laws and regulations for the cloud scenario that are already known

Table 12.8 Indirect stakeholder template—updated version

Name	See direct stakeholder template
Description	See direct stakeholder template
Relations to other stakeholders	For each relation from this stakeholder to another direct or indirect stakeholder (no line explicitly shown), name the kind of relation between the stakeholders, e.g., protected by, controlled by law, implement laws
Motivation	State the motivation of the stakeholder for having any reason of considering the cloud for its work or the motivation for having any kind of relation to stakeholders of the direct or indirect environment, e.g., protect privacy of citizens or implement concrete laws of an economic community
Compliance	Identify relevant laws as well as regulations based on the indirect stakeholders. Specify and identify the ones relevant for the stakeholder at hand, e.g., the U.S. Health Insurance Portability and Accountability Act (HIPAA)

is no strict rule on the kind of diagrams to include or their scope. It depends on the size and technology involved in the cloud. Hence, our method scales, because we can attach diagrams for different decomposition levels of the cloud or views to the *Cloud System Analysis Pattern*. In our running example, we are interested in adding behavioral information about business processes to the pattern. Hence, we use UML (UML Revision Task Force 2010) activity diagrams to show this particular view of the cloud system.

12.6.2 Instantiate the Extended Cloud Pattern with Our Running Example

We instantiate our cloud pattern using the example of the online banking service introduced previously. We consider the use cases introduced in Sect. 12.6 and in particular the financial institute (bank) and its bank customers. The financial institute is located in Germany. The bank plans to hire an internal software development unit to develop software for online banking in the cloud and a customized OS for the developed online banking software. Hence, the bank plans to outsource the affected IT processes to the cloud to reduce costs and scale up their system for a larger number of customers. Customer data such as the transaction log history are stored in the cloud database, and transactions such as money transfers are processed in the cloud.

The cloud developer creates SaaS systems for the bank institute via developing in a PaaS environment and customizing an OS for a given IaaS offer. The bank authorizes its internal software department to design and build the cloud-specific software according to the interface and platform specification of the cloud provider. The main goal of the cloud provider, in our example a company called *Hulda*, is to maximize profit by maximizing the workload of the cloud. Therefore, subgoals are to increase the number of customers and their usage of the cloud, i.e., the amount of data as well as the number and frequency of calculation activities they outsource into the cloud. Fulfilling security requirements is only an indirect goal to acquire customers and convince them to increase the set of processes they outsource. The bank customer is a person, juristic or natural, who has an account at the bank, which enables him or her to do financial transactions via the banking offers. In our scenario, these financial transactions can be conducted via the web service the bank offers using the cloud (Fig. 12.7).

Basically, the online banking cloud service is embedded in an environment consisting of two parts, namely the *Direct System Environment* and the *Indirect System Environment*. The *Direct System Environment* contains stakeholders and other systems that directly interact with the cloud through associations, e.g., the *Bank Customer*. Moreover, associations between stakeholders in the *Direct* and *Indirect System Environment* exist, but not between stakeholders in the *Indirect System Environment* and the cloud. For example, the *Legislator Germany* is part of the *Indirect System Environment*. Typically, the *Indirect System Environment* is a significant source for compliance and privacy requirements .

We derive the indirect stakeholders required for this scenario based on the instantiation of the *Direct System Environment*. The *Cloud* is located in Germany and the USA. This is the reason for the *indirect stakeholders Legislator* Germany and

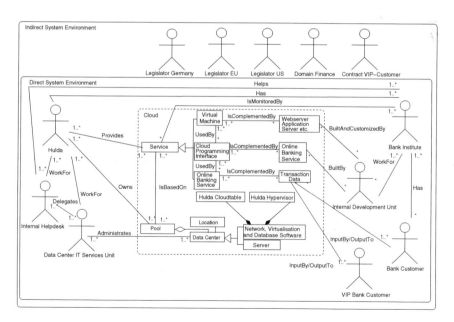

Fig. 12.7 Example instantiation of our extended cloud system analysis pattern

Legislator US. Germany is a member of the European Union, resulting in an additional set of regulations. They are described by the *Legislator* EU that represents a set of EU regulations. The financial institute has also several contractual obligations, one of which is the *Contract VIP Customer* that represents a contract between the bank institute and *VIP Bank Customer* that defines the 24/7 availability of the on-line banking system. As examples, we present one stakeholder template instance for an indirect stakeholder (see Table 12.9 cf. Beckers et al. 2011) and one for a direct stakeholder (see Table 12.10 cf. Beckers et al. 2011).

The *VIP Bank Customer* relates to the cloud in a similar manner as the *Bank Customer*, because the pattern focuses on structural information. We show an activity diagram for conducting a financial transaction in Fig. 12.8, which illustrates the difference in behavior between those customers. We have two different kinds of end customers, the *Bank Customer* and the *VIP Bank Customer*. The *VIP Bank Customer* is entitled to a 24/7 financial transaction service, while the *Bank Customer* is only entitled to a normal financial transaction service. The difference is that the normal transaction services gives only very limited guarantees to the availability of the service. The 24/7 transaction service provides the guarantee that a transaction can be conducted at 99.9999 % of the time and that any occurring problem is fixed within 5 min.

The process depicted in Fig. 12.8 begins with either the *Bank Customer* or the *VIP Bank Customer* initiating a financial transaction. We focus on the transaction of the *Bank Customer*. The transaction request is sent to the cloud and executed if sufficient resources exist. The *Financial Institute* rented only a limited amount of resources in the cloud, and scaling these resources causes an increase in costs. Thus, these increases in resources will not happen for single *Bank Customers*. Only if sufficient number of *Bank Customers* are using the *Online Banking Service*, the *Financial Institute* will increase the resources. This is fundamentally different from the *VIP Bank Customers*, who pay for the scaling of resources. Hence, if resources are not sufficient for conducting a financial transaction of a *Bank Customer*, the system notifies the *Bank Customer* that the transaction is not possible at this time

Table 12.9 Indirect stakeholder template: Legislator Germany

Name	*Legislator Germany*
Description	The *Legislator Germany* represents all German laws relevant for this cloud scenario
Motivation	The German laws try to control the risks of companies (*Hulda* and *Bank Institute*) and to protect the privacy of the *Bank Customers* by regulating disclosure of personal data
Relations to other stakeholders	Controlled by law: The laws have to be obeyed by all stakeholders of the *Direct System Environment*
Compliance	The following regulations might be considered: • Privacy protection: e.g., BDSG • Risk management: e.g., AktG

Table 12.10 Direct stakeholder template: Bank customer

Name	*Bank Customer*
Description	The *Bank Customer* uses the online banking service of the *Bank Institute*. The *Bank Customer* is not entitled to a 24/7 transaction service
Motivation	The *Bank Customer* wants low cost and secure financial transactions via the bank's cloud computing offer
Relations to the cloud	*InputBy/OutputTo*: *InputBy financial data, data related to a person*, which is required for billing of the *Bank Institute* and maintenance of the cloud
Cloud deployment scenarios	The bank considers a public cloud, because it offers significant savings in terms of money
Location	The *Bank Customer* is located in Germany
Relations to other direct stakeholders	*Has*: *Bank Institute* as SaaS provider
Assets	Financial data and all data related to the person
Compliance	The following laws might be of relevance: • Privacy protection: BDSG Sections 3, 4, 9 and 11 • Risk management: AktG Sections 91 and 93

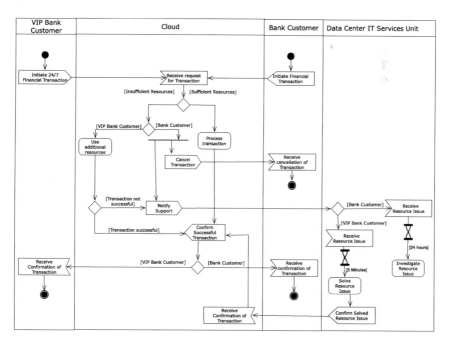

Fig. 12.8 Activity diagram describing the process of conducting a financial transaction

and suggests to try again later. In addition, the cloud sends a message to the *Data Center IT Services Unit*. The unit investigates within 24 h if the cloud resources need to be increased, due to a significant number of requests. Otherwise, the unit does nothing.

Should there not be sufficient resources to conduct the financial transaction of the *VIP Bank Customer*, the cloud service automatically uses further resources to conduct the transaction. In addition, should any problem occur during the transmission, the *Data Center IT Services Unit* has to solve the problem within 5 min. The *VIP Bank Customer* is informed about the successful transaction afterwards.

This concludes our ISMS scope description in our example. We propose to use multiple processes and further refinements in scope descriptions, but for space reasons we limit ourselves to one. In addition, process descriptions can also benefit from documentation standards for IT Services, e.g., ITIL (HM Government 2012). These provide example processes for typical tasks regarding IT management. We show how to relate the ITIL and ISO 27001 standards using the cloud pattern in detail in Beckers et al. (2013b).

12.7 PACTS Step 3: Identify Assets

The ISO 27001 standard lists the protection of assets with adequate security goals already on its first page. In Table 12.11 we state several references of the standard toward asset identification. These references occur in particular during the scope definition, policy definition, and risk estimation phases of the standard (Fig. 12.9).

Table 12.11 Demands for asset identification of the ISO 27001 standard

Significance for the ISO 27001 standard	The design goal of the ISO 27001 ISMS is to protect assets with adequate security controls and this is stated already on p. 1 of the standard
Related section(s) of the standard	Section 4 describes the information security management system and in particular in *Section 4.2—Establishing and managing the ISMS* states the scope definition. Section 4.2.1a demands the definition of assets. Section 4.2.1b concerns the definition of ISMS security policies and it demands that the policy shall consider assets. Section 4.2.1d that concerns risk identification uses the scope definition to identify assets, to analyze threats to assets, and to analyze the impacts of losses of these assets. Section 4.2.1e concerns risk analysis, which also clearly defines to analyze assets and to conduct a vulnerability analysis regarding assets in light of the controls currently implemented

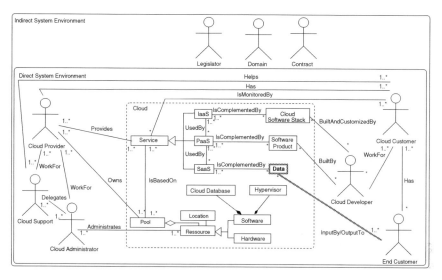

Fig. 12.9 Asset identification for the end customer

Changes for ISO 27001:2013 Compliance

ISO27001:2013 does not consider asset identification as mandatory (ISO/IEC 2013, p. 4). However, the risk identification in Section 6.1.2 in the standard could be supported by identifying assets and their vulnerabilities and threats. Some controls in Appendix A even refer to assets such as control *A8—Asset management*.

This section is inspired by the idea of using the relations in the cloud system analysis pattern to identify assets presented in Beckers et al. (2011). We improved the content of the aforementioned publication with a structured method in this work. We enhance the asset identification by already identifying assets simply using the cloud pattern. These assets can be reused for different projects. In addition, we check the instantiated cloud pattern for further assets. Hence, the contribution of this section is a structured method for asset identification using the cloud system analysis pattern and its instantiation.

Figure 12.10 depicts an overview of this method. We explain its steps in the following.

Instantiate Asset Template—The ISO 27001 standard defines an asset (ISO/IEC 2005b, p. 2) as follows: "anything that has value to the organization." The organization in our case is either the *cloud provider* or the *cloud customer*.

We identify assets in the cloud pattern by analyzing the associations (the lines) from all stakeholders toward the cloud. We check if the cloud elements at the end of

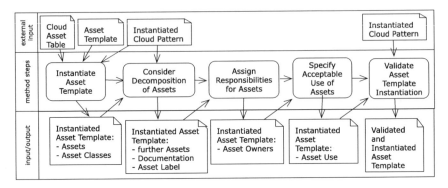

Fig. 12.10 A method for cloud-specific asset identification

the associations potentially have value to the stakeholders and, thus, are assets. If they are assets, we check if associations between these and further cloud elements lead to further assets. Let us take the end customer to illustrate this step (see Fig. 12.9). The association from the stakeholder to the cloud leads to *Data*. This data has potential value to the end customer, because the *Data* should be processed or stored by the cloud. It is not likely that the end customer invests money into processing or storing *Data*, which has no value to him or her. Hence, we identify *Data* as an asset. The relation between *Data* and *SaaS* is also investigated. The *SaaS* is used by the end customer, but not of particular value. We assume that the *End Customer* can also use other offers.

The identified assets are listed in Table 12.12. In the following we explain the technique for identifying assets using the cloud pattern. The benefit of identifying assets in the pattern is that the identified assets can be reused for each instantiation of the pattern. The pattern also contains the information about a so-called *asset provider*. This is a stakeholder that either owns an asset or creates an asset. For example, the *cloud developer* built the *software product*. The last column of the table contains a reasoning that the asset has value to a stakeholder and harm to it would affect this stakeholder. The stakeholder could, e.g., suffer financial loss in case his or her assets are financial data. In addition, the data could also contain personal information, and leaking it could also harm the stakeholder.

An example for considering not only direct relations to the *cloud* is the *cloud provider* (see Fig. 12.11), who owns the *pool*. Harm to the *pool* could result in the bankruptcy of the *cloud provider*. The pool consists of *resources*, which are also assets. We include the *Hardware, Software, Cloud Database*, and *Hypervisor* in a similar manner.

We use an asset template in our method, shown in Table 12.13. Our asset template collects all the information required for assets by the ISO 27001 standard (ISO/IEC 2005b, p. 15).

Table 12.12 Cloud asset table

Asset	Asset provider	Asset reasoning
Cloud software stack	Cloud developer	The *Cloud Software Stack* is the basis for the *software product* of the *cloud customer*. Harm to it can affect the functionality of the *software product* and cause financial harm to, as well as harm to the reputation of the *cloud customer*
Software product	Cloud developer	The *software product* is essential to the business of the *cloud developer*, and harm to it can cause financial harm to the *cloud customer*. Harm to this asset can also cause harm to the reputation of the *cloud customer*
Data	End customer	Harm to the asset can possibly cause financial loss and privacy violations to the *end customer*. The harm depends on the kind of data
Resources	Cloud provider	The resources are the essential infrastructure of the *cloud*, and harm to these can cause bankruptcy of the *cloud provider*
Hardware	Cloud provider	The resources are the essential infrastructure of the *cloud*, and harm to these can cause bankruptcy of the *cloud provider*
Software	Cloud provider	The resources are the essential infrastructure of the *cloud*, and harm to these can cause bankruptcy of the *cloud provider*
Hypervisor	Cloud provider	The resources are the essential infrastructure of the *cloud*, and harm to these can cause bankruptcy of the *cloud provider*
Cloud database	Cloud provider	The resources are the essential infrastructure of the *cloud*, and harm to these can cause bankruptcy of the *cloud provider*

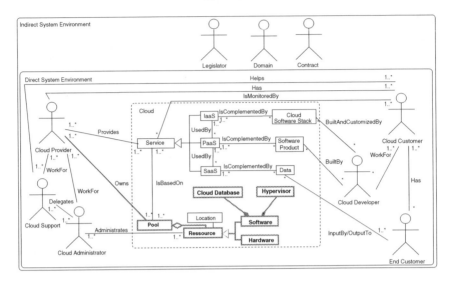

Fig. 12.11 Asset identification for the cloud provider

Table 12.13 Instantiated asset template

Asset	Asset owner	Asset use	Asset class	Asset label
Webserver, application server, etc.	Mr. Smith	External activity diagram	Software	AS_SO_100
Online banking service	Mr. Jones	External activity diagram	Software	AS_SO_110
Transaction data	Mr. Jones	See Fig. 12.15	Data	AS_DA_120
Data center	Mr. Mintz	External activity diagram	Physical	AS_PH_100
Network, virtualisation and database software	Ms. Lock	External activity diagram	Software	AS_SO_130
Hulda cloudtable	Mr. Lock	External activity diagram	Software	AS_SO_140
Hulda hypervisor	Mr. Lock	External activity diagram	Software	AS_SO_150
Server	Ms. Mintz	External activity diagram	Hardware	AS_HA_100

In the first step, we instantiate our cloud asset table (see Table 12.12) and enter the names of the assets into the first column of our asset template. We explain the instantiation of the remaining columns in the following.

Consider Decomposition of Assets—The second step of our method is the decomposition of the assets, which are already listed in the instantiated asset template. Figure 12.12 shows the decomposition of the *transaction data* of the *VIP Bank Customer*. As an example we present the details of the *transaction data* as a UML class

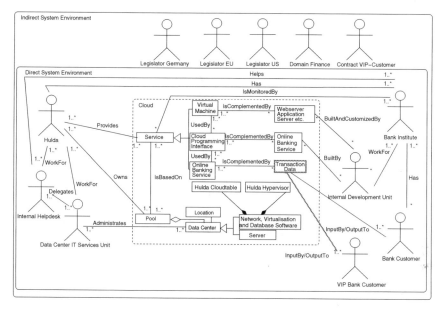

Fig. 12.12 Asset identification of the VIP bank customer

Transaction Data
Originator Name: String
Originator Account: String
Originator City State Zip: String
Originator Country: String
Receiver Name: String
Receiver Address: String
Receiver City State Zip: String
Receiver Country: String
Amount: String
ValueDate: DateTime
Message: String

Fig. 12.13 Decomposition of the asset transaction data

diagram, (see Fig. 12.13). This details can help to classify the assets in categories. For example, which elements of the transaction data contain personal information. Assets should be decomposed if the decomposition reveals further information for classifying or describing the assets. For example, the *Webserver, Application Server, etc.* is decomposed into two web servers for redundancy, and also several different types of application servers. All of these have to be listed in the asset template.

Assign Responsibilities for Assets—This step of our method concerns the assignment of responsibilities for assets and relates to the second column of our asset template that lists so-called *asset owners*. The standard defines the term as follows: "The term 'owner' identifies an individual or entity that has approved management responsibility for controlling the production, development, maintenance, use and

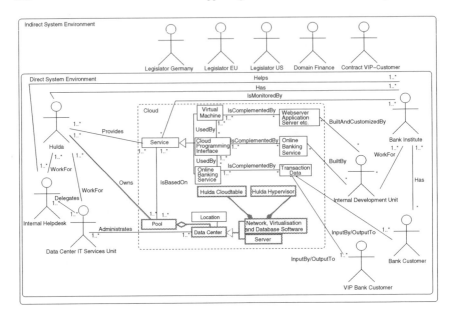

Fig. 12.14 Asset identification for the cloud provider Hulda

security of the assets. The term 'owner' does not mean that the person actually has property rights to the asset" (ISO/IEC 2005b, p. 15) (Fig. 12.14).

Specify Acceptable Use of Assets—The third column states the acceptable use of that asset, because using it outside of the specification can result in harm to that asset and stakeholders, respectively. We propose to specify the acceptable use of assets in UML activity diagrams. For example, Fig. 12.15 specifies the acceptable use of the asset *transaction data*. The data is sent to the cloud, where it is processed and stored for 30 days. Thirty days is the time the data is stored for billing purposes. Afterwards it is deleted. If the deletion fails, the *Data Center IT Services Unit* gets a message from the cloud and executes the deletion of the data. Deletion in the cloud could fail, e.g., because the data is distributed and replicated over many computers and one of these replications might not delete the data. A possible reason is that due to a malfunction of the communication, one of the computers that stores a replication does not receive the delete command. Assets also have to be classified in order have a unique identifier per asset. We propose a classification into *Hardware, Software, Data,* and *Physical.* The difference between *Hardware* and *Physical* is that *Physical* refers to items that are buildings in which a data center is located, while *Hardware* is exclusively parts of a computer. The fourth column states the classification of an asset and the last column its unique label. We propose a labeling schema that uses the first two letters of the word asset, followed by the first two letters of the asset class and an increasing number.

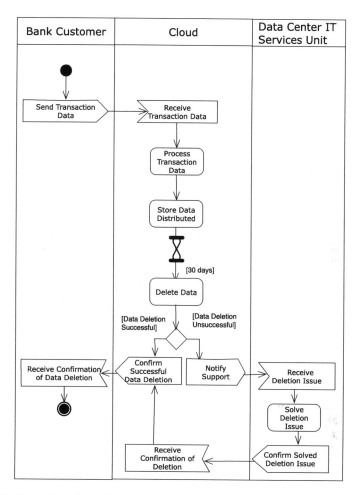

Fig. 12.15 Description of use of transaction data

A Procedure to Check the Asset Template Instantiation—We propose to check the instantiation of the asset template via the following procedure. We propose the following checks:

- Check the instantiated cloud pattern for assets. For example, we present the asset identification for *Hulda* in Fig. 12.14. This is in particular relevant if the cloud pattern has been extended with further cloud elements or stakeholders.
- Check all cloud elements for not yet considered assets. This validation condition should check for completeness.
- Check if the instantiated asset template has empty fields. After the termination of this step all fields of the instantiated asset template should be filled. If this is not the case, the information for the missing fields has to be elicited.

- Check if assets are considered more than once. The asset template does not contain a check for duplicate entires. The security expert shall check if two assets are in fact the same and remove the duplicates.
- Do some assets require further refinement? This check answers the question if the decomposition is complete. The security expert shall determine if an asset contains further assets or if the decomposition is complete.

12.8 PACTS Step 4: Analyze Threats

The ISO 27001 standard demands a threat analysis in order to determine and analyze risks to identified assets (see Table 12.14). The background of this section are the cloud security issues from CSA and Gartner. We combine these threats with our cloud pattern and use the results of misuse cases to elicit security requirements. Our contribution in this section is a cloud-specific threat analysis method.

Table 12.14 Threat analysis demands within the ISO 27001 standard

Significance for the ISO 27001 standard	The ISO 27001 standard concerns threat analysis in several sections for determining the risks to assets
Related Section(s) of the standard	Sect. 4.2.1d demands a threat analysis for assets for the purpose of identifying risks and the vulnerabilities that might be exploited by those threats. Section 4.2.1e concerns risk analysis and evaluation and demands to determine likelihoods and consequences for threats. Section 4.2.4d concerns the review process of the ISMS and also demands a threat identification. Section 7.2 that concerns the management review of the ISMS also demands a threat analysis

---

Changes for ISO 27001:2013 Compliance

Threat analysis is not a mandatory step in ISO27001:2013, as well (ISO/IEC 2013, p. 4). However, the risk identification, analysis, evaluation, and treatment steps in the standard could be supported by identifying vulnerabilities and threats. However, controls in Appendix A refer to vulnerabilities and threats such as control *A 12.6—Technical vulnerability management*.

12.8.1 Cloud Security Alliance—Top Threats to Cloud Computing

The Cloud Security Alliance presents a list of seven threats for clouds and their relations to IaaS, PaaS, and SaaS (Cloud Security Alliance (CSA) 2010). We use this particular list of cloud threats, because it summarizes the experience in the field of cloud computing from the point of view of a large industrial consortium. In the following, we present a short summary of these threats:

Abuse and Nefarious Use of Cloud Computing—Nefarious can mean criminal or treacherous.[15] This threat describes the abuse of the scalable cloud resources, e.g., storage or network capacity. For example, the resources can be used by spammers or malicious code authors.

This threat refers to the cloud service models: IaaS and PaaS.

Insecure Interfaces and APIs—Clouds provide interfaces for provisioning, management, orchestration of services. Security functions, e.g., authentication, access control, and encryption rely upon these. Hence, malicious use of these interfaces has to be prevented. An example for the malicious usage of interfaces is the eavesdropping during clear-text transmission of content.

This threat refers to the cloud service models: IaaS, PaaS, and SaaS

Malicious Insiders—The cloud provider controls access to the cloud. A cloud customer or end customer has very limited transparency considering data access permissions provided to cloud employees. Hence, the threat of malicious insiders, which are employees of the cloud provider, scales with the resources and offered services in the cloud. An example for a specific problem is policy compliance. Cloud customers or end customers have no influence on the hiring or monitoring of the cloud providers' employees.

This threat refers to the cloud service models: IaaS, PaaS, and SaaS.

Shared Technology Issues—The different stakeholders in the cloud use the same physical resources, e.g., Central processing Units (CPU)s and Graphics Processing Units (GPU)s. These are shared using so-called *Hypervisors*, which provide isolation properties for these physical resources. Side channel attacks on these *Hypervisors* can provide a stakeholder with inappropriate levels of control of the underlying cloud infrastructure.

This threat refers to the cloud service models: IaaS.

Data Loss or Leakage—The threats to data in a cloud scales with the amount of data stored in it. Deletion or alteration of data without a backup is an example. Moreover, cloud databases store data in a distributed way. The links to records in these cloud databases can be destroyed, which results in unrecoverable data. In addition, the loss of an encoding key renders data useless.

This threat refers to the cloud service models: IaaS, PaaS, and SaaS.

Account or Service Hijacking—Clouds provide numerous services and credentials, and passwords are often reused. Thus, compromised credentials provide access

---

[15]According to http://thesaurus.com/browse/Nefarious?s=t.

to a large set of data about activities and transactions of stakeholders. Thus, the attacker can exploit the reputation of a cloud customer and launch a large-scale attack on its end customers. The cloud customer's reputation can lead to directed phishing and farming attacks toward its end customers.

This threat refers to the cloud service models: IaaS, PaaS, and SaaS.

Unknown Risk Profile—Cloud customers and end customers do not own cloud resources. Hence, cloud providers can apply the so-called *security by obscurity* policy. Thus, the cloud customers and end customers do not know the exact specifications of the security mechanisms used in the cloud. This results in an unknown exposure of assets and increases the difficulty of creating a risk profile for a cloud scenario. This threat refers to the cloud service models: IaaS, PaaS, and SaaS.

12.8.2 Gartner's Cloud Security Risks Assessment

Gartner assessed the security risks of cloud computing and defined a list of recommendations for cloud customers that help to evaluate cloud providers. The difference between risk and threats is that high level risks cause loss to a stakeholder, while threats exploit vulnerabilities and can be used to realize attacks (Fabian et al. 2010). Hence, in order to map risks to threats we have to link them to possible vulnerabilities. In the following, we present Gartner's evaluation criteria and relate each criterion to the CSA threats. In addition, we formulate a new threat for each criterion that could not be mapped to an original CSA threat.

Privileged User Access—Sensitive data processing outside the organization or by nonemployees leads to an uncertain level of risk, because the security controls of the organization are bypassed. The cloud customer depends on the controls of the cloud provider for upholding security assurances for sensitive data. The cloud provider should release information about hiring and an oversight of all staff that has access to the sensitive data and controls concerning access to the data.

Relation to CSA threats: Malicious Insiders, Unknown Risk Profile.

Compliance—Regulations hold the cloud customer responsible for the security of their organization's and customer's data. Hence, cloud customers should demand security certifications, which include documentation of controls, as well as security audits.

Relation to CSA threat: Unknown Risk Profile.

Data Location—Several privacy regulations demand that personal information stays in certain geographical regions. The cloud customer has to know if the cloud provider upholds privacy regulations and can restrict personal information from flowing into restricted geographical regions.

Relation to new threat: Unrestricted Flow of Personal Information.

Data Segregation—Data has to be transferred to the cloud and stored in it. Encryption is one solution for protecting the data. In transit most cloud providers use SSL, but for storing data in the cloud, not all cloud providers use encryption. The cloud customer has to check, which kind of encryption is used and who tested and analyzed its implementation. In addition, the cloud provider has to provide the information of who has the key for decrypting the data.

Relation to CSA threats: Data Loss or Leakage, Insecure Interfaces and APIs, Malicious Insiders.

Availability—The cloud customer has to check the commitments regarding availability of the cloud provider. These commitments have to be in contractual form of service level agreements. The contracts have to contain written penalties for the cloud provider.

Relation to new threat: Insufficient Service Level Agreements.

Recovery—The cloud customer requires insurances for data recovery in case of total disaster to the cloud. The cloud provider has to provide the specification of the backup systems and describe in detail if data replication is conducted and if a complete or partial data recovery after disaster is possible.

Relation to CSA threat: Data Loss or Leakage.

Investigative Support—Logging in clouds is difficult, because of changing hosts and data centers. Hence, the cloud customer can have difficulties to prove wrongdoings or even to conduct investigations. The cloud customer requires the cloud providers' written commitment to provide the means for investigations and evidence storing.

Relation to new threat: Impossible Investigations

Viability—The long-term viability of the cloud has to be evaluated. What happens to the data of the cloud customer if the cloud provider becomes insolvent or is acquired? The cloud provider has to provide assurance that the data and applications can still be accessed after these events.

Relation to CSA threat: Data Loss or Leakage.

Support in Reducing Risk—The cloud customer has to evaluate the level of information and support provided by the cloud provider to safely and reliably use the cloud. Does the cloud provider support definitions of policies and attack prevention?

Relation to CSA threat: Unknown Risk Profile.

We propose to add the following threats to the CSA list based upon the missing relations between Gartner's cloud security risks and CSA threats.

Unrestricted Flow of Personal Information—Personal information is protected by regulations in several countries. If the cloud has no means to control the flow of personal information and to determine the location of it, these laws can be violated.

Associated with Gartner Risk: Data Location.

Insufficient Service Level Agreements—Missing availability of the cloud and with it the cloud customer's data and application can cause financial loss and even put the cloud customer out of business. The reason is that the financial loss of the cloud customer, which results from the missing availability of the cloud, is not compensated by the cloud provider.

Associated with Gartner Risk: Availability.

Impossible Investigations—Logging in clouds is difficult to implement, because of the complexity of clouds. Hence, the cloud customer can be without the means to investigate possible wrongdoings or prove crimes in the cloud.

Associated with Gartner Risk: Investigative Support.

In the following section we show how to relate these threats to our cloud pattern as a preparation to use both in a structured method. In the future, we will extend the list of considered cloud security threats, e.g., with the cloud security issues listed in Beckers and Jürjens (2010).

12.8.3 Relations Between Threats and the Cloud Pattern

We analyze the threats to clouds using the threats introduced in the previous sections (Sects. 12.8.1 and 12.8.2). First, we map the threats to the security goals: *Confidentiality*, *Integrity*, *Availability*, *Authorization*, and *Non-Repudiation*. We also consider *Privacy* in this work, as discussed in Sect. 12.4. The results are illustrated in Table 12.15. Note that we abbreviated the following threats: *Abuse and Nefarious Use of Cloud Computing* to *Abuse of Cloud Computing*, *Unrestricted Flow of Personal Information* to *Unrestricted Flow of PI*, and *Insufficient Service Level Agreements* to *Insufficient SLAs* for the remainder of this work. The relation between a threat and a security goal and privacy is either a *cause* or an *effect* relation. The entry "cause" means that a violation of this security goal is essential for the threat to occure. For example, the threat *Abuse of Cloud Computing* is caused by not proper *Authorization*, because attackers should not be authorized to use the cloud for conducting attacks. The entry "effect" states that a threat has an impact on one or more security goal(s), e.g., the threat *Insecure Interfaces and API* has an effect on the security goals *Confidentiality*, *Integrity*, *Availability*, and *Non-Repudiation*. For example, an insecure interface could result in an integrity checking mechanism of files, which can be disabled by an attacker. One cause for this threat is a lack of authorization in the APIs and one effect is that the integrity of the data can be compromised.

Second, we propose to relate these threats to our cloud pattern (Sect. 12.6). We present a mapping from the threats to the stakeholders described in the cloud pattern in Table 12.16. The mapping from the threats to the technical elements in the cloud pattern are depicted in Table 12.17. We use also a cause relationship in these tables, but we do not consider an effect relationship. Instead, we use an "x" that indicates that an element of the pattern is somehow affected by the respective threat, e.g., the *Cloud Provider* is somehow affected by the threat *Abuse of Cloud Computing* (see Table 12.16). This relationship is not as determined as an effect relationship and to determine what exactly is affected, it is necessary to conduct a further examination on the threatened element. Some cells in Tables 12.16 and 12.17 do not have any of the above-mentioned entries assigned. Nevertheless, a free cell does not imply that there is no relation between threats and elements in the cloud pattern. It simply means that we did not come across any relation so far.

The tables provide an initial overview from which stakeholder or technical element a threat originates. This provides a beginning of the threat analysis. For example, we want to analyze origins for the threat *Shared Technology Issues*. We consider Table 12.15 first. The threat *Shared Technology Issues* can be caused by violations of security goals confidentiality, integrity, and authorization. Next, we check Table 12.16 and see that the stakeholders *Cloud Developer* and *Cloud Administrator* can cause

Table 12.15 Relations between cloud threats and security/privacy goals

Threats	Security goals					
	Confidentiality	Integrity	Availability	Authorization	Non-Repudiation	Privacy
Abuse of cloud computing	effect	effect	effect	**cause**	effect	effect
Insecure interfaces and API	**cause**	**cause**	effect	**cause**	effect	effect
Malicious insiders	effect	effect	effect	**cause**	effect	effect
Shared technology issues	**cause**	**cause**	effect	**cause**	effect	effect
Data loss or leakage	**cause**	**cause**	**cause**	effect	effect	effect
Account or service Hijacking	**cause**	effect	effect	**cause**	effect	effect
Unknown risk profile	effect	effect	effect	**cause**	**cause**	effect
Unrestricted flow of PI	effect	effect	effect	effect	effect	**cause**
Insufficient SLAs	effect	**cause**	**cause**	effect	effect	effect
Impossible investigations	effect	effect	effect	effect	**cause**	effect

PI Personal Information, *SLA* Service Level Agreement

Table 12.16 Cloud threats: direct system environment view

Threats/cloud pattern	Cloud provider	Cloud customer	Cloud developer	End customer	Cloud administrator	Cloud support
Abuse of cloud computing	x	cause	cause	x	x	x
Insecure interfaces and API	x	x	cause	x		
Malicious insiders	x/cause	x	cause	x	cause	cause
Shared technology issues	x	x	cause	x	cause	
Data loss or leakage	cause	x	cause	x	cause	cause
Account or service hijacking		x	cause	x	cause	cause
Unknown risk profile	cause	x	cause	x	cause	
Unrestricted flow of PI		x	cause	x	cause	cause
Insufficient SLAs	cause	x		x		
Impossible investigations	x/cause	x	cause	x	cause	

PI Personal Information, *SLA* Service Level Agreement

the threat. Finally, Table 12.17 reveals that cloud elements *Cloud Software Stack* and *Software* can cause the threats. All these cloud elements and stakeholders would have to be analyzed at first for a threat analysis.

12.8.4 Cloud Threat Patterns

Our threat analysis is twofold: First, we execute the threat analysis on the (not in-stantiated) cloud pattern in this section. Second, we consider the instantiated cloud pattern in a structured method in the next section (Sect. 12.8.5), which relies on the results presented in this section. The idea is to conduct a threat analysis once on the cloud pattern and reuse the results for all possible instantiations of the pattern. Hence, security engineers can reuse threat analysis results that are relevant for all instantiations and focus on the particular threats that only occur in the instantiation.

In particular, we contribute a *cloud threat table* in this section. The table lists the stakeholders and cloud elements that cause each threat in Table 12.18 and we also

Table 12.17 Cloud threats: cloud view

Threats/cloud pattern	IaaS	PaaS	SaaS	CloudSoftwareStack	Software product	Data	Resources	Hardware	Software
Abuse of cloud computing	x	x	x	x	x	x^a			
Insecure interfaces and API	x/cause	x/cause	x/cause	x/cause	x/cause				x/cause
Malicious insiders	x	x	x	x/cause		x	x	x	x
Shared technology issues	x	x	x	x/cause	x		x	x	x/cause
Data loss or leakage	x	x	x	x/cause	x/cause	x			x/cause
Account or service hijacking	x	x	x	x/cause	x/cause	x			
Unknown risk profile	x	x	x	x/cause	x/cause	x	x/cause	x	x/cause
Unrestricted flow of PI	x	x	x	x/cause	x/cause	x	x/cause	x	x/cause
Insufficient SLAs	x	x	x	x	x	x			
Impossible investigations	x	x	x	x/cause	x/cause	x	x/cause	x	x

[a] Considering browser based attacks. *PI* Personal Information, *SLA* Service Level Agreement

Table 12.18 Cloud threat table

Threat	Cloud element	Stakeholder	Threat actions
Abuse of cloud computing	–	cloud developer, cloud administrator, cloud support, cloud customer	Conduct cybercrime, execute treacherous IT attack
Insecure interfaces and API	IaaS, PaaS, SaaS, cloud software stack, software product	Cloud developer	Ignore security functions, corrupt interface, create backdoor
Malicious insiders	Cloud software stack, resource, software, hypervisor, software product	Cloud developer, cloud provider, cloud administrator, end customer, cloud customer, cloud support	Neglect employee monitoring, missing background checks, hide information about employees
Shared technology issues	Cloud software stack, IaaS, resource, software, hardware, hypervisor	Cloud developer, Cloud administrator	Side channel attacks, misconfiguration
Data loss or leakage	Cloud software stack, resource, software, cloud satabase, software product, IaaS, PaaS	Cloud provider, cloud developer, cloud administrator, cloud support	change data, loose data, not conducting backups
Account or service hijacking	Cloud software stack, resource, software, cloud database, software product, IaaS, PaaS, SaaS	cloud developer, cloud administrator, cloud support	Attack end customer, steal credentials
Unknown risk profile	Cloud software stack, resource, software, hardware, cloud database, hypervisor, software product, IaaS, PaaS, SaaS	cloud provider, cloud developer, cloud administrator	restrict security information, incomplete information gathering
Unrestricted flow of PI	Cloud software stack, resource, software, cloud database, software product, IaaS, PaaS, SaaS	Cloud developer, cloud administrator	Ignore personal information, conduct global data distribution
Insufficient SLAs	–	Cloud provider	Write incomplete SLAs, write insignificant penalties
Impossible investigations	Cloud software stack, software product, resources, hardware, software, cloud database, hypervisor	Cloud provider, cloud developer, cloud administrator	Implement incomplete logging, erase logging data

identified possible misuse actions. These are actions an attacker uses to conduct an attack with regard to a certain threat. We created the table by a thorough analysis of all the threats mentioned in Sects. 12.8.1 and 12.8.2, as well as the relations to the cloud pattern (Sect. 12.8.3). Only after one threat has been successfully investigated, the analysis of the next one begins. To investigate the threats caused by cloud elements, we use Table 12.17. We select the entries in the table that contain a *cause*-relationship and highlight the corresponding cloud elements in the cloud pattern. The investigation of the threats caused by cloud stakeholders is based on Table 12.16. Similar to the cloud elements, we also highlight the cloud stakeholders that cause the threat in the cloud pattern. We check if further cloud elements or stakeholders exist, which can also give rise to the threat. For this purpose, we explore the relations between cloud elements. We start with the highlighted cloud elements and reason about the cloud elements that have relations (lines) to it. If a cloud element contributes to a threat, it is also highlighted in the cloud pattern. This is possible because we assume that cloud threats are caused by elements that have direct relations to each other. This continues until a cloud element does not contribute to a threat or no further cloud elements exist. We show exemplarily for one threat how we identified the relevant entries for our cloud threat table. We choose the threat *Shared Technology Issues* and we use Table 12.17 to identify cloud components that cause the threat. For our selected threat we have two *cause*-relationships, namely the *Cloud Software Stack* and *Software*. Table 12.16 states that the *Cloud Developer* and the *Cloud Administrator* are a possible cause for this threat. We mark the found elements in bold and red, (see Fig. 12.16). We investigate the relationships from the highlighted stakeholders and cloud elements. Following the relations from the *Cloud Software Stack* to other cloud elements, we get to the *Virtual Machine* leading us to the *Service* (see Fig. 12.16).

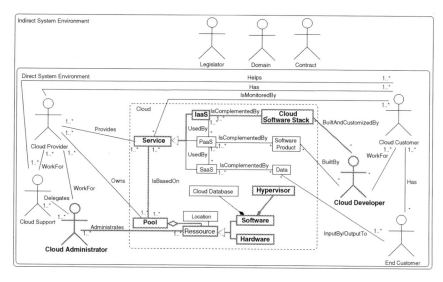

Fig. 12.16 Pattern-based cloud computing threat analysis for shared resources

Both cloud elements support the sharing of resources. Hence, both cloud elements give rise to the threat. The *PaaS* cloud element only uses shared resources, because the sharing of resources is managed in the *IaaS* layer or at the *Pool*. Hence, it is not highlighted (see Fig. 12.16). The threat we investigate is caused by managing the sharing of IT resources, e.g., the slicing of a physical hard drive into sections that *Virtual Machines* can use. The *Cloud Programming Interface* (Sect. 10.4) uses shared resources, but it does not participate in managing the sharing of IT resources. This cloud element just uses the IT resources it gets assigned by the *IaaS* layer or its instance of a *Virtual Machine*. Thus, we also exclude the *Cloud Programming Interface* as a cause for this particular threat. The *Virtual Machine* has also a relation to the *Service*, which has in turn a relation to the *Pool* that contains *Resources* and *Locations*. In this case, *Locations* do not give rise to the *Shared Technology Threat*, because the location of the technology is not related to the process of sharing IT resources. However, the *Resource* is relevant to the threat, because it contains the technology that allows the sharing of IT resources in the cloud. This technology is further refined into *Software* and *Hardware*. The *Hardware* is relevant, because it is the *Resource* that is shared, and the *Software* is also shared and orchestrates the sharing, as well. The *Hypervisor* is a particular software for sharing resources and thus also highlighted. The *cloud database* is not relevant, because even though it provides resources, it is not involved in the sharing of the resources. The *Cloud Customer* is not involved in the technical realization of the *software product*. Thus, the stakeholder is not marked. The same argument holds for the *Cloud Provider*.

12.8.5 A Method for Pattern-Based Threat Analysis for Clouds

We present our method in Fig. 12.17 and explain each step in the following:

Create Misuse Cases—We instantiate the columns *stakeholder* and *cloud element* of the cloud threat table. For example, the *cloud provider* is instantiated with *Hulda*. We use the misuse case notation (Opdahl and Sindre 2009) for the first step and in particular a textual representation of misuse cases as introduced by

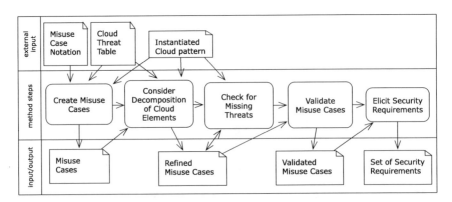

Fig. 12.17 A method for pattern-based threat analysis for clouds

Deng et al. (2011). We iterate over all threats in our instantiated cloud threat table (see Table 12.18) and write misuse cases using the instantiated cloud pattern. We also consider the identified assets (Sect. 12.7) in the misuse cases. For example, we consider the following misuse case for the *Malicious Insiders* threat and the asset *online banking service*:

Misuse Case 1.   The *bank institute* neglects to monitor the *internal development unit* during the development of the *online banking service*. The *bank institute* also did not conduct background checks of the team hired for the development. This results in multiple blackmail attacks on bank customers conducted by members of the *internal development unit*.

Consider Decomposition of Cloud Elements—We propose to analyze the initial set of misuse cases for further threats or threats that require refinement. If cloud components are involved in the threats, these should be decomposed to provide the information for a more detailed threat analysis. For example, we consider the misuse case:

Misuse Case 2.   Integrity of *Transaction Data* might be compromised by a shared technology attack exploiting a vulnerability.

We decompose the *Hulda Hypervisor*, depicted in Fig. 12.18, because it organizes the sharing of resources. Hence, a threat analysis of this element is of particular interest. In our example, the *Hulda Hypervisor* separates a virtual machine running the operating system *Hulda OS* and another virtual machine that runs *Other OS*. The *Hulda Hypervisor* arranges that the *Hulda OS* can only access the *Hulda Storage* and the *Other OS* can access only the *Other Storage*. A possible threat is that, due to, e.g., a configuration mistake of the *Hulda Hypervisor*, the *Other OS* can access the *Hulda Storage*. In particular, the transaction data of the *bank customer* and the *VIP bank customer* are threatened.

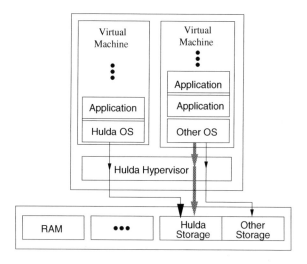

Fig. 12.18 Hypervisor threat analysis

The *Data Center IT Services Unit* can be the stakeholder that causes a mistake in the configuration of the *Hulda Hypervisor*.

This information is considered in a refinement of the initial Misuse Case 2 into two separate Misuse Cases:

Misuse Case 2.   Integrity of *Transaction Data* might be compromised by a shared technology attack exploiting a vulnerability on the *Hulda Hypervisor* in order to gain access from one OS to another OS.

Misuse Case 3.   The *Data Center IT Services Unit* does not configure the *Hulda Hypervisor* to ensure isolation of *Bank Customers* and *VIP Bank Customers* using the *Online Banking Service*.

Check for Missing Threats—The instantiated cloud pattern has to be analyzed for missing or incomplete threats. We use the marked cloud pattern introduced in Sect. 12.8.4 and check if the information in the instantiated cloud pattern leads to further cloud elements that have to be considered. For example, *Hulda* uses the *Hulda Cloudtable* and *Hulda Hypervisor*. These components are implemented in such a way that the *cloud table* optimizes itself using the configuration information in the *Hulda Hypervisor* and also has the ability to adapt the *Hulda Hypervisor* configuration regarding sharing of database resources. Hence, we have to include the *Hulda Cloudtable* in the threat analysis for the shared resources threat (Fig. 12.19).

Validate Misuse Cases—We propose to check the misuse cases via several validation conditions. We identified the following:

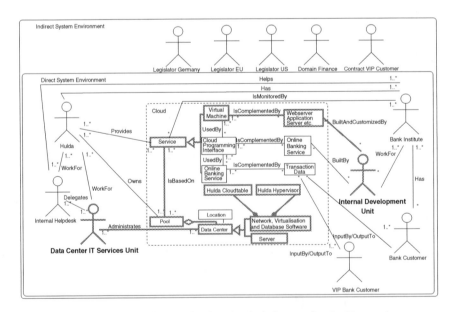

Fig. 12.19 Pattern-based cloud computing threat analysis for an online banking service

- Each misuse case has to address at least one element of the instantiated cloud pattern.
- Each misuse case has to refer to at least one asset.
- The decomposition of cloud components has to be conducted to a sufficient degree.
- Each misuse case has to consider at least one cloud component and one stakeholder.
- Each threat in the Cloud Threat Table (see Table 12.18) has to be considered in at least one misuse case.

Elicit Security Requirements—We use the information about threats collected in the previous steps and create at least one misuse case for each threat. Afterwards, the threat is used as a basis for eliciting a security requirement.

According to Fabian et al. (2010), a security requirement is typically a confidentiality, integrity, or availability requirement. It refers to a particular piece of information, the *asset* that should be protected, and it indicates the *counter-stakeholder* against whom the requirement is directed. A *stakeholder* is an individual, a group, or an organization that has an interest in the system under construction. Furthermore, the *circumstances* of a security requirement describe application conditions of functionality, temporal, spatial aspects, or the social relationships between stakeholders. Hence, circumstances have relations to functional requirements, stakeholders, etc., which shall be considered in the system-to-be. We use the elicited threats as inputs for misuse cases. We use them to derive security requirements and check for missing threats. We propose a table as introduced by Deng et al. (2011) that lists misuse cases and their corresponding security requirement for this step. In contrast to the work of Deng et al., we do not consider solutions in this step. We discuss them in Sect. 12.10 in relation to ISO 27001 security controls. We present exemplary misuse cases and security requirements in Table 12.19.

Table 12.19 From misuse cases to security requirements for our running example

Misuse case	Security requirement
1. The *bank institute* neglects to monitor the *internal development unit* during the development of the *online banking service*. The *bank institute* also did not conduct background checks of the team hired for the development. This results in multiple black mail attacks on bank customers perpetrated by members of the *internal development unit*	Conduct background checks of the members of the *internal development unit* and hire external auditors to monitor the work of the *internal development unit*
2. Integrity of *Transaction Data* might be compromised by a shared technology attack exploiting a vulnerability on the *Hulda Hypervisor* in order to gain access from one OS to another OS	Ensure the integrity of the *Transaction Data* is not harmed by side channel attacks caused by the *Hulda Hypervisor*
3. The *Data Center IT Services Unit* does not configure the *Hulda Hypervisor* to ensure isolation of *Bank Customers* and *VIP Bank Customers* using the *Online Banking Service*	The *Hulda Hypervisor* has to be configured such that isolation of all users of the *Online Banking Service* is ensured
...	...

The elicitation of security requirements concludes our threat analysis. Security requirements are of importance, because they allow a statement if they are fulfilled or not. Hence, if all security requirements of a cloud scenario can be fulfilled, we can state that the security level of a cloud system is sufficient.

The security requirements elicited are used for the establishment of an ISMS, which is a process. Several security requirements should lead to implementations in software, e.g., the integrity checks of transactional data in the second security requirements in Table 12.19. These requirements need to be refined with technical details, e.g., the software that conducts the integrity checks.

12.9 PACTS Step 5: Conduct Risk Assessment

Risk assessment is mentioned in numerous sections of the ISO 27001 standard. In our PACTS method, risk is used to asses if an asset requires an additional control or not. We provide a list of every mentioning in the standard in Table 12.20. In addition, we use the CORAS method (Sect. 2.5.2) for risk management in our PACTS method, because it is asset-based and we are able to integrate it in our work. Hence, CORAS is background and the integration into our work is the contribution of this section, which we explain in the following.

Preparation for the Analysis—We consider the description of use cases presented in Sect. 12.5 in this step. These are the initial descriptions of cloud use cases that also describe the target of the risk management.

Customer Presentation of Target—We use our cloud analysis pattern, templates, and business processes as presented in Sect. 12.6 as customer target description that considers the documentation of the previous step as an input. We show in detail how to conduct risk identification and analysis in Hüning et al. (2010) and Gsell et al. (2010).

Table 12.20 Risk assessment demands of the ISO 27001 standard

Significance for the ISO 27001 standard	The ISO 27001 standard states that managing risk by implementing security controls as a main goal of the process the standard creates. The standard mentions this already on p. 1 and risk assessment is a part of risk management
Related section(s) of the standard	Section 4.2.1b states that the ISMS policy has to align with the risk management. Section 4.2.1c demands a risk assessment that includes criteria for accepting risks and identifying the acceptable risk levels. Section 4.2.1d concerns risk identification and Sect. 4.2.1e demands risk analysis and evaluation. Section 4.2.1h demands management approval for acceptable levels of risk

Refine the Target Description using Asset Diagrams—The asset identification presented in Sect. 12.7 is used as an input for the CORAS asset diagrams. Moreover, the high-level security goals presented in Sect. 12.5 in policy patterns (see Sect. 12.10) can be input for high-level risk analysis, which are derived from these security goals.

Approval of Target Description—Risk evaluation criteria and likelihood scales have to be defined using CORAS. We explain in the following how to set up likelihood and consequences scales. Risk assessment can be conducted either quantitatively or qualitatively. Quantitative risk assessment demands that the likelihood and consequences scales contain numeric values. These have to express in which time frame a risk is likely to occur and what the consequences are in, e.g., the number of affected assets. Should these numbers not be available, likelihood and consequences tables can contain a qualitative scale that does not contain numbers. This qualitative scale is a starting point for risk assessment. We present an example of a qualitative likelihood scale in Table 12.21. For each of the direct assets we define a separate consequences table. We provide one example of a consequence scale for the *Pool* (see Fig. 12.19) considering availability in Table 12.22.

Table 12.21 Qualitative likelihood scale for clouds

Likelihood value	Description
Certain	A high number of similar incidents have been recorded; has been experienced a very high number of times by several users
Likely	A significant number of similar incidents have been recorded; has been experienced a significant number of times by several users
Possible	Several similar incidents on record; has been experienced more than once by the same actor
Unlikely	Only very few similar events on record; has been experienced by few actors
Rare	Never experienced by most actors throughout the total lifetime of the Cloud

---

Changes for ISO 27001:2013 Compliance

ISO 27001:2013 Section 6.1.2 concerns information security risk assessment. Section 6.1.2a shall establish risk acceptance criteria and criteria for performing risk assessment. Section 6.1.2b ensures that risk assessment produces comparable results. Risk identification is demanded in Section 6.1.2c, risk analysis in Section 6.1.2d, and risk evaluation in Section 6.1.2e. Section 6.1.3 concerns information security risk treatment. ISO 27001:2013 demands in Section 6.1.3a that controls from all possible sources can be selected, but the standard demands in Section 6.1.3b that these controls have to be mapped to controls in Annex A.

Table 12.22 Qualitative consequence scale for the cloud's pool

Consequence	Generic interpretation
Catastrophic	Permanent blackout (e.g., by loss of electricity); Can put the cloud provider *Hulda* out of business
Major	Large-scale outage for a short time; Can cause significant monetary reparations for the cloud provider *Hulda*
Moderate	Several outages of small scale for a short Time; Can cause monetary reparations for the cloud provider *Hulda*
Minor	Few outages for a short time; Tolerable if easy to recover from and if very rare
Insignificant	A single outage; No impact on operations of the cloud provider *Hulda*

Table 12.23 Risk evaluation matrix for the cloud's pool

		Consequence				
		Insignificant	Minor	Moderate	Major	Catastrophic
Frequency	Rare					
	Unlikely					
	Possible					
	Likely					
	Certain					

We present a risk evaluation matrix in Table 12.23. The red (or dark gray) parts indicate combinations of consequence and likelihood that result in unacceptable risks. The green (or light gray) combinations result in an acceptable risk.

Risk Identification using Threat Diagrams—We use our threat analysis presented in Sect. 12.8 as an input for CORAS threat diagrams. Our threats and misuse cases can be used to model the unwanted incidents and threat scenarios in CORAS.

Risk Estimation using Threat Diagrams—For all threats to assets relations the risks have to be evaluated using the tables created in the *Approval of Target Description* step of CORAS.

Risk Evaluation using Risk Diagrams—The risks are modeled in risk diagrams, which explicitly state the risks that are acceptable and which are unacceptable. This estimation is based on the risk evaluation matrix we created specifically for our cloud example (see Table 12.23). The step also considers discussions with the customer if the assets have acceptable risks or not.

Risk Treatment using Treatment Diagrams—For all assets that have unacceptable risk, controls have to be implemented or improved to create acceptable risk levels for these assets. The approach presented in Sect. 12.10, which instantiates policy patterns and selects appropriate security controls, can support this step.

12.10 PACTS Step 6: Create Security Policies and Reason About Controls

An important part of assembling an ISMS is to define security policies (see Table 12.24). They define security goals and state high-level solutions. Security policies are the basis for selecting and defining ISO 27001 controls, which are solutions for security goals.

A security policy has to be in accordance with business requirements and relevant laws and regulations as well as with contractual obligations regarding security. The policy has to be in written form and contain management approval. The policy document is also published and communicated to all stakeholders who the policy concerns. The policy shall also be reviewed at planned intervals and adapted to changes in the organization and its environment (ISO/IEC 2005b, p. 13).

The ISO 27001 standard defines also ISMS policies that are: "considered as a superset of the information security policy. These policies can be described in one document" (ISO/IEC 2005b, p. 4). In particular, the ISMS policy contains an alignment with risk management, risk evaluation criteria, and management approval. The security goals and high-level actions have to be aligned with these activities.

Table 12.24 Security policy demands of the ISO 27001 standard

Significance for the ISO 27001 standard	The ISO 27001 standard concerns high level ISMS policies during the establishment of the ISMS to guide the focus of security and security policies as controls that define in detail what a specific security control should achieve
Related section(s) of the standard	Section 4.2.1b concerns the definition of ISMS policies

---

Changes for ISO 27001:2013 Compliance

ISO 27001:2013 allows to select any control to treat a risk, but the controls have to be mapped to controls in Annex A of the standard (ISO/IEC 2013, p. 4). The controls in Annex A differ in ISO 27001 and ISO 27001:2013 slightly. We provide a mapping between the controls of ISO 27001 and ISO 27001:2013 in Appendix C that illustrate the differences.

Table 12.25 Controls of the ISO 27001 standard

Control name	Control objective	Important demands
A.5 Security policy	Provide directions for information security	Documentation and review requirements
A.6 Organization of information security	Manage security within the organization and with external parties	Clear management commitment, responsibilities, coordination, and independent consultation and review
A.7 Asset management	Achieve and ensure appropriate protection levels for assets	Identify assets, assign responsibilities for assets, classify assets, define and document rules for treatment of assets
A.8 Human resources security	Provide security training for employees, communicate responsibilities, provide structured exit procedures	Specify role and terms of employment, define responsibilities and provide security education and training, define disciplinary process, define termination responsibilities, return of assets and removal of rights
A.9 Physical and environmental security	Prevent unauthorized physical access, damage and interference to secure areas and equipment	Establish security parameter, physical controls for access to secure rooms. Equipment shall be protected, e.g., from power failure and the support for the equipment shall be ensured, e.g., by protecting cable connection from interference
A.10 Communications and operations management	Ensure secure operations of information processing, especially for service delivery from third parties, ensure availability, integrity, and confidentiality of information processing	Guidelines for processes, e.g., segregation of duties, and specific demand that ensure the goals, e.g., back up and monitoring of processes
A.11 Access control	Control the access to information	Ensure access control on information systems, networks, operating systems etc.
A.12 Information systems acquisition, development and maintenance	Embed security in information systems and prevent misuse of information	Specific measures are demanded, e.g., security requirements analysis, input/output data validation, use of cryptography, prevent information leakage, etc.
A.13 Information security incident management	Identify security events and weaknesses associated with information security and provide timely corrective action, ensure a consistent and effective approach	Ensure a reporting for security events and security weaknesses, learn from information security incidents

(continued)

Table 12.25 (continued)

Control name	Control objective	Important demands
A.14 Business continuity management	Protect critical business processes from effects of information system failures and ensure their timely resumption	Include security and risk management in the business continuity management process, reassess and test the business continuity plans
A.15 Compliance	Ensure compliance with laws, regulations, contractual obligations, security requirements, organizational security policies, and standards, consider system audits	Identify relevant laws, regulations, contractual obligations, etc., and also data and privacy protection measures, check the compliance to these laws, regulations, contractual obligations, etc., and use also audits to check compliance

We provide a pattern-based method to formulate security policies and select ISO 27001 controls to address the security concerns in the policies. Our approach also addresses changes of policies and an integration with risk management.

12.10.1   Controls in the ISO 27001 Standard

Annex A of the ISO 27001 standard describes the normative controls of the standard. We present a short overview of these in Table 12.25. The numbering of the controls starts with A.5 and ends with A.15. The reason for not starting the numbering with A.1 is that the control numbering shall align with the controls listed in the ISO 27002 standard. This standard provides guidelines on how to implement the controls and also further controls, but it is not normative.

12.10.2   A Method for Establishing ISO 27001 Policies

We propose a method for establishing ISO 27001 security policies and reasoning about ISO 27001 security controls (see Fig. 12.20). In the following, we describe its different steps.

Check Assets needs for Security Controls—We check the risk levels of each asset in this step. During the risk assessment, acceptable risk levels have been defined. If the risk level of an asset is above this threshold, a security control is required to decrease the risk level. The output of this step is a list of all assets that do not have an acceptable risk level.

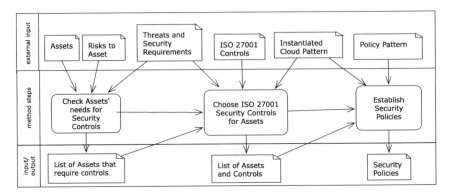

Fig. 12.20   A method for establishing ISO 27001 policies

Choose ISO 27001 Security Controls for Assets—Security controls shall reduce the risk that threats can harm assets. We use the list of assets created in the previous step and iterate over all the assets. We have elicited threats and security requirements for assets in Sect. 12.8. We consider the security requirements for an asset and iterate over them as well. We reason which of the ISO 27001 controls can fulfill a security requirement.

In order to ease the effort for selecting relevant ISO 27001 controls, we contribute an overview of the controls in Table 12.25. The first column contains the *Control Names*, the second column contains the *Control Objectives*. These are taken from the ISO 27001 standard. The last column states *Important Demands*, which we have taken partially from the standard and included also demands that we consider useful. Beside the *Important Demands*, we consider the column *Control Objective* contains relevant terms, which we can try to relate to the security requirement. We also reason why the remaining controls do not fulfill this requirement. This results in a list that explains, which controls are relevant for each security requirement and which are not. We use this list as an ISO 27001 specific document, the so-called *Statement of Applicability*. The *Statement of Applicability* has to contain a reasoning about the selection of ISO 27001 controls.

Establish Security Policies—We specify security policies using our policy pattern (see Fig. 12.21), which we explain in the following. The pattern describes the structure of a security policy considering ISO 27001 controls and our cloud pattern. The pattern is composed of four main parts, described in the following.

The *Policy Scope* contains at least one *Addressed Cloud Stakeholder*, who is a *Cloud Stakeholder* from the cloud pattern, who is addressed in the policy. The *Addressed Cloud Stakeholders take part in* an *Activity* that *uses Data* or *Cloud Elements*. Data can be any data, while a cloud element refers to an element of the cloud in the cloud system analysis pattern (Sect. 12.6). The *Addressed Cloud Stakeholders* also *use or own Data or Cloud Elements*. The pattern can also contain

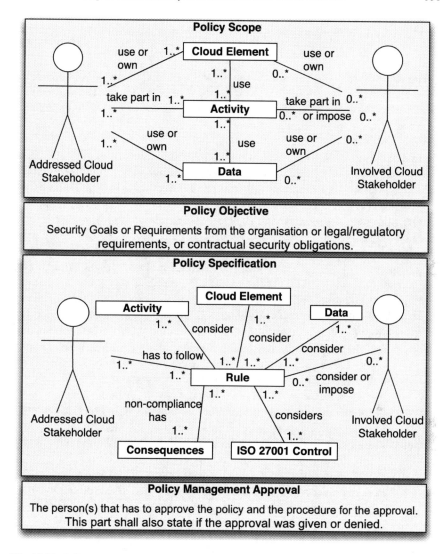

Fig. 12.21 Policy pattern

Involved Cloud Stakeholders that can *take part in* or *use or own* a *Data or Cloud Element*. The security expert has to check if a *Cloud Stakeholder* is an *Addressed Cloud Stakeholder* or an *Involved Cloud Stakeholder*. The policy pattern has to be instantiated for each scenario using the instantiated cloud pattern, e.g., Fig. 12.7.

The *Policy Objective* states the goal and security requirements of the policy. Security requirements are refinements of security goals (Fabian et al. 2010). These can originate from organizational goals, laws and regulations, or business contracts.

These goals are formulated in natural language and shall only refer to the elements of the *Policy Scope*. The goals have to be in alignment with the goals of the ISMS policy (Sect. 12.5).

The *Policy Specification* addresses the *Addressed Cloud Stakeholder* who *has to follow Rules*. These *Rules consider* at least one *ISO 27001 Control*. The *noncompliance* of these *Rules* has *Consequences*. *Rules* also *consider* at least one *Activity* and *Data* and *Cloud Element*. A *Rule* can also consider *Involved Cloud Stakeholders*. The elements *Activity, Cloud Element, Data* and *Involved Cloud Stakeholder* have to occur in the *Policy Scope*.

The *Policy Management Approval* considers the demands of the ISO 27001 standard for management commitment (e.g., ISO 27001 Sect. 4.2.1 b). The *Policy Management Approval* contains a statement of who is responsible for approving the policy as well as the procedure for approval. The *Policy Management Approval* also contains the date of the approval.

We analyzed the controls in Table 12.25 for relations to the cloud pattern and in order to ease the identification of possible stakeholders for the instantiation of our policy pattern.

We present the results in Table 12.26. The column *Control Name* states the name of each control, the column *Addressed Stakeholder* lists possible stakeholders as candidates from the cloud system analysis pattern for *Addressed Stakeholders* in our policy pattern. The column *Action* lists activities related to a control. These can help to instantiate the *Activity* element in the policy pattern. The column *Cloud Element* lists potentially relevant cloud elements for these controls.

The policy patterns ease the effort of writing policies. For example, the provided elements in the patterns help not to miss an important element of a policy. Moreover, several parts of the policy pattern use elements of the cloud system analysis pattern and descriptions created in previous sections. This should improve the consistency between the different documents for the ISO 27001 standard.

Moreover, the instantiated patterns are a vital part of the ISMS specification, because they define the decisions for the protection of each asset within the cloud scenario. They are also the basis for refining high level ISO 27001 controls into concrete security solutions.

12.10.3 Cloud Security Alliance: Cloud Controls Matrix

The Cloud Security Alliance (CSA) provides an MS Excel spreadsheet called Cloud Controls Matrix (CCM)[16] (see Fig. 12.22)[17], which provides security controls the

---

[16]Cloud Security Alliance-Cloud Controls Matrix Version 3.0.1: https://cloudsecurityalliance.org/research/ccm/.

[17]Reproduced by permission of the Cloud Security Alliance. Please contact the Cloud Security Alliance (Europe), 34 Melville Street, Edinburgh, EH3 7HA, United Kingdom for the definitive version of this document.

Table 12.26 Controls of the ISO 27001 standard and their relations to the cloud system analysis pattern

Control name	Addressed stakeholder	Action	Cloud element
A.5 Security policy	All	All	All
A.6 Organization of information security	All	Security management activity e.g. clear management commitment	None
A.7 Asset management	Cloud provider, cloud customer, end customer	Activities regarding identify, classify and protect assets	Data, software product, cloud software stack, etc.
A.8 Human resources security	Cloud support, cloud administrator, cloud developer	Activities regarding training, responsibility assignment, designing and implementing exit procedures etc.	None
A.9 Physical and environmental security	Cloud provider, cloud administrator	Activities regarding concerning physical access and prevention of damage/ interference of hardware	All that are physical e.g. hardware
A.10 Communications and operations management	Cloud provider, cloud administrator, cloud support, cloud developer, cloud customer	Activities regarding guidelines for processes, e.g., segregation of duties	All
A.11 Access control	Cloud support, cloud developer	Activities regarding implement and monitor access to information	All that are software
A.12 Information systems acquisition, development and maintenance	Cloud administrator, cloud software developer	Activities regarding eliciting of security requirements and vulnerability detection e.g. penetration testing and specific measures, e.g., cryptography	All that are software
A.13 Information security incident management	cloud administrator, cloud support, cloud developer	Activities regarding reporting security events and issues, ensuring a consistent and effective response, learning from security incidents	All

(continued)

Table 12.26 (continued)

Control name	Addressed stakeholder	Action	Cloud element
A.14 Business continuity management	Cloud administrator, cloud support, cloud developer	Activities regarding business continuity management for business processes, e.g., security and risk management	All
A.15 Compliance	cloud administrator, cloud support, cloud developer cloud provider, cloud customer, cloud end customer	Activities regarding identifying laws, regulations and contractual obligations. privacy protection, monitor compliance to the laws regulations and contractual obligations, compliance audits	All

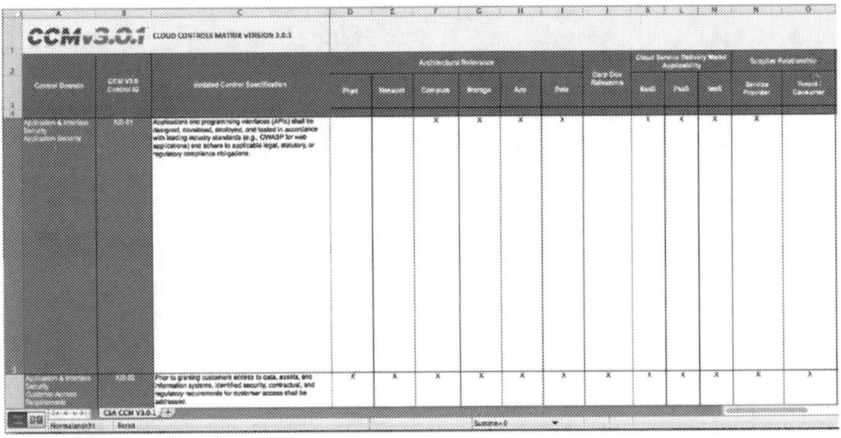

Fig. 12.22 CSA-CCM screenshot (see footnote 17)

CSA defined such as *Application &Interface Security Data Integrity*, which have a so-called control specification that explains the details of the control. For example, the mentioned control is specified as follows in version 3.0.1 of CCM "Data input and output integrity routines (i.e., reconciliation and edit checks) shall be implemented for application interfaces and databases to prevent manual or systematic processing errors, corruption of data, or misuse."

The controls are mapped to several cloud system attributes. First, to one of the following parts of computer architectures: Physical, Network, Compute, Storage, App, and Data. Secondly, the controls are mapped to cloud service models: SaaS, PaaS, and IaaS. Thirdly, a mapping exists to supplier relationships: service provider and tenant/consumer. In contrast, the CSA-CCM does not explain relations between these elements in a comprehensive model, such as our cloud system analysis pattern.

Furthermore, the CSA controls are mapped to numerous standards and regulations such as ISO 27001, ISO27001:2013, COBIT, IT-Grundschutz, etc. These mappings can be used within our PACTS method to find refinements or similar controls to the ones selected in ISO 27001 or ISO 27001:2013. For example, the control *Application and Interface Security Data Integrity* in CCM is mapped to ISO 27001 control: *A.11.5.6* (and other ISO 27001 controls), as well as to COBIT 5.0 control APO09.03 (among other COBIT 5.0 controls) (Table 12.27).

We present the following mapping from the cloud architectural, service, and supplier mappings to our cloud system analysis pattern (see Table 12.28). This mapping can be used to identify the cloud pattern's parts or stakeholders to which a particular CCM control refers.

12.10.4 Application of Our ISO 27001 Policy Method to Our Running Example

Check Assets' Needs for Security Controls—We have conducted the risk analysis of the asset *Hulda Hypervisor*, and the resulting risk level for the *Hulda Hypervisor* is unacceptable. This is the resulting of the risk assessment in alignment with the customer. In our example, the unacceptable risk level is caused by the *Shared Technology* threat (Sect. 12.8). We consider the following security requirement in our example, which addresses this threat (see Table 12.19): The *Data Center IT Services Unit* does not configure the *Hulda Hypervisor* to ensure isolation of *Bank Customers* and *VIP Bank Customers* using the *Online Banking Service*.

Choose ISO 27001 Security Controls for Assets—We consider all the controls listed in Table 12.25 for our security requirement (see Table 12.19): The *Hulda Hypervisor* has to be configured such that isolation of all users of the *Online Banking Service* is ensured. We argue that all controls are relevant and finish this step (see Table 12.27). If we had found out that a control is not relevant, a solid argument would have to be presented why this is the case.

Establish Security Policies—We instantiate our policy pattern (see Fig. 12.21) for each of the security requirements identified during the threat analysis. We show an example instantiation of our policy pattern in Fig. 12.23. We focus on the *Data Center IT Services Unit*. In this policy the involved stakeholders are the *Bank Customer* and the *VIP Bank Customer*. The policy addresses the requirement that the configuration of the *Hulda Hypervisor* has to be protected. The policy specification states that only the *Data Center IT Services Unit* is allowed to change the configuration of the *Hulda Hypervisor*. In our example, a reconfiguration of the *Hulda Hypervisor* is done on behalf of the *VIP Bank Customer*. Our example policy considers the control A5 in particular, therefore it is printed in bold. The other controls have to be implemented in subsequent steps. The policy was approved by Mr. Jones, who is responsible for establishing the ISMS (Sect. 12.5).

Table 12.27 Control reasoning for the ISO 27001 standard for our running example

Control name	Control reasoning
A.5 Security policy	We require a clearly defined policy for access to the *Hulda Hypervisor*'s configuration data. We address this control by instantiating a policy pattern
A.6 Organization of information security	Organizational demands for defining processes of how change requests regarding the *Hulda Hypervisor* have to be satisfied. For example, an end customer might wish to join two online banking accounts and the hypervisor shall merge the related resources. In this case, the end customer needs a way to request these changes
A.7 Asset management	We have already addressed this control in Sect. 12.7 and the asset *Hulda Hypervisor* is documented and responsibilities have been assigned to it
A.8 Human resources security	We require security training for the *Data Center IT Services Unit*, which should be the only stakeholder allowed to change the configuration data of the *Hulda Hypervisor*. The *Data Center IT Services Unit* has to be trained in the access control mechanisms of the *Hulda Hypervisor* and how to communicate with other cloud stakeholders about change requests
A.9 Physical and environmental security	The *Hulda Hypervisor* is software, but the software runs on a physical hardware server. We have to protect this server and need a control for it. For example, the server room shall be in a closed room with an emergency power supply. The emergency power supply shall keep the servers working during power shortages. In addition, the room has to be locked in order to prevent unauthorized access to the server
A.10 Communications and operations management	We have to define processes for accessing the *Hulda Hypervisor* as well as maintaining and repairing it
A.11 Access control	The used mechanism to ensure access control has to be chosen and described in detail. For example, the mechanisms defined in the ANSI standard for role-based access control (?, ?) could be used for this purpose. Moreover, for clouds the XACML (?, ?) standard for extensible and XML-based access control can be relevant

(continued)

Table 12.27   (continued)

Control name	Control reasoning
A.12 Information systems acquisition, development and maintenance	We use this control to check the implementation of our chosen security mechanisms for access control, e.g., via penetration testing (?, ?) of a particular software solution
A.13 Information security incident management	For each possible security incident a process has to be defined, so that the *Data Center IT Services Unit* can react accordingly. The possible incidents can be derived from the threats and security requirements elicited in Sect. 12.8
A.14 Business continuity management	The selection of the cloud provider *Hulda* was done carefully by a bidding process among different providers. The availability values and response times for incidents were compared to multiple providers. *Hulda* offered the best values and could convince with a qualified 24x7 support team in place, which the customer can call when there is a major incident or outage. Furthermore, *Hulda* could provide disaster recovery on the hardware side, e.g., VMware HA (?, ?) and VMware Vmotion (?, ?). *Hulda* also provided proof of the existence of a fallback data center
A.15 Compliance	We have to identify relevant laws and regulations using the method presented in Sect. 12.12

12.10.5 Consistency Checks

We propose to check instantiated policy patterns via several validation conditions. We identified the following so far:

- All cloud stakeholders that are referred to in the policy pattern have to occur in the cloud pattern.
- All controls in the policy pattern have to occur in ISO 27001 Annex A, or a reason has to be given why none of them is applicable.
- All cloud elements that are referred to in the policy pattern have to occur in the cloud pattern.
- Each policy has to refer to at least one asset, and the asset has to be referred to in at least one threat.
- The policy scope, policy objective, and policy specification have to refer to the same asset and the same stakeholders.

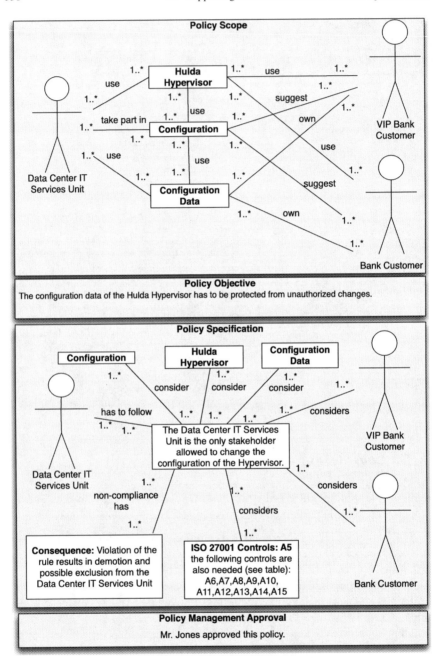

Fig. 12.23 Example instance of our policy pattern

Table 12.28   Mapping cloud pattern to CSA-CCM

Cloud pattern	CCM terminology
Indirect environment	
Legislator	–
Domain	–
Contract	Corp Gov relevance
Direct environment	
Cloud provider	–
Cloud support	–
Cloud administrator	–
Cloud customer	Tenant/Consumer
Cloud developer	Service provider
End customer	Tenant/Consumer
Cloud: Service	
IaaS	IaaS
PaaS	PaaS
SaaS	SaaS
Cloud software stack	–
Software product	App
Data	Data
Cloud: Pool	
Resource	–
Location	–
Software	Compute, network
Hardware	Phys, network
Hypervisor	
Cloud database	Storage
Stakeholder templates	
Name	–
Description	–
Motivation	–
Relations to other stakeholders	–
Relations to the cloud	–
Relations to other direct stakeholders	–
Assets	–
Compliance and privacy	–

12.10.6 Policy Change Pattern

We also present our policy change pattern (see Fig. 12.24), which allows stakeholders to propose changes for an existing policy pattern. The change pattern has several parts. The *Change Scope* describes the scope of the policy after the change. *Reasons*

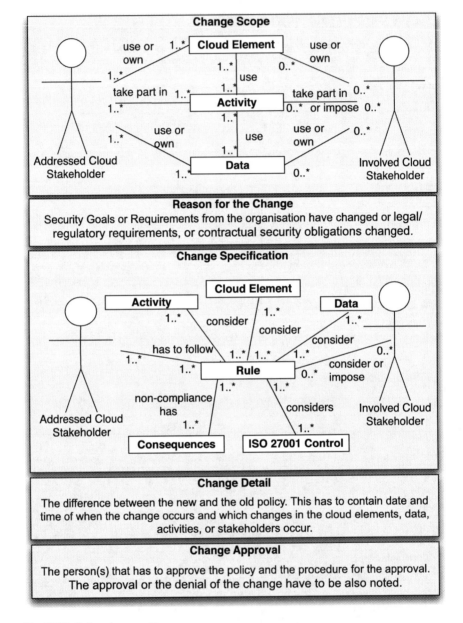

Fig. 12.24 Policy change pattern

for the Change define the security goals or requirements that cause the change. The *Change Specification* describes the security policy specification after the change, and the *Change Detail* describes the difference between the new and the old policy. The *Change Approval* states who has to approve the change and how it has to be approved.

12.11 PACTS Step 7: Design ISMS Specification

The ISO 27001 standard demands a documentation of the ISMS (see Table 12.29). Section 4.3.1 demands a documentation of several parts of the ISMS. A list is provided in the column *ISO 27001 Documentation* of Table 12.30. These "documents and records may be in any form or type of medium". (ISO/IEC 2005b p. 8). Hence, we propose a mapping in Table 12.30 of the generated artifacts from our method to the documentation demands of the ISO 27001 standard. The column *Support from our Method* states the artifact that relates to a specific part of the ISMS. The column *Sections* states the section of our work that describes how to create the artifacts mentioned in the column *Support from our Method*.

We describe our mapping in the following (see Table 12.30). We use our policy pattern and the attached management commitment templates to document the ISMS policies and objectives. The scope and boundaries of the ISMS are documented using our cloud system analysis pattern, and the procedures and controls are documented as part of our chosen security controls. The risk methodology is our described risk approach, and the risk assessment report uses the asset identification, threat analysis, and risk analysis approaches. The risk treatment plan contains the risk estimation for each asset and the established controls to reduce the risk to acceptable levels.

The information security procedures are documented using our policy patterns. The control and protection of records is the documentation of the selected security solution for the control *A.10.7.4 Security of system documentation*, which describes the protection of the ISMS documentation against unauthorized access. This control has to be referenced in one of our policy pattern instances and a solution has to be implemented for it (Table 12.31).

Table 12.29 Documentation demands of the ISO 27001 standard

Significance for the ISO 27001 standard	The ISO 27001 standard requires documentation in any form or medium to ensure the ISMS produces satisfying results
Related section(s) of the standard	Section 4.3 lists documentation demands of the standard. The section contains subsections for general documentation concerns, control of documents and control of records. Records provide evidence for conformity with the ISMS requirements and proof of the effective operation of the ISMS

Changes for ISO 27001:2013 Compliance

The documentation demands for ISO 27001:2013 can be found in Table 12.31.

Table 12.30 Support of our method for ISO 27001 documentation demands

ISO 27001 documentation	Support from our method	Sections
ISMS policies and objectives	Management approval and policy pattern	12.5
Scope and boundaries of the ISMS	Cloud system analysis pattern	12.6
Procedures and controls	Documentation of security controls	12.10
The risk assessment methodology	Our risk methodology	12.9
Risk assessment report	Asset identification, threat analysis, and risk assessment	12.7, 12.8 and 12.9
Risk treatment plan	Risk assessment and control selection	12.9 and 12.10
Information security procedures	Policy pattern	12.10
Control and protection of records	Security solution concerning the control *A.10.7.4 Security of system documentation*	12.10
Statement of applicability	Reasoning about controls	12.10

The *Statement of Applicability* defines the implemented controls for each asset. It also contains a reasoning for the controls not selected for this asset (Sect. 12.10). The results of the legal compliance (Sect. 12.12) analysis and privacy analysis (Sect. 12.13) are cross-cutting concerns and considered in all documents.

12.12 Considering Legal Compliance in the PACTS Method

The ISO 27001 standard mentions the importance of considering law already on page 1 of the standard. We listed all sections that demand legal consideration in Table 12.32. The section on how to establish an ISMS explicitly states that legal obligations have to be considered during the definition of the ISMS policy and the selection of controls. The standard also has an explicit control for legal compliance. The controls for security policy and human resource security state the explicit consideration of laws when applying the control. In a previous work (Beckers et al. 2012), we developed a pattern-based approach for identifying relevant laws for a software engineering project. We presented the *law pattern* for structuring laws. We structured requirements in a similar pattern, the *law identification pattern*. This allowed us to use a matching algorithm from requirements to laws, because requirements and laws were documented in a similar structure. In a subsequent work (Beckers et al. 2012), we integrated this work with our cloud pattern. We used the information in the pattern to instantiate the law identification pattern. This work (Beckers et al. 2012) also contains a detailed description of how to map laws to requirements as part of a structured method for identifying relevant laws.

Table 12.31 Support for ISO 27001:2013 documented information demands

ISO 27001:2013 documented information	Support from our method	Sections
Information security policy	Management approval and policy pattern	12.5
Information security objectives	Management approval and policy pattern	12.5
Scope of the ISMS	Cloud system analysis pattern	12.6
Risk assessment and treatment methodology	Our risk methodology	12.9
Risk assessment results	Asset identification[a], Threat analysis[a], and risk assessment	12.7, 12.8 and 12.9
Risk treatment results	Risk assessment and control selection	12.9 and 12.10
Statement of applicability	Reasoning about controls	12.10
Evidence of corrective actions	Documentation of security controls	12.10
Procedures to the effective planning	Policy pattern	12.10
Competence records	Management approval template	12.5
Monitoring and measuring results	Documentation of security controls	12.10
Audit programme and results	Documentation of security controls	12.10
Management review results	Documentation of the management review activities and their results	12.5

[a] Note that these steps are not mandatory for ISO27001:2013, but recommend to support security reasoning

Table 12.32 Legal compliance in the ISO 27001 standard

Significance for the ISO 27001 standard	The consideration of legal obligations is already mentioned in the first paragraph on page 1 of the standard. It is stated that the compliance to the standard does not confer immunity for legal obligation. Hence, legal obligations have to be known in order to be able to follow this demand
Related section(s) of the standard	Section 4.2.1b concerns the definition of an ISMS policy that includes the consideration of legal compliance. Section 4.2.1g concerns the selection of controls, which demand also legal and regulatory compliance

---

Changes for ISO 27001:2013 Compliance

ISO27001:2013 states in Section 4.2 *Understanding the needs and expectations of interested parties* that the organizsation shall determine the requirements of interested parties, which may include legal and regulatory requirements. Moreover, the control *A 18.1—Compliance with legal and contractual requirements* concerns the consideration of legal requirements.

We conducted a literature review regarding clouds and laws. Section 12.12.1 presents the results of this review and shows in particular the difficulty of identifying relevant laws. We describe how our law identification process works in Sect. 10.7. We also propose a novel modification of the law identification process, which integrates our law identification method with our PACTS method. In Sect. 12.12.3 our novel process is applied to the running example. Section 12.12.4 describes how to derive controls from the results of the law identification process.

12.12.1 Overview on Compliance Issues of Clouds

A PriceWaterhouseCoopers study from 2010[18] reveals that identifying compliance requirements is a significant challenge for compliance management in clouds. Compliance requirements are requirements derived from relevant laws or regulations. In particular, we identified the following cloud compliance issues:

Identification of relevant Laws—The identification of relevant laws and the elicitation of legal requirements for a software system is essential in order to be *compliant*. This is considered to be difficult, because it is a cross-disciplinary task in laws and software and systems engineering (Biagioli et al. 1987). This task has a significant complexity for clouds. Due to the number of different stakeholders, functionalities and locations, a high number of different laws have to be considered.

Data Location and Deletion—Cloud providers are often not able to provide detailed information on the location of their customers' data (Jansen 2011). This is relevant, e.g., to obey privacy laws. In some cases the information is available, but not disclosed to the cloud customer (Jansen 2011). Another open question is how a cloud provider can prove that data has been deleted (Chow et al. 2009), which is also relevant for compliance to privacy laws.

Choosing a Legislation—Cloud providers and customers are often located in different countries. In this case, the laws of the cloud provider's country are usually relevant. However, cloud providers and customers can agree on using the laws of one country for their cloud business. Furthermore, contracts have to fill the gap between the chosen law and the law that is not chosen (Duisberg 2011). The chosen law is

---

[18] http://www.pwc.de/en/prozessoptimierung/trotz-einiger-bedenken-der-virtuellen-datenverarbe itung-gehoert-die-zukunft.jhtml.

binding to all cloud stakeholders. Thus, understanding the laws of the different countries and making an informed choice is essential and presents a significant challenge.

Contractual Obligations—Contracts are also used to define the ramifications of violations of the clouds' SLAs. In addition, contracts have to fill the gap between the agreed law for a cloud system and national law of the cloud stakeholders (Chow et al. 2009). Detecting these gaps completely is a challenge, because of the lack of methods to support this task.

Subcontractor Issues—The previous issues multiply in complexity when the cloud provider can use subcontractors, e.g., from another country. Moreover, it is hardly possible for cloud customers to detect that their data has been processed by a third party (Chow et al. 2009).

Audibility—The use of distributed computing environments, spread all over the globe, provides a challenge for auditing demands (Chow et al. 2009).

The cloud pattern can be used to elicit relevant information for *Contractual Obligations*, *Subcontractor Issues*, *Audibility*, *Data Location and Deletion*, because the relevant direct and indirect cloud stakeholders and cloud elements including their location are part of our extended cloud pattern. Moreover, our pattern is extensible and further attributes, stakeholders, or cloud elements can be included with little effort.

In this work, we focus on identifying relevant laws and regulations that affect IT risk management and privacy. Hence, we tackle in particular the *Identification of relevant Laws* issue. The *Choosing a Legislation* issue can also be supported by our main focus, because one can identify relevant laws for a cloud project using our method. The resulting requirements from these laws can provide the means for an informed decision regarding a legislation.

In our running example, we chose the German law as the binding law. However, we believe that our law identification and analysis method is also valid for laws of other nations. In order to give an idea of the number of laws, regulations, and standards that would have to be considered, we present the following list, which could be extended even further:

> Law on Monitoring and Transparency in Businesses (KonTraG), Stock Corporation Act, (AktG) German banking act (KWG), Securities Trading Act (WpHG), Minimum Requirements for Risk Management (MaRisk), Commercial Code (HGB), Tax Code (AO), State Data Protection Acts (LDSG), Telemedia Act (TMG), Federal Data Protection Act (BDSG).

From this short survey one can recognize that even for our small running example a huge number of laws might become relevant. This fact emphasizes the need for an engineering method for the identification of relevant laws and their analysis. We focus on relevant compliance regulations for privacy in our running example. We only explain the laws and regulations that we use in the example.

In 1995, the European Union (EU) adopted the *Directive 95/46/EC* on the processing of personal data that represents the minimum privacy standards that have to be included in every national law. Germany implements the European Privacy Directive in the *Federal Data Protection Act (BDSG)*. According to *Section 1 BDSG* all private and public bodies that process, store, and use personal data have to comply with the BDSG. IT systems have increased the feasibility of unwanted disclosure, because storage capacity and speed of computers allow to store, search and correlate data. *Section 9 Sentence 1 BDSG* states different requirements that have to be fulfilled by technical and organizational measurements for protecting personal data, e.g., physical and virtual access control to data and the separation of storing and processing data collected for different purposes. Furthermore, it must be verifiable whether personal data has been deleted, and by whom and that data has only been processed with the permission of the customer.

Moreover, the EU law as well as *Section 4b BDSG* forbid sharing data with companies or governments in countries that have weaker privacy laws. For exchange with companies in the United States (US), there exists the *Safe Harbor* agreement. But under the *US Patriot Act*, officials could access information about citizens of other countries if that information is physically located within the US or accessible electronically. The priority of the Patriot Act has never been explicitly tested in court, but is a risk for bringing privacy-critical data into the cloud as data centers can be technically distributed worldwide. As cloud computing is considered as contracted data processing, the cloud customer is responsible to adhere to the complete BDSG, according to *Section 11 BDSG*. The law further defines the contract between customer and outsourcing provider. For example, after ending the contract all data has to be deleted.

12.12.2 PACTS Step 8: Identify Relevant Laws and Regulations

In the context of our PACTS method, the overall requirements engineering process (cf. Beckers et al. 2012) is not of central relevance. Hence, we present here only a brief description of the steps of our law identification method. We propose to store laws in a specific structure, which helps us to search for them. The structure is our so-called *law pattern*. The structure is derived from legal literature and discussed in detail in Sect. 10.7. We also introduced law patterns and a method of how to use them, which we apply in this part of the PACTS method. We modify the method for a better integration with PACTS.

The original law identification method is designed to be embedded into a full requirements engineering process (Sect. 10.7). At this level, using this method results in a detailed set of relevant law sections and their relation to requirements. Hence, the impact of laws on the system-to-be can be easily derived and reflected. But there are scenarios in which using an RE process is not feasible and, hence, detailed requirements are not available. For example, when only a preliminary ISO 27001 complaint ISMS documentation is the aim, or legacy systems are part of the ISMS.

In this section, we present a modified version of our law identification method that relies on the artifacts created in our PACTS method. Note that the only step modified

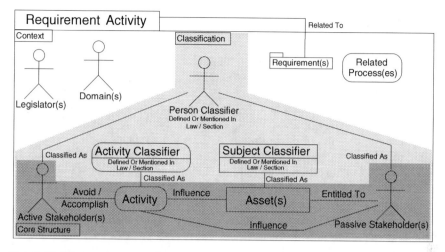

Fig. 12.25 Law identification pattern

is the *Instantiation of Law Identification Pattern (Core)*, which we describe in the following. The remaining steps have to be executed as described in Sect. 10.7.

Instantiation of Law Identification Pattern (Core)—We instantiate the core part of our law identification pattern, which is the dark gray area in Fig. 12.25. We use the instances of the following artifacts of our PACTS method for the instantiation of the core part of our law identification pattern:

- our cloud pattern (Sect. 12.6)
- our cloud asset table (Sect. 12.7)
- the modeled processes (Sects. 12.6 and 12.7).

We consider all direct stakeholders of our cloud pattern and select one at a time. The next part is an iteration over each of their assets by considering our asset templates. We select all processes in which an asset appears, as well. For each process, we select those activities that are related to the asset and also executed by the selected stakeholder. We use this information to instantiate the core part of a law identification pattern for each activity.

We use the following mapping between elements of our law identification pattern and the artifacts of our PACTS method:

Active Stakeholder The active stakeholder of core structure corresponds to the selected direct stakeholder.

Activity The activity of the core structure corresponds to the currently selected activity.

Asset The asset of the core structure corresponds to the currently selected asset.

Passive Stakeholder The passive stakeholder of the core structure corresponds to the asset provider of the currently selected asset.

12.12.3 Example

We present an example of our modified law identification method and apply the method to our running example in the following.

Instantiation of Law Pattern—We illustrate our approach by an example based on Section 4b of the German *Federal Data Protection Act (BDSG)*. Hence, we fill our database with all laws of the BDSG in order to be able to discover dependent laws. Then, we use our *law identification pattern* for our running example.

The resulting law pattern instance is depicted in Fig. 12.26. The light gray words close to an element of an instance refer to the type of the element in the original pattern. We consider the transfer of data outside of Europe. For example, in Fig. 12.26 the light gray words *Activity Classifier* near *Abroad Transfer* indicate that the *Abroad Transfer* element is an instantiation of the *Activity Classifier* element in the original pattern.

We instantiate BDSG Sect. 4b, which refers to further sections of the BDSG, e.g., *BDSG Sect. 1*, which we also instantiate. This is noted in the *Context* part of our pattern (white area in Fig. 12.26, cf. Beckers et al. 2012). The *Legislator(s)* and *Domain(s)* can be instantiated according to the considered legislators (e.g., *Germany* and *General Public* in the *Context* part.). We instantiate *Activity* with *Abroad Transfer*. *Addressee*, *Target Subject*, and *Target Person* are instantiated using the related Sect. 1 BDSG. Finally, we instantiate the *Classification* part (light gray area in Fig. 12.26). We discuss the hierarchies for the law in Beckers et al. (2012). For example, one of the hierarchies states that a *Natural Person* is a specialization of *Authority*. The hierarchies in Fig. 12.27 (cf. Beckers et al. 2012) are updated with *Transfer*, defined in Sect. 3 BDSG, with a specialization *Abroad Transfer*.

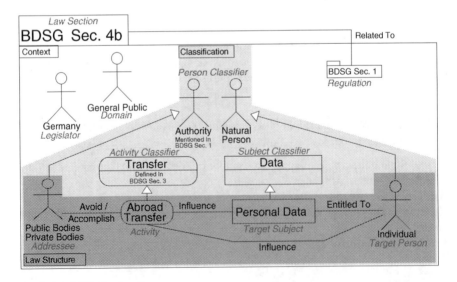

Fig. 12.26 BDSG Section 4b

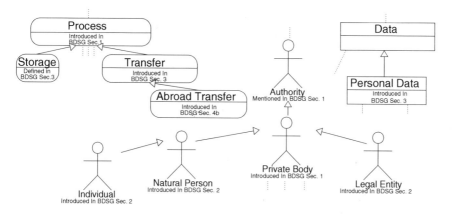

Fig. 12.27   Hierarchies for person (*bottom*), subject (*upper right*), and activity (*upper left*)

Table 12.33   Asset template instance (excerpt from Table 12.13)

Asset	Asset owner	Asset use	Asset class	Asset label
Transaction data	Mr. Jones	... Fig. 12.15 ...	Data	AS_DA_120

Instantiation of Law Identification Pattern (Core)—After instantiating the BDSG law, we consider the information of our ISMS. For our example, we select *Hulda* as *direct stakeholder* from the cloud analysis pattern instance (see Fig. 12.7). For Hulda, we select the corresponding *asset template instances*, which are described in Sect. 12.7. From these instances, we select the *Transaction Data* entry (see Table 12.33). From the entry we derive that the *asset* is *Transaction Data*. For this asset, we select *VIP Bank Customer, and Bank Customer* as *Asset Providers* from the cloud asset table (see Table 12.34). The related *process* is described as activity diagram (see Fig. 12.15). This process describes several *activities*, which are executed by the cloud offered by *Hulda*, and which are related to the transaction data. One of those activities is *Store Data Distributed*. We select this activity and instantiate the law identification pattern core (see Fig. 12.28).

Considering our mapping in Fig. 12.28, we instantiate the *Active Stakeholder* with our direct stakeholder *Hulda*. For the *Activity* we instantiate *Store Data Distributed*. The *Asset* is the *Transaction Data*. And the *Passive Stakeholders* are our asset providers *VIP Bank Customer* and *Bank Customer*.

Full Instantiation of Law Identification Pattern—We use processes, activities and assets documented as described in Sect. 10.7 for fully instantiating the law identification pattern. Figure 12.28 presents an example. We instantiate the legislators *Germany, US, EU*, the domain *Finance*, as well as the process *Offering Transaction Data Processing*. The activity *Store Data Distributed* is classified as *Abroad*

Table 12.34 Cloud asset table (excerpt from Table 12.12) Instance

Asset	Asset provider	Asset reasoning
	...	
Transaction data	VIP bank customer, bank customer	Harm to the asset can possibly cause financial loss and privacy violation to the *VIP Bank Customer, Bank Customer*. The harm depends on the kind of data
	...	

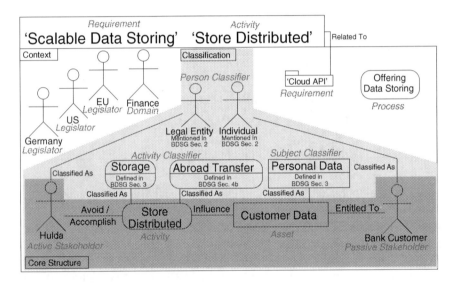

Fig. 12.28 Law identification pattern instance

Transfer and *Storage, Transaction Data* is classified as *Personal Data, (VIP) Bank Customer* is classified as *Individual*, and *Hulda* is classified as *Legal Entity*, based on a discussion between legal experts and software engineers.

Pattern Matching—The matching of the classification area of the law identification pattern instance (light gray area) and the law structure and classification part of the law pattern instance (light and dark gray areas) reveal relevant laws. This law identification uses the previously documented hierarchies. The matching is successful for *Abroad Transfer, Personal Data*, and *Individual*. Hence, we have a match between our law identification pattern instance shown in Fig. 12.28 and our law pattern instance of Sect. 4b BDSG, depicted in Fig. 12.26. The matching for *Hulda*, who is classified as *Legal Entity*, uses the hierarchy depicted in Fig. 12.27. This reveals that *Legal Entity* is a specialization of *Private Bodies*, which results in identifying Sect. 4b BDSG as relevant.

Law Revision—The resulting set of relevant laws has to be validated by lawyers. The lawyers confirm that the BDSG is indeed relevant for the scenario.

12.12.4 PACTS Step 9: Define Compliance Controls

Finally, we have to integrate results of our PACTS Step 8 to refine the ISO 27001 *A.15.1 Compliance with Legal Requirements*. This step of our PACTS method describes how to map the instructions from laws to requirements. We distinct between functional requirements and nonfunctional requirements. Functional requirements describe "what the system does" (Summerville 2007, p. 119) and nonfunctional requirements describe global requirements, e.g., reliability and maintainability on the system-to-be (Summerville 2007). In our view, security requirements are nonfunctional requirements. Software engineers refine these requirements into software specifications. These are implementable requirements.

The content of laws can be translated into functional and nonfunctional requirements or a restriction to a functional requirement. For example, the appendix to BDSG Section 9 demands specific methods, e.g., access control. These have to be part of the software specifications, while other laws shall be transformed into nonfunctional requirements, e.g., Section 17 TKG demands confidentiality of information. This would have to be transformed into a security requirement. Further laws simply demand a specific functionality that shall be transformed into a functional requirement. For example, the appendix to BDSG Section 9 also demands that the passing on of personal information has to be controlled. This leads to a functional requirement that states that all transmissions of personal information have to be documented. In addition, a law might provide different options to deal with a situation. These would have to be considered as well.

However, the restrictions that a law imposes changes the envisioned system. The software engineer or his/her employer has to decide if the changed system, or at least a functionality of it, is still useful. This might lead to the decision to stop using the system or functionality.

We have to distinguish between the different kinds of requirements or specifications that instructions from laws in order to achieve a seamless integration of the instructions from laws into a given software engineering process. Hence, we propose a method that can help decide how to translate the instructions from laws into requirements when they are captured in law patterns. We assume that a significant number of demands from laws have to be translated into security requirements. That is the reason why we focus on these in our method.

According to Fabian et al. (2010), a security requirement is typically a confidentiality, integrity, or availability requirement. It refers to a particular piece of information, the *asset*, that should be protected, and it indicates the *counter-stakeholder* against whom the requirement is directed. A *stakeholder* is an individual, a group, or an organization that has an interest in the system under construction. Furthermore, the *circumstances* of a security requirement describe application conditions of functionality, temporal, spatial aspects, or the social relationships between stakeholders. Hence, circumstances have relations to functional requirements, stakeholders, etc., which shall be considered in the system-to-be.

In order to determine if the instruction from the law can be transformed, we propose to try to instantiate the instruction as a security requirement. We define preconditions for each of the steps of the instantiation and give advise of how to check if these preconditions are fulfilled. The method is iterative and if one precondition fails, the method terminates and the following preconditions are not checked anymore. If one of the preconditions fail, the instruction cannot be a security requirement. Afterwards we present a method for determining if the instruction is a further functional requirement or a technical measure that has to be integrated into the software specification.

For each instantiated requirement activity pattern matching at least one instance of a law paragraph pattern do:

Instantiate stakeholder

Precondition: The stakeholder has to be a stakeholder in the sense of security requirements.
Determination: Check if the stakeholder has an interest in the system under construction.
Activity: Describe the stakeholder and his/her interest to the system. Use the descriptions from the law patterns, cloud system analysis pattern and the templates. Consider the information in the instantiated law paragraph and requirement activity patterns into which the direct and indirect stakeholders can be distinguished. The addressee is a direct or an indirect stakeholder, and the target person is a direct stakeholder. The legislator and the domain are indirect stakeholders.

Instantiate asset

Precondition: The asset has to be an asset in the sense of security requirements and it has to be owned by the stakeholder.
Determination: Check if the asset is some piece of information or hardware or software of the stakeholder and if it should be protected with respect to confidentiality, integrity, or availability.
Activity: Describe the asset of the stakeholder and the protection requirements in terms of confidentiality, integrity, or availability.

Instantiate counter-stakeholder

Precondition: A counter-stakeholder has to exist.
Determination: Check if a counter-stakeholder exists who threatens the confidentiality, integrity, or availability of the asset.
Activity: Describe the counter-stakeholder and the threat he/she presents to the asset in terms of confidentiality, integrity, or availability. In contrast to stakeholders, laws, and therefore the instantiated law paragraph and requirement activity patterns, do not define counter-stakeholders. It has to be reasoned if they exist or not.

Instantiate circumstance

Precondition: The circumstances of the security requirement have to be related to functionalities, stakeholders, or aspects of the system.

Determination: Check if stakeholder, asset, counter-stakeholder are related to the system-to-be.

Activity: Describe the relations which the stakeholder, asset, counter-stakeholder have to existing functional requirements or to other stakeholders, assets, counter-stakeholders, etc. Typically, security requirements are considered in the context of functional requirements. Therefore, the functional requirement that is the source of the activity the instantiated requirement activity pattern refers to is the basis for specifying the security requirement.

When all the preconditions are true, the instruction from the law we were looking at can be translated into a security requirement. If this is not the case, we have to determine if the instruction in the law has to be translated into a functional requirement or has to be integrated into the specification of the software. We propose a method of exclusion. Hence, we have to decide if the law prescribes a mechanism or a requirement for the system-to-be. The difference is that a requirement just states a problem that the system shall address, e.g., the system has to store some information. In this case, the software engineer has to find a mechanism that implements the requirement. In contrast, if a law demands access control to certain types of data, it is already a mechanism that solves a problem, e.g., the problem that the data has to be kept confidential. Finally, we have to create a policy pattern as described in PACTS Step 6 for the functional or security requirement.

12.12.5 Example

We execute the process in the following for the Law Identification Pattern Instance presented in Fig. 10.18 and the previous section. We analyze if the law can also be translated into a security requirement from the point of view from the stakeholder Hulda.

Instantiate stakeholder

Hulda is a stakeholder of the cloud online banking system, because he/she accomplishes distributed storage in the cloud computing system.

Instantiate asset

The customer data is not an asset in the sense of a security requirement to Hulda, because he/she stores data for a fee in the cloud, and the integrity, confidentiality and availability of the data is not of primary concern to the stakeholder. The reason is that he/she is not the owner of the data. These demands could be part of an contractual obligation of the stakeholder. However, we do not consider these here. The restricted activity in question is the abroad transfer of the distributed storage. The method terminates at this step, because the asset is not an asset in the sense of a security requirement.

The law identification pattern instance cannot be translated into a security requirement. The next part of the method is to decide if the abroad transfer can be translated into a functional requirement or has to be integrated into the software specification. Abroad transfer is not a specific mechanism. It is rather a restriction on a functional requirement that the cloud software system should store data distributed. The requirement has to be modified into: The cloud software system should store data distributed, but the distribution has to be restricted to European Union. At this point the security engineer or his/her employer has to decide if the system-to-be is still useful with this restriction and a policy pattern as described (see PACTS Step 6).

We also investigate a further part of the law identification pattern instance in Fig. 12.28. We analyze via our method if the abroad transfer for the bank customer is translated into a security requirement, because we can list one example for each of the required instantiations.

Instantiate stakeholder

The bank customer is a stakeholder of the cloud online banking system, because the bank customer uses the online banking software in the cloud.

Instantiate asset

The cloud costumer transfers data to the cloud and the cloud stores the data distributed. This data is an asset to the bank customer and he/she wants the data to be confidential from, e.g., other users of the online bank. The bank customer also wants the data to be available and he/she also wants its integrity preserved.

Instantiate counter-stakeholder

A possible counter-stakeholder are the other customers of the cloud banking system. He/she could get access to the data of the cloud customer, because they both have access to the cloud data storage. Hence, the confidentiality of the data can be violated.

Instantiate circumstance

The store distributed functional requirement is related to the bank customer's data stored in the cloud.

We include the resulting security requirement and create a policy pattern in accordance with PACTS Step 6.

Table 12.35 Privacy in the ISO 27001 standard

Significance for the ISO 27001 standard	Relation to Section 4.2.1 of the Plan Phase and several phases of the do, check, act phases of the standard according to Annex B
Related section(s) of the standard	Annex B-OECD principles

Changes for ISO 27001:2013 Compliance

Privacy is mentioned in ISO27001:2013 in control *A.18.14—Privacy and protection of personally identifiable information.* The risk identification in Section 6.1.2 of ISO27001:2013 allows to identify privacy risks e.g. the loss of personal information, which can in turn lead to the selection of the previously mentioned privacy related control.

12.13 Considering Privacy in the PACTS Method

The ISO 27001 considers privacy in Annex B (see Table 12.35), which, in turn, refers to Section 4.2.1 of the plan phase of the standard. Annex B references *The Fair Information Practice Principles* (-or short FIPs) of the OECD (1980). These are widely accepted privacy regulations. They state that a person's informed consent is required for the data that is collected, collection should be limited for the task it is required for and erased as soon as this is not the case anymore. The collector of the data shall keep the data secure and shall be held accountable for any violation of these principles.

In previous work, we developed textual patterns to formulate privacy requirements (Beckers and Heisel 2012) and proposed a computer-aided privacy threat analysis (Beckers et al. 2014). Both methods are based on problem frames, which we introduce in Sect. 2.5.3. These methods were developed based on our experience with the building of privacy-preserving Voice-over-IP systems (Hofbauer et al. 2012c, a, b). We compared privacy requirements methods via an extension of the conceptual framework for security requirements engineering (Sect. 2.4) and analyzed the privacy goals these methods support (Beckers 2012). We chose a subset of these goals for our method that most methods had in common.

We combine our previous work into one structured method for privacy requirements and threat analysis. We also adapt this method to integrate it with our PACTS method in Sects. 12.13.1–12.13.3. Section 12.13.4 contains an example of the privacy steps of our PACTS method.

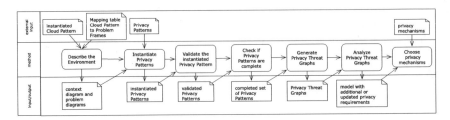

Fig. 12.29 A method for considering privacy using our cloud system analysis pattern and the problem frames notation

12.13.1 A Method for Considering Privacy in an ISMS

We refine the Steps 10–12 of PACTS (see Fig. 12.4) and combine them into a privacy method. We describe the steps of our privacy method for privacy requirements elicitation and threat analysis in the following. We state the steps of PACTS (see Fig. 12.4) and underneath the steps of our privacy method (see Fig. 12.29) that refine them (Sect. 12.4). The Problem Frames (Jackson 2001) notation uses an abstraction of the system-to-be and models the environment of the system around it. Privacy analysis concerns the flow of personal information in a system. Our choice is to analyze privacy concerns using problem frames, because we can use the abstraction of the system-to-be and its environment to model relevant information flows of personal information and elicit privacy requirements for the system-to-be.

12.13.2 Pacts Step 10: Instantiate Privacy Patterns

Describe the environment—We use the information in the cloud pattern as an input for our privacy method. We reuse previous research (Beckers and Heisel 2012; Beckers et al. 2014) regarding privacy in this method. Both of these works use the problem frame notation (see Sect. 2.5.3). Hence, we present a mapping of our cloud pattern and the stakeholder templates to the problem frame notation in Table 12.36. We first map the stakeholders of the indirect environment to the problem frame notation. These are part of the indirect environment, which are considered as requirements in problem diagrams. For example, the legislator stakeholder contains a set of laws, which are mapped to functional and nonfunctional requirements (see Sect. 12.12.4). The stakeholders of the direct environment are all considered as biddable domains in the context diagram and subsequent problem diagrams. We divided the cloud into cloud elements that are part of the service and elements that are part of the pool. The IaaS, PaaS, and SaaS layers are considered as an integral part of the cloud and, hence, only used and not part of the machine to be built, while the cloud software stack and the software product can be part of the machine to be built. Data maps to a lexical domain. All elements that are part of the pool can be either a causal domain or part of the machine, except for the location(s) of the resource. We combined the

Table 12.36 Mapping cloud pattern to problem frame notation

Cloud pattern	Problem frame notation
Indirect Environment	
Legislator	Requirement
Domain	Requirement
Contract	Requirement
Direct environment	
Cloud provider	Biddable domain
Cloud support	Biddable domain
Cloud administrator	Biddable domain
Cloud customer	Biddable domain
Cloud developer	Biddable domain
End customer	Biddable domain
Cloud: Service	
IaaS	Causal domain
PaaS	Causal domain
SaaS	Causal domain
Cloud software stack	Causal domain or machine domain
Software product	Causal domain or machine domain
Data	Lexical domain
Cloud: Pool	
Resource	Causal domain or machine domain
Location	Part of the description of a Causal Domain
Software	Causal domain or machine domain
Hardware	Causal domain or machine domain
Hypervisor	Causal domain or machine domain
Cloud database	Causal domain or machine domain
Stakeholder templates	
Name	Name of the biddable domain
Description	Used in the description of the domain
Motivation	Used in the description of the domain
Relations to other stakeholders	Basis for one or more phenomena
Relations to the cloud	Basis for one or more phenomena
Relations to other direct stakeholders	Basis for one or more phenomena
Assets	Causal domain or lexical domain
Compliance and privacy	Considered as requirements in problem diagrams

elements of the templates for direct and indirect stakeholders. The name, description, and motivation are used in the attributes of domains. Relations between stakeholders, or between stakeholders and the cloud, give rise to phenomena between domains. Assets are lexical domains if they are only data. If they also have a physical representation, these are causal domains. For example, an address would be a lexical domain and a hard drive containing an address is a causal domain. Compliance and privacy demands are mapped to requirements in problem diagrams.

The first step of our method uses the mapping table (see Table 12.36) and an instantiated cloud pattern as an input. This results in a context diagram and several problem diagrams. These are accompanied by textual requirements.

Instantiate privacy patterns—In the second step, we use textual patterns for privacy requirements introduced in previous work (Beckers and Heisel 2012). Privacy requirements are difficult to elicit for any given software engineering project that processes personal information. The problem is that these systems require personal data in order to achieve their functional requirements and privacy mechanisms that constrain the processing of personal information in such a way that the requirement still states a useful functionality.

We present privacy patterns for anonymity, pseudonymity, unlinkability and unobservability in accordance with the definitions of the ISO 15408 standard - Common Criteria for Information Technology Security Evaluation (or short CC) (ISO/IEC 2012). Our analysis of related work regarding privacy requirements engineering methods in Beckers (2012) also showed that all methods we analyzed consider at least the privacy goals anonymity, pseudonymity, unlinkability and unobservability. Thus, we focus on these privacy goals. Our privacy patterns have a textual representation that can be instantiated using the problem frame notation. We also show predicate patterns that can validate the instantiation of our privacy patterns. We presented an exhaustive discussion about privacy terminology in Beckers and Heisel (2012) and present only the Common Criteria definition of these terms in this work.

The privacy specification in the CC defines four privacy goals. These goals can be refined into privacy requirements for a given software system. *Anonymity* means that a subject is not identifiable within a set of subjects, the anonymity set. *Unlinkability* of two or more items of interest (IOI) means that within a system the attacker cannot sufficiently distinguish whether these IOIs are related or not. *Unobservability* of an IOI means that an IOI is not detectable by any subject uninvolved in it and anonymity of the subject(s) involved in the IOI even is given, even against other subject(s) involved in that IOI. A pseudonym is an identifier of a subject other than one of the subject's real names. Using pseudonyms means *pseudonymity*.

We explain specific privacy domain types, which we use in the remainder of our method:

- A *Stakeholder* is a *BiddableDomain* (and in some special cases also a *CausalDomain*) with some relation to stored or transmitted personal information. It is not necessary that a stakeholder has an interface to the machine.
- A *CounterStakeholder* is a *BiddableDomain* that describes all subjects (with their equipment) who can compromise the privacy of a *Stakeholder* at the machine. We do not use the term attacker here, because the word attacker indicates a malicious intent. Privacy of stakeholders can also be violated by accident.
- *PersonalInformation* is a *CausalDomain* or *LexicalDomain* that represents personal information about a *Stakeholder*. The difference between these domains is

that a *LexicalDomain* describes just the stored information, while a *CausalDomain* also includes the physical medium the data is stored upon, e.g., a hard drive.

- *StoredPersonalInformation* is *PersonalInformation*, which is stored in a fixed physical location, e.g., a hard drive in the U.S.
- *TransmittedPersonalInformation* is *PersonalInformation*, which is transmitted in-between physical locations, e.g., data in a network that spans from Germany to the U.S.
- *InformationAboutPersonalInformation* is a *CausalDomain* or *LexicalDomain* that represents information about *PersonalInformation*, e.g., the physical location of the name and address of a stakeholder.

We introduce four textual privacy patterns in the following. These texts contain "[ ]" that contain in turn textual descriptions of domains that can be instantiated and their domain types. In particular, the terms in *italic* inside the square brackets can be instantiated. The type of the domain is stated after the ":" and the type can be a single domain or a set of domains. We use the symbol "$\mathbb{P}$" in front of the domain type for types that allow sets of domains. These patterns are a subset of the patterns presented in Beckers and Heisel (2012).

Anonymity—

> Preserve anonymity of [*Stakeholders* : BiddableDomain] and prevent disclosure of their identity by [*CounterStakeholders* : $\mathbb{P}$ BiddableDomain]

Pseudonymity—A *Pseudonym* is a *LexicalDomain* used as an identifier of a *Stakeholder* without revealing *PersonalInformation*. *Authorized Users* are *Stakeholders* who are allowed to know the identity of the *Stakeholder* the *Pseudonym* belongs to.

> Preserve pseudonymity of [*Stakeholders* : BiddableDomain] via preventing
> [*CounterStakeholders* : $\mathbb{P}$ BiddableDomain] from relating
> [*Pseudonyms* : LexicalDomain] to their [*Stakeholders* : BiddableDomain].
> The following [*Authorized Users* : $\mathbb{P}$ BiddableDomain] exist.

Unlinkability—A *ConstrainedDomain* is a *CausalDomain* or a *ConnectionDomain* that is constrained by a functional or privacy requirement.

> Preserve unlinkability of two or more [*ConstrainedDomains* : $\mathbb{P}$ CausalDomain] for
> [*Stakeholders* : BiddableDomain] and prevent
> [*CounterStakeholders* : $\mathbb{P}$ BiddableDomain] from disclosing that the
> [*ConstrainedDomains* $\mathbb{P}$ CausalDomain] have a relation to the
> [*Stakeholder* : BiddableDomain].

Unobservability—

> Preserve unobservability of a [*ConstrainedDomains* : $\mathbb{P}$ CausalDomain] that is used by
> [*Stakeholders* : BiddableDomain] and prevent
> [*CounterStakeholders* : BiddableDomain] from recognizing that the
> [*ConstrainedDomains* : $\mathbb{P}$ CausalDomain] exist.

Validate the instantiated privacy patterns—In addition, for each privacy pattern a predicate pattern exists that can be used to validate its instantiation. This validation is based upon problem frame models, and it validates that the pattern is instantiated with the correct privacy domain types. We introduce our predicate patterns in the following.

Validate the instances of the anonymity pattern with:

$$anon_{cs}: BiddableDomain \times \mathbb{P}\,BiddableDomain \rightarrow Bool$$

The suffix "cs" indicates that this predicate describes a requirement considering a certain **CounterStakeholder**. The definition of anonymity by ? (?) states that a stakeholder shall not be identifiable from a set of stakeholders. This is the so-called *anonymity set*, which is represented in our pattern by a $\mathbb{P}\,Biddable\,Domain$.

Validate the instances of the pseudonymity pattern with:

$$pseudo_{cs}: LexicalDomain \times BiddableDomain \times \mathbb{P}\,BiddableDomain \times$$
$$LexicalDomain \times BiddableDomain \times \mathbb{P}\,BiddableDomain \rightarrow Bool$$

Validate the instances of the unlinkability pattern with:

$$unlink_{cs}: \mathbb{P}\,CausalDomain \times BiddableDomain \times \mathbb{P}\,BiddableDomain \times$$
$$\mathbb{P}\,CausalDomain \times BiddableDomain \rightarrow Bool$$

Validate the instances of the unobservability pattern with:

$$unobserv_{cs}: \mathbb{P}\,CausalDomain \times BiddableDomain \times \mathbb{P}\,BiddableDomain \times$$
$$\mathbb{P}\,CausalDomain \rightarrow Bool$$

Check if privacy patterns are complete—The predicate patterns can also be used to identify incomplete privacy requirements. In incomplete privacy patterns instantiation domain types are missing, which results in incomplete requirements. Hence, we check for each requirement that all the textual gaps are instantiated. In addition, we check for each gap that has to be instantiated with sets of domains that no domain is missing in the set. For example, requirements that have to be instantiated with a *Stakeholder* have to name an instance of a *Biddable Domain* from the context diagram, e.g., *Bank Customer*. Several privacy patterns require instantiation with sets of *Biddable Domains*. For example, a privacy pattern might not only be directed toward *Bank Customers*, but also toward the *Bank Institute*. In this case, we can reason for all biddable domains in the context diagram if they are a *CounterStakeholder* or not.

In addition, a requirement might be missing, e.g., because of an incomplete threat analysis. In order to execute this check, all personal information in the cloud has to be elicited. For each *CausalDomain* or *LexicalDomain* we have to check if these are *StoredPersonalInformation*, *TransmittedPersonalInformation* or *InformationAbout-PersonalInformation*. If this is the case, we check if these were considered in the privacy threat analysis. If this is not the case, we have to redo the privacy threat analysis and start our process from the beginning.

12.13.3 Pacts Step 11: Analyze Privacy Threats

Generate privacy threat graphs—The Problem-Based Privacy Analysis (ProPAn) (Beckers et al. 2014) is a method including tool support[19] for identifying privacy threats during the requirements analysis of software systems using the problem frame notation. Our approach does not rely entirely on the privacy analyst to detect privacy threats, but allows a computer-aided privacy threat identification that is derived from the relations between stakeholders, technology, and personal information in the system-to-be. We defined a UML-profile for privacy requirements and a reasoning technique that identifies stakeholders, whose personal information is stored or transmitted in the system-to-be, and stakeholders from whom we have to protect this personal information. For this purpose, we have tool support that uses problem diagrams to create a privacy threat graph. This graph uses the information from the instantiated privacy patterns and documents the information flow of personal information in the system-to-be. In our method, graphs are labeled and directed and the set of vertices is a subset of the domains occurring in the model. The edges are annotated with problem diagrams and point from one domain to another. Hence, the graph can be used to evaluate the information flow between the domains. A formal definition is provided in previous work of which Rene Meis is the main author and the author of this book provided support for integrating of privacy goals into the analysis (Beckers et al. 2014).

Analyze privacy threat graphs—The analysis of the privacy threat graph reveals if privacy requirements need to be added or updated in the model. The results of the analysis lead to a refined set of privacy requirements.

Choose privacy mechanism—The last step of our method is to *choose a privacy mechanism* that solves the problem. For example, to achieve pseudonymity a privacy-enhancing identity management system (Clauß et al. 2005) can be chosen.

12.13.4 Example of Our Privacy Method

We used the mapping in Table 12.36 to create a context diagram based on our instantiated cloud pattern. The context diagram is depicted in Fig. 12.30. The *Data Center* contains *Server*s and *Network, Virtualisation and Database Software* in particular

---

[19]http://www.uni-due.de/swe/apf12.shtml.

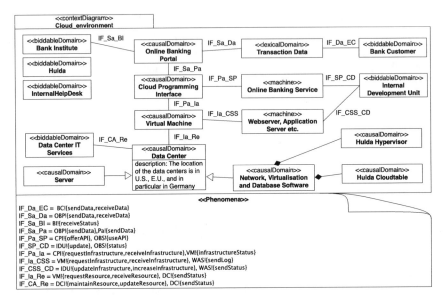

Fig. 12.30 Context diagram of the instantiated cloud analysis pattern

the *Hulda Hypervisor* and the *Hulda Cloudtable*. We added a number of phenomena between the domains after the mapping. The *Data Center IT Services Unit* maintains and updates resources of the *Data Center* and the *Data Center* can send a status message to the *Data Center IT Services Unit*. *Hulda* and the *InternalHelpDesk* do not communicate with any technical element of the cloud. The *Virtual Machine* can request and receive resources from the *Data Center* and the *Data Center* can send a status back to the *Virtual Machine*, e.g., if the requested resources can be provided. The *Webserver, Application Server, etc.* can request and receive IT infrastructure resources from the *Virtual Machine* and the *Virtual Machine* can send a log back to the *Webserver, Application Server, etc.*. The *Cloud Programming Interface* can also request infrastructure resources from the *Virtual Machine* and also receive a status message back. The *Online Banking Service* uses the API of the *Cloud Programming Interface*. In order to acquire knowledge about the API, the *Cloud Programming Interface* can send a description to the *Online Banking Service*. The *Internal Development Unit* can update and increase the infrastructure provided by the *Webserver, Application Server, etc.* and receive a status back. The *Internal Development Unit* can also update the *Online Banking Service* and receive a status message back that states if the update was installed successfully. The *Bank Institute* can receive a status message from the *Online Banking Service* to check if it is working properly. The *Online Banking Service* can also receive *Transaction Data* from a *Bank Customer* and conduct a financial transaction.

We consider three textual requirements in our example, which we illustrate in the following.

- **R1** The *Bank Customer* can store *Transaction Data* using the *Online Banking Service* via the *Online Banking Portal*.
- **R2** All *Transaction Data* from the *Online Banking Service* is stored in the *Hulda Cloudtable*.
- **R3** The *Data Center IT Services Unit* ensures the availability of the *Hulda Cloudtable* and the *Online Banking Service*.

We draw a problem diagram for each requirement in order to refine it. We present a problem diagram for **R1** in Fig. 12.31. We consider the domains *Internal Development Unit*, *Online Banking Service*, *Cloud Programming Interface*, the *Online Banking Portal*, and the *TansactionData* in this problem diagram. The descriptions of the domains and the phenomena between them are the same as in the context diagram. We refined the *TansactionData* with specific data such as the *Name*, *Address*, and *Transaction* information. The requirement *R1* ≪refersTo≫ the *Bank Customer* and the *TransactionData* and ≪constrains≫ the *Online Banking Service* in such a way that only *Bank Customers* can conduct financial transactions. The requirement also ≪constrains≫ the *Online Banking Portal* in a way that enables the *Bank Customer* to enter *TransactionData*, which is forwarded to the *Online Banking Service*.

Instantiate privacy patterns—The *Bank Institute* wants to protect the privacy of the *Bank Customer* that uses the *Online Banking Portal*. The decision makers of the *Bank Institute* discuss our privacy patterns and select the anonymity privacy pattern, because anonymity eliminates all information about the identity of a *Bank Customer*. The decision makers of the *Bank Institute* do not consider the problem in detail and take the decision after a short discussion. Our experience is that the

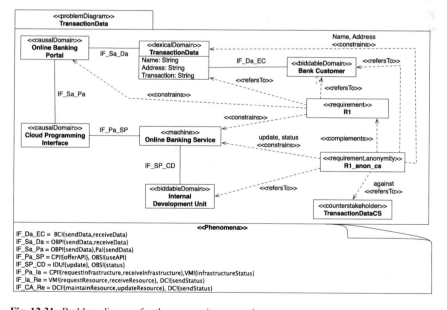

Fig. 12.31 Problem diagram for the anonymity constrain

decision about privacy requirements is often taken in a simplistic manner after only short discussions. In the following, we illustrate the instantiated privacy pattern for our example.

> Preserve anonymity of *Bank Customers* and prevent disclosure of their identity by the *Internal Development Unit*.

The anonymity requirement is included in the problem diagram shown in Fig. 12.31. The anonymity requirement is labeled *R1_anon_cs* in the diagram. The *R1_anon_cs* requirement ≪complements≫ the requirement **R1** with an anonymity demand. The diagram illustrates that the *R1_anon_cs* requirement ≪constrains≫ the *TransactionData*. In particular, the *Name* and *Address* attributes of it should be anonymized. The requirement protects against a generic counterstakeholder, which we label *TransactionDataCS*. In addition, a possible concrete counterstakeholder is the *Internal Development Unit*. The reason is that the *Internal Development Unit* shares the phenomenon *IF_SP_CD* with the *Online Banking Service*. This phenomenon states that the *Online Banking Service* sends status messages to the *Internal Development Unit*. These messages might contain parts of the *TransactionData*. Hence, the *R1_anon_cs* requirement ≪constrains≫ this update message that the *Internal Development Unit* sends.

Validate the instantiated privacy patterns—We execute the validation check for the anonymity pattern. The result is true, because the instantiated stakeholder and the counterstakeholder are both instantiated and are biddable domains, which is the correct domain type.

Check if privacy patterns are complete—The iteration over all the possible biddable domains reveals that the biddable domain *Data Center IT Services Unit* is a possible counter-stakeholder and it is also integrated into the privacy pattern and the problem diagram. The reason is that the *Data Center IT Services Unit* can identify the identity of *Bank Customers*.

Generate privacy threat graphs—We use the ProPAn tool to generate a privacy threat graph for our scenario to visualize, which domains have a relation to each other. The relations are labeled with the requirements that introduce them. We do not show the entire privacy threat graph, but for our scenario we expect at least 30 requirements and at least 90 relations between domains from our previous experience when using ProPAn. The effort for modeling the 30 problem diagrams can be calculated by 1 hour per problem diagram. This results in at least 30 person hours.

Analyze privacy threat graphs—We present a privacy threat graph for the *Bank Customer* considering the requirements **R1, R2,** and **R3** in Fig. 12.32 and the counterstakeholder *Data Center IT Services Unit*. The edges from the counterstakeholder are red (light graey), bold edges with white arrowheads.

We have two possibilities to solve the privacy threats identified by the threat graph:

- We can consider the edges starting from the counter-stakeholder and restrict the information the counter-stakeholder can access. In our example, we have to restrict the information that the *Data Center IT Services Unit* can access. The relevant edge concerns *R3* and we can restrict the information flow that is caused by this requirement.

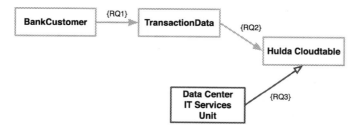

Fig. 12.32 Privacy threat graph

- We consider the other edges of the threat graph, as well. This results in a restriction of the information flow between the domains. For example, we could restrict the information flow between the *TransactionData* and the *Hulda Cloudtable*, which is related to *R2*.

We can see that the counter-stakeholder *Data Center IT Services Unit* may gain information from the lexical domain *Transaction Data*, while it is stored in the *Hulda Cloudtable*. In order to fulfill the anonymity requirement, the *Name* and *Address* attributes have to be erased. If these are not erased, anonymity cannot be achieved. However, this data is required for the *Bank Institute* to conduct banking business with the *Bank Customer*. A possible solution is to apply pseudonymity instead of anonymity. In this case, the *Name* and *Address* attributes are replaced with other values and a later mapping is possible.

Choose privacy mechanism—After identifying privacy threats we have to reason, which technical privacy enhancement technique to use. We refer to the work of Deng et al. (2011) that represents an extensive list of these mechanisms.

12.14 Related Work

We present related work regarding the ISO 27001 standard and clouds as well as legal compliance and privacy.

ISO 27001 and Cloud Computing

Calder (2009) and Kersten et al. (2011) provide advice for an ISO 27001 realization. In addition, Klipper (2010) focuses on risk management according to ISO 27005. The author also includes an overview of the ISO 27000 series of standards. However, none of these works consider to use security requirements engineering methods.

Cheremushkin and Lyubimov (2010), Lyubimov et al. (2011) present a UML-based meta-model for several terms of the ISO 27000, e.g., assets. This meta-models can be instantiated and, thus, support the refinement of these terms. However, the authors do not present a holistic approach to information security. The work mostly constructs models around specific terms in isolation. The CF of Fabian et al. (2010), on the other hand, presents a holistic framework for information security.

Mondetino et al. investigate possible automation of controls that are listed in ISO 27001 and ISO 27002 (Montesino and Fenz 2011). Their work can complement our own.

Fenz et al. (2007) introduce an ontology-based framework for preparing ISO/IEC 27001 audits. They provide a rule-based engine which uses a security ontology to determine if security requirements of a company are fulfilled.

Auty et al. (2010) base their work on the controls in ISO 27001 (ISO/IEC 2005b) and 27002 (ISO/IEC 2005a) standards. The authors discuss if these controls are adequate for cloud computing or if adjustments have to be considered. The discussion ranges from social to technical threats. This approach can complement our own by proposing updates for the ISO 27001 and ISO 27002 controls based upon the findings of Auty et al.

Shaikh and Haider (2011) map the existing research on cloud security threats to the categories context, problem description, technique used, and proposed models or tools. The authors conclude that privacy and data loss are the threats that cause the most concerns. This approach differs from our own, because the authors aim to identify the most severe threats rather than providing a threat analysis method.

Greenwood and Sommerville (2011) propose to use responsibility modeling to identify threats for cloud computing. The method operates on the same abstraction levels as goal based notations such as i*. The method models agents in the system, resources, responsibilities and the relations among them. Resources can be information or physical systems. The threat analysis investigates the responsibilities and describes conditions in which case a threat to this responsibility can occur. Our method is based upon a cloud pattern, which can be reused for different projects. The work of Greenwood and Summerville requires a new model for each project.

Grobauer et al. (2011) investigate risks for cloud providers and users. The authors base their work upon the ISO 27005 (ISO/IEC, 2008) standards for high level risk criteria, and they use the risk taxonomy of the Open Group (Open Group 2009) for a refinement of these criteria. The authors map these risks to a cloud architecture description from IBM. The authors describe risks, e.g., for the cloud management interface. The work can complement our own by integrating the risks into our threat analysis.

Legal Compliance and Privacy

Breaux et al. (2006) and Breaux and Antón (2008) present a framework that covers analyzing the structure of laws using a natural language pattern. This pattern helps to translate laws into a more structured and restricted natural language and then into a first-order logic. The idea of using first-order logic in the context of regulations is not a new one. For example Bench-Capon et al. (1987) made use of first-order logic to model regulations and related matters. In contrast to our work, the authors of those approaches assume that the relevant laws are already known and thus do not support identifying legal texts.

Siena et al. (2008) describe the differences between legal concepts and requirements. They model the regulations using an ontology, which is quite similar to the natural language patterns described in the approaches mentioned before. The ontology is based on the Hohfeld taxonomy (Hohfeld 1917), which describes the means and relations between the different means of legal texts in a very generic way. Thus Hohfeld does not structure a certain law at all but aims at the different meanings of laws. Hence, the resulting process in Siena et al. (2008) to align legal concepts to requirements and the given concepts are quite high level and cannot directly applied to a scenario. In a second work Siena et al. (2009) try to bridge the gap between the requirements engineering process and compliance using a goal-oriented approach. In contrast to our approach, they do not identify relevant laws and do not intertwine compliance regulations with already elicited requirements.

Álvarez et al. (2002) describe reusable legal requirements in natural language, which is based on the Spanish adaption of the EU directive 95/46/CE concerning personal data protection. We believe that the work by Álvarez et al. complements our work, i.e., applying our law identification method can precede using their security requirements templates.

Deng et al. (2011) present a threat tree for privacy based upon the threat categories: linkability, identifiablitiy, non-repudiation, detectability, information disclosure, content unawareness, and policy/consent noncompliance. These threats are modeled for the elements of an information flow model, which includes the components data flow, data store, processes, and entities. Privacy threats are described for each of these components. Hence, privacy threat identification for an existing data flow model is simplified, because for each data flow element in a model only the threats shown in the tree need to be considered. The work differs from our own, because the privacy threat identification has to be carried out manually.

The PriS method (Kalloniatis et al. 2008) elicits privacy requirements in the software design phase. Privacy requirements are modeled as organizational goals. Furthermore, privacy process patterns are used to identify system architectures, which support the privacy requirements. The PriS method starts with a conceptual model, which also considers enterprise goals, stakeholders, privacy goals, and processes. It is based upon a goal-oriented requirements engineering approach, while our work uses a problem-based approach as a foundation. The difference is that our work focuses on a description of the environment as a foundation for the privacy analysis, while the PriS method uses organizational goals as a starting point. In addition, the PriS method has to be carried out manually.

Hafiz (2006) describes four privacy design patterns for the network level of software systems. These patterns solely focus on anonymity and unlinkability of senders and receivers of network messages from protocols, e.g., http. The patterns are specified in several categories. Among them are intent, motivation, context, problem and solution, as well as forces, design issues and consequences. This work focuses on privacy issues on the network layer and can complement our work in this area.

12.15 Summary

The decision whether a cloud service is chosen by a costumer relies, amongst other reasons, on the trustworthiness of the cloud system. One way to establish this trust is to demonstrate that security, privacy, and compliance are taken seriously by the cloud provider. This is usually achieved by providing certified services. A well-known standard for such a certification is the ISO 27001 standard. However, establishing an ISMS as required by this standard is a nontrivial task. Furthermore, the standard does not take the special needs of cloud computing into consideration yet. With the work presented in this chapter we intend to close the aforementioned gap. We do so by providing a structured pattern-based method to establish an Information Security Management System (ISMS) according to the ISO 27001 standard. It has been tailored to suit the demands of the cloud computing domain. We introduce specific patterns for clouds to elicit the context of the envisioned ISMS. The approach further allows us to refine the initially elicited context with behavior descriptions. It also provides the means for documenting management commitment, threat and risk analysis, as well as a pattern-based definition of security policies compliant to the ISO 27001 standard. We enhance the approach by providing validation conditions that can be used to check the instantiated context as well as policy patterns. It is, for example, possible to check whether a given responsible stakeholder in the policy pattern is also present in the context-pattern. Moreover, we take the standard's demand to consider legal compliance and privacy into account.

In summary, the benefits of our approach are:

- A structured method for establishing a cloud-specific ISMS compliant to ISO 27001.
- Detailed steps for asset identification, threat analysis, risk management and security reasoning.
- The pattern-based method provides the means for consistency checks, e.g., for the instantiation of the pattern.
- Consideration of legal compliance via steps for identifying laws and regulations.
- Support for formulating and validating privacy requirements and conducting a privacy threat analysis.
- A systematic support to generate the required ISMS documentation in compliance with the standard.
- Integration of proven existing methods, e.g., CORAS and Misuse Cases.
- Integrating requirements engineering for security, legal compliance, and privacy to construct a holistic ISMS.

To sum up, we analyzed the ISO 27001 demands for system development and documentation with regard to cloud computing systems. Based on these insights, we provide a method that relies upon existing requirements engineering methods. Moreover, our method relies on patterns for several security tasks, e.g., context descriptions, threat analysis, and policy definition. Our method can ease the effort of establishing a cloud-specific ISMS and can produce the necessary documentation for an ISO 27001 compliant ISMS.

References

Álvarez, J. A. T., Olmos, A., & Piattini, M. (2002). Legal requirements reuse: A critical success factor for requirements quality and personal data protection. In *Proceedings of the International Conference on Requirements Engineering (RE)* (pp. 95–103). IEEE Computer Society.

Armbrust, M., Fox, A., Griffith, R., Joseph, A. D., Katz, R. H., Konwinski, A., et al., (2009). Above the clouds: A Berkeley view of cloud computing. Technical report. San Francisco, U.S.: EECS Department, University of California, Berkeley.

Auty, M., Creese, S., Goldsmith, M., & Hopkins, P. (2010). Inadequacies of current risk controls for the cloud. In *Proceedings of the 2010 IEEE Second International Conference on Cloud Computing Technology and Science* (pp. 659–666). IEEE Computer Society.

Beckers, K. (2012). Comparing privacy requirements engineering approaches. In *Proceedings of the International Conference on Availability, Reliability and Security (ARES)—6th International Workshop on Secure Software Engineering (SecSE 2012)* (pp. 574–581). IEEE Computer Society.

Beckers, K., & Jürjens, J. (2010). Security and compliance in clouds. In *ISSE 2010 Securing Electronic Business Processes: Highlights of the Information Security Solutions Europe 2010 Conference* (pp. 91–101). Vieweg + Teubner.

Beckers, K., & Heisel, M. (2012). A foundation for requirements analysis of privacy preserving software. In *Proceedings of the International Cross Domain Conference and Workshop (CD-ARES 2012)*

Beckers, K., Küster, J.-C., Faßbender, S., & Schmidt, H. (2011). Pattern-based support for context establishment and asset identification of the ISO 27000 in the field of cloud computing. In *Proceedings of the International Conference on Availability, Reliability and Security (ARES)* (pp. 327–333). IEEE Computer Society.

Beckers, K., Faßbender, S., & Schmidt, H. (2012). An integrated method for pattern-based elicitation of legal requirements applied to a cloud computing example. In *Proceedings of the International Conference on Availability, Reliability and Security (ARES)—2nd International Workshop on Resilience and It-Risk in Social Infrastructures(RISI 2012)* (pp. 463–472). IEEE Computer Society.

Beckers, K., Côté, I., Faßbender, S., Heisel, M., & Hofbauer, S. (2013a). A pattern-based method for establishing a cloud-specific information security management system. *Requirements Engineering, 18*(4), 1–53.

Beckers, K., Hofbauer, S., Quirchmayr, G., & Wills, C. C. (2013b). A mapping between ITIL and ISO 27001 processes for use by a high availability video conference service provider. In *Proceedings of the International Cross Domain Conference and Workshop (CD-ARES 2013)* (pp. 224–239). Springer.

Beckers, K., Faßbender, S., Heisel, M., & Meis, R. (2014). A problem-based approach for computer aided privacy threat identification. In *Proceedings of the Annual Privacy Forum APF 2012* (pp. 1–16). Springer.

Bench-Capon, T., Robinson, G., Routen, T., & Sergot, M. (1987). Logic programming for large scale applications in law: A formalization of supplementary benefit legislation. In *International Conference on Artificial Intelligence & Law (ICAIL)* (pp. 190–198). ACM.

Biagioli, C., Mariani, P., & Tiscornia, D. (1987). ESPLEX: A rule and conceptual model for representing statutes. In *Proceedings of the 1st International Conference on Artificial Intelligence and Law (ICAIL)* (pp. 240–251). ACM.

Breaux, T. D., & Antón, A. I. (2008). Analyzing regulatory rules for privacy and security requirements. *IEEE Transactions on Software Engineering, 34*(1), 5–20.

Breaux, T. D., Vail, M. W., & Antón, A. I. (2006). Towards regulatory compliance: Extracting rights and obligations to align requirements with regulations. In *RE* (pp. 46–55). IEEE Computer Society.

Buyya, R., Ranjan, R., & Calheiros, R. N. (2009). Modeling and simulation of scalable cloud computing environments and the cloudSim toolkit: Challenges and opportunities. In *Proceedings of the International Conference Von High Performance Computing and Simulation (HPCS)*. IEEE Computer Society.

Calder, A. (2009). *Implementing information security based on ISO 27001/ISO 27002: A management guide*. Zaltbommel: Van Haren Publishing.

Cheremushkin, D. V., & Lyubimov, A. V. (2010). An application of integral engineering technique to information security standards analysis and refinement. In *Proceedings of the International Conference on Security of Information and Networks* (pp. 12–18). ACM.

Chow, R., Golle, P., Jakobsson, M., Shi, E., Staddon, J., Masuoka, R. et al., (2009). Controlling data in the cloud: Outsourcing computation without outsourcing control. In *Cloud Computing Security Workshop (CCSW)* (pp. 85–90). ACM.

Clauß, S., Kesdogan, D., & Kölsch, T. (2005). Privacy enhancing identity management: protection against re-identification and profiling. In *Proceedings of the 2005 Workshop on Digital Identity Management* (pp. 84–93). ACM.

Cloud Security Alliance (CSA). (2010). Top threats to cloud computing v1.0. https://cloudsecurityalliance.org/topthreats/csathreats.v1.0.pdf.

Deng, M., Wuyts, K., Scandariato, R., Preneel, B., & Joosen, W. (2011). A privacy threat analysis framework: supporting the elicitation and fulfillment of privacy requirements. *Requirements Engineering, 16*, 3–32.

Duisberg, A. (2011). Gelöste und ungelöste Rechtsfragen im IT-outsourcing und cloud computing. In *Trust in IT* (pp. 49–70). Springer.

Fabian, B., Gürses, S., Heisel, M., Santen, T., & Schmidt, H. (2010). A comparison of security requirements engineering methods. *Requirements Engineering—Special Issue on Security Requirements Engineering, 15*(1), 7–40.

Fay Chang, S. G., & Dean, J. (2006). Bigtable: A distributed storage system for structured data (Technical report). Google. Retrieved from http://labs.google.com/papers/bigtable-osdi06.pdf.

Fenz, S., Goluch, G., Ekelhart, A., Riedl, B., & Weippl, E. (2007). Information security fortification by ontological mapping of the ISO/IEC 27001 standard. In *Proceedings of the International Symposium on Dependable Computing* (pp. 381–388). IEEE Computer Society.

Gartner. (2008). Assessing the security risks of cloud computing. (http://www.gartner.com/id=685308).

Greenwood, D., & Sommerville, I. (2011). Responsibility modeling for identifying sociotechnical threats to the dependability of coalitions of systems. In *2011 6th International Conference on System of Systems Engineering* (SOSE), (pp. 173–178).

Grobauer, B., Walloschek, T., & Stocker, E. (2011). Understanding cloud computing vulnerabilities. *Security Privacy, IEEE, 9*(2), 50–57.

Gsell, H., Weißenberg, N., Beckers, K., & Hirsch, M. (2010). Process intelligence in der finanzwirtschaft. In *Software as a Service, Cloud Computing und Mobile Technologien* (pp. 63–75). GITO Verlag.

Hafiz, M. (2006). A collection of privacy design patterns. In *Proceedings of the 2006 Conference on Pattern Languages of Programs* (pp. 1–13). ACM.

HM Government. (2012). IT Infrastructure Library (ITIL). (http://www.itil-officialsite.com/home/home.aspx).

Hofbauer, S., Beckers, K., & Quirchmayr, G. (2012a). Conducting a privacy impact analysis for the analysis of communication records. In *Proceedings of the International Conferences on Perspectives in Business Informatics Research (BIR 2012)* (pp. 148–161). Springer.

Hofbauer, S., Beckers, K., & Quirchmayr, G. (2012b). A privacy preserving approach to call detail records analysis in VoIP systems. In *Proceedings of the International Conference on Availability, Reliability and Security (ARES)—7th International Workshop on Frontiers in Availability, Reliabilityand Security (FARES 2012)* (pp. 307–316). IEEE Computer Society.

Hofbauer, S., Beckers, K., Quirchmayr, G., & Sorge, C. (2012c). A lightweight privacy preserving approach for analysing communication records to prevent VoIP attacks using toll fraud as an example. In *Proceedings of the Second International Symposium on Anonymity and Communication Systems (ACS-2012)* (pp. 992–997). IEEE Computer Society.

Hohfeld, W. N. (1917). Fundamental legal conceptions as applied in judicial reasoning. *The Yale Law Journal, 26*(8), 710–770.

Hüning, N., Gsell, H., Weißenberg, N., Hirsch, M., & Beckers, K. (2010). Process compliance und process intelligence in der finanzwirtschaft. *ERP Management, 2*, 39–42.

IETF. (1997). Hmac: Keyed-hashing for message authentication (IETF RFC 2104). Fremont, U.S.: Internet Engineering Task Force (IETF). Retrieved from http://tools.ietf.org/rfc/rfc2104.txt

I. G. (2007). Cobit 4.1. IT Governance Institute (I. G.) ISA.

ISO. (2008). Quality management systems—Requirements (ISO/IEC 9001). Geneva, Switzerland: International Organization for Standardization (ISO).

ISO/IEC. (2005a). Information technology—Security techniques—Code of practice for information security management (ISO/IEC 27002). Geneva, Switzerland: International Organization for Standardization (ISO) and International Electrotechnical Commission (IEC).

ISO/IEC. (2005b). Information technology—Security techniques—Information security management systems—Requirements (ISO/IEC 27001). Geneva, Switzerland: International Organization for Standardization (ISO) and International Electrotechnical Commission (IEC).

ISO/IEC. (2008). *Information technology—security techniques—information security risk management* (ISO/IEC 27005). Geneva, Switzerland: International Organization for Standardization (ISO) and International Electrotechnical Commission (IEC).

ISO/IEC. (2009). Information technology—Security techniques—Information security management systems—Overview and Vocabulary (ISO/IEC 27000). Geneva, Switzerland: International Organization for Standardization (ISO) and International Electrotechnical Commission (IEC).

ISO/IEC. (2012). Common Criteria for Information Technology Security Evaluation (ISO/IEC 15408). Geneva, Switzerland: International Organization for Standardization (ISO) and International Electrotechnical Commission (IEC).

ISO/IEC. (2013). Information technology—Security techniques—Information security management systems—Requirements (ISO/IEC 27001). Geneva, Switzerland: International Organization for Standardization (ISO) and International Electrotechnical Commission (IEC).

Jackson, M. (2001). *Problem frames. Analyzing and structuring software development problems.* Boston: Addison-Wesley.

Jansen, W. A. (2011). Cloud hooks: Security and privacy issues in cloud computing. In *Hawaii International Conference on System Sciences (HICSS)* (pp. 1–10). IEEE Computer Society.

Kalloniatis, C., Kavakli, E., & Gritzalis, S. (2008). Addressing privacy requirements in system design: The PriS method. *Requirements Engineering, 13*, 241–255.

Kersten, H., Reuter, J., & Schröder, K.-W. (2011). IT-Sicherheitsmanagement nach ISO 27001 und Grundschutz. Vieweg+Teubner.

Klipper, S. (2010). Information Security Risk Management mit ISO/IEC 27005: Risikomanagement mit ISO/IEC 27001, 27005 und 31010. Vieweg+Teubner.

Lyubimov, A., Cheremushkin, D., Andreeva, N., & Shustikov, S. (2011). Information security integral engineering technique and its application in isms design. In *Proceedings of the International Conference on Availability, Reliability and Security (ARES)* (pp. 585–590). IEEE Computer Society.

Mell, P., & Grance, T. (2009). The NIST definition of cloud computing. Working Paper of the National Institute of Standards and Technology (NIST).

Moeller, R. (2007). *Coso enterprise risk management: Understanding the new integrated ERM framework.* New York: Wiley.

Montesino, R., & Fenz, S. (2011). Information security automation: how far can we go? In *Proceedings of the International Conference on Availability, Reliability and Security (ARES)* (pp. 280–285). IEEE Computer Society.

Moyano, F., Beckers, K., & Fernandez-Gago, C. (2014). Trust-aware decision-making methodology for cloud sourcing. In *Proceedings of the 26th International Conference on Advanced Information Systems Engineering (CAiSE)*. (pp. 136–149). Springer.

OECD. (1980). OECD Guidelines on the Protection of Privacy and Transborder Flows of Personal Data (Technical report). Paris, France Organisation for Economic Co-operation and Development (OECD). Retrieved from http://www.oecd.org/document/18/0,3746,en_2649_34255_1815186_1_1_1_1,00&&en-USS_01DBC.html.

Opdahl, A. L., & Sindre, G. (2009). Experimental comparison of attack trees and misuse cases for security threat identification. *Information and Software Technology, 51*, 916–932.

Open Group. (2009). The open group's risk taxonomy. Technical report. Berkshire, United Kingdom: Author. (http://pubs.opengroup.org/onlinepubs/9699919899/toc.pdf).

Scarfone, K. A., Souppaya, M. P., & Hoffman, P. (2011). Sp 800-125. Guide to security for full virtualization technologies. Technical report. Gaithersburg, MD, United States NIST.

Shaikh, F., & Haider, S. (2011). Security threats in cloud computing. In *2011 International Conference for Internet Technology and Secured Transactions* (ICITST), (pp. 214–219).

Siena, A., Perini, A., & Susi, A. (2008). From laws to requirements. In *RELAW* (pp. 6–10). IEEE Computer Society.

Siena, A., Perini, A., Susi, A., & Mylopoulos, J. (2009). A meta-model for modelling law-compliant requirements. In *Proceedings of the International Workshop on Requirements Engineering and Law* (RELAW) (pp. 45–51). IEEE Computer Society.

Streitberger, W., & Ruppel, A. (2009). Cloud-Computing Sicherheit—Schutzziele. Taxonomie. Marktübersicht, Fraunhofer Institute for Secure Information Technology (SIT). Technical report. Darmstadt, Germany: Fraunhofer Institute for Secure Information Technology (SIT).

Summerville, I. (2007). *Software engineering* (8th ed.). Boston: Addison-Wesley.

UML Revision Task Force. (2010). OMG unified modeling language: Superstructure.

Vaquero, L. M., Rodero-Merino, L., Caceres, J., & Lindner, M. (2008). A break in the clouds: Towards a cloud definition. Special interest group on data communication (SIGCOMM). *Computer Communication Review, 39*(1), 50–55.

Chapter 13
Validation and Extension of Our Context-Pattern Approach

Abstract The previous chapter introduced our PACTS methodology that supports the establishment of a cloud-specific information security management system (ISMS) compliant to the ISO 27001 standard. In this chapter, we present the results of our collaboration with the industrial partners of the ClouDAT project. The Clou-DAT project develops a method including tool support to help small and medium enterprises active in the cloud computing domain to establish an ISO 27001 ISMS. The members of the ClouDAT project decided to base their method on our PACTS method and evaluated our approach. The results of their validation offered valuable insights, which we discuss in this chapter. In addition, the ClouDAT project members were missing a simpler way to formulate security requirements and our collaboration resulted in an extension of PACTS for this purpose. In particular, our joint work focused on the creation of a textual pattern for security requirements, which can be instantiated with elements such as stakeholders from our cloud system analysis pattern.

13.1 Introduction

Security and privacy concerns are essential in cloud computing scenarios because cloud customers and end customers have to trust the cloud provider with their critical business data and even their IT infrastructure. In projects, these are often addressed late in the software development life cycle because they are difficult to elicit in cloud scenarios, due to the large amount of stakeholders and technologies involved. We contribute a catalog of security and privacy requirements patterns that support software engineers in eliciting these requirements, as requirement patterns provide artifacts that support reusing requirements. Our work shows how these requirements can be classified according to cloud security and privacy goals. Furthermore, we provide a structured method on how to elicit the right requirements for a given scenario. We mined these requirements patterns from existing security analyses of public organizations such as European Union Agency for Network and Information Security

© Springer International Publishing Switzerland 2015
K. Beckers, *Pattern and Security Requirements*,
DOI 10.1007/978-3-319-16664-3_13

(ENISA)[1] and the Cloud Security Alliance (CSA),[2] from the practical experience of the industrial partners in the ClouDAT project, and from our previous research in cloud security (Chap. 12).

Software requirements that appear in documents (short: SRD) can be classified into categories, namely functional, quality (Fabian et al. 2010), and nonfunctional requirements:

- **Functional requirements** describe the functionality of a system that satisfies the needs of its stakeholders. Based on these requirements, we can derive the activities that describe how requirements fulfill the needs of its stakeholders. Subsequently, we can identify the assets bound to the functional requirements and activities. A functional requirement could be *scalable data storing*. One of the activities associated with this requirement is the activity *store distributed* that refers to an asset *customer data* in our example.
- **Security requirements** are typically confidentiality, integrity, or availability requirements. They refer to a particular piece of information, the *asset*, that should be protected, and they indicate the *counter-stakeholder* against whom the requirement is directed. A *stakeholder* is an individual, a group, or an organization that has an interest in the system under construction. Furthermore, the *circumstances* of a security requirement describe application conditions of functionality, temporal, spatial aspects, or the social relationships between stakeholders.
- **Other Nonfunctional Requirements** are requirements that express needs of stakeholders, which usually cannot be verified, for example, the usage of the system has to be intuitive.

In our recent research (Beckers et al. 2014a), we analyzed a number of SRD from small and medium enterprises (SME) with regard to the consideration of security. We observed that a big percentage of SMEs had problems eliciting and describing their security needs. The documents contained only very vague statements toward security, e.g., the system must not have any security issues, while functional requirements, in contrast, were given in more detail. This is the case because the SMEs were able to describe what their required functionalities were. Moreover, the relevant security requirements were often very similar. Hence, we propose reusable textual requirement patterns. We previously analyzed the activities required to describe the context of cloud computing systems and to elicit security and privacy requirements. We used analysis patterns of clouds to support these activities (Beckers et al. 2013). We illustrated how to describe security textually from this information (Beckers et al. 2013), which should be in compliance with the security standard ISO 27001 (ISO/IEC 2005).

Our contribution is a pattern catalog for security requirements with regard to cloud computing systems. We base our catalog on the work of Withall (2007) from Microsoft, who provides guidelines and examples for formulating software requirements. Other researchers have also proposed catalogs for textual requirements.

---

[1] ENISA: http://www.enisa.europa.eu.
[2] Cloud Security Alliance (CSA): https://cloudsecurityalliance.org.

Palomares et al. (2013) present a catalog for functional software requirements for content management systems. Liu et al. (2008) present requirements patterns for networked software systems. Our work differs from these approaches because we focus on providing reusable requirements specifically for security in the context of the cloud computing domain.

The SMEs also had problems regarding the category "other nonfunctional requirements". However, this category is not within the scope of this work. Therefore, it will not be considered further.

This chapter is based on the publications (Beckers et al. 2013, 2014a, b) in which we create a process for pattern mining, conducting the mining of threats in existing documentation, and defining validation criteria for instantiating the textual security requirements pattern. Ludger Goeke and Isabelle Côté described the security requirements patterns and created a mapping between these requirements and the ISO 27001 standard. Our work will be integrated with the PACTS method in the future in Step 4 *Analyze Threats* and Step 6 *Create Security Policies and Reason about Controls*. Step 4 will be enhanced with a mapping between the cloud threats described in PACTS and the security requirements patterns. This has the effect that when a certain threat is selected, the security requirements mapped to this threat are selected, as well. Step 6 will benefit from the mappings of the security requirements to the ISO 27001 controls. This mapping will be used to consider the selected patterns in Step 4 and automatically suggest ISO 27001 controls to address the instantiated security requirements patterns.

13.2 The ClouDAT Framework

The ClouDAT framework is a result of the ClouDAT project.[3] The project, in turn, is a collaboration of the SMEs *ITESYS GmbH* and *LinogistiX* as well as the *University of Duisburg-Essen (UDE)*. The project has several associated partners: *paluno—the Ruhr institute for software technology*, the *Fraunhofer Institute of Software and Systems Engineering (ISST)*, and the *TÜV NORD GROUP*. This project is funded by the EU and the Ministry of Innovation, Science, Research and Technology of the German State of North Rhine-Westphalia. The SMEs involved in this projects make sure that the obtained results, frameworks, tools, and methods meet the needs shared by most SMEs.

The developed framework is going to be published as open source. This allows interested parties to try out and use the framework free of charge.

The framework's goal is to provide a means for an SME to create a cloud-specific information security management system (ISMS) compliant to the ISO 27001 (ISO/IEC 2005) standard. An ISMS is a process that ensures the security of an organization or parts there of. Currently, the framework includes:

---

[3]The ClouDAT project homepage: http://ti.uni-due.de/ti/clouddat/de/.

- A structural metamodel of a cloud system and a corresponding context-pattern and templates to elicit all relevant information of a cloud scenario (Chap. 10). The context-pattern is part of a larger pattern language that provides the means to also support different types of systems like service-oriented architectures (Chap. 11).
- A simple method that describes how to conduct a security analysis and to create a cloud-specific ISMS (Chap. 12).
- Tool support for eliciting and analyzing the required information and also to create the required documentation for an ISO 27001 certification (Chap. 10).
- A catalog of security requirements patterns for cloud systems.
- A mapping of security requirements patterns to mandatory controls in Annex A of the ISO 27001 standard.

Changes for ISO 27001:2013 Compliance

ISO 27001:2013 allows to select any control to treat a risk, but the controls have to be mapped to controls in ANNEX A of the standard (ISO/IEC 2013, p. 4). But it is possible to use only the controls in ANNEX A, as well. Hence, our approach is still applicable. Just note that the controls in ANNEX A differ in ISO 27001 and ISO 27001:2013 slightly. We provide a mapping between the controls of ISO 27001 and ISO 27001:2013 in Appendix C that illustrate the differences.

13.3 A Catalog for Cloud Security Requirements Patterns

We present a catalog of security requirements patterns for the domain of cloud computing. We build our patterns from state-of-the-art threat and risk catalogs, which we map to security requirements. In particular, we consider the threat and risk catalogs of the Cloud Security Alliance (2010), Gartner (2008), the European Network and Information Security Agency (2009), the German Federal Office for Information Security (BSI) (BSI 2010; Essoh 2010), Eurocloud (2010), BITKOM (2009), and Fraunhofer SIT (Streitberger and Ruppel 2009), as well as the book by Mather et al. (2009), and the works of Mell and Grance of the U.S. National Institute of Standards and Technology (Mell and Grance 2009).

The current version of the catalog contains a total of 78 security requirements. This set of security requirements does not claim to be complete. Each user of the catalog can extend the set of requirements, if necessary. We constructed the structured ClouDAT method for mining security requirements patterns. This ClouDAT method is depicted in Fig. 13.1 and contains the following steps.

1. *Analysis*: We analyze all the listed threat and risk catalogs (see above in the beginning of this subsection) for risks, threats, security, and privacy related texts. We marked the identified parts of the texts and stated to which cloud stakeholders

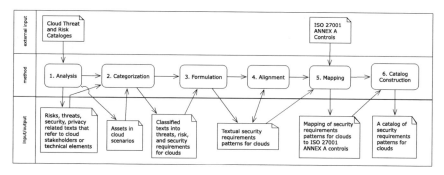

Fig. 13.1   Our catalog construction method

or technical elements they refer to. We also mined the identified assets in the cloud scenarios. The term asset refers to anything a stakeholder places value upon.

2. *Categorization*: We classify the identified texts into threats, risks, and security requirements. Threats refer to, e.g., attackers that may exploit vulnerabilities in the cloud system. Risks focus on unwanted incidents that can cause a threat to occur. A security requirement states the confidentiality, integrity, and availability needs for the protection of an asset.

3. *Formulation*: We transform each threat and risk into security requirements. This is achieved by stating the needs for protection of assets being referred to. In this step, we check if security requirements appear multiple times. We formulate a new textual security requirement pattern for each requirement that appears at least twice.

4. *Alignment*: We check for each of the selected requirement patterns if it refers to a cloud element or a stakeholder. We replace these parts of the textual requirements patterns with placeholders of a certain type. Such a type refers to an element of our cloud system analysis pattern (Chap. 10), e.g., an IaaS service.

5. *Mapping*: We map the security requirements patterns to security controls of the ISO 27001 standard. We base our mappings on the content of the patterns and past experiences. If an ISO 27001 control is a possible solution to a security requirement pattern according to our judgement or our previous experience, we created a mapping and documented the argumentation.

6. *Catalog Construction*: We categorize the resulting security requirements into protection needs. An example for such a need is authenticity. Thus, we obtain authenticity requirements patterns. We adopted the means of pattern representation proposed by Withall (2007) to our approach (Sect. 13.4).

13.4 Representing Cloud Security Requirements Patterns

Withall (2007) provides guidelines and examples for formulating software requirements based upon project experience. The author explains the need for documenting requirements including assumptions, glossary, document history and references.
Withall's work aims at writing textual requirements, which also consider domain
knowledge in the form of assumptions to these requirements. Our work differs
from Withall's, because we do not only provide patterns for requirements. Our work
focuses on the elicitation of domain knowledge with our cloud system analysis pattern (Chap. 10), as well. The requirement patterns mentioned in Withall (2007) are
ordered by the domains

- fundamental,
- information,
- data entity,
- user function,
- performance,
- flexibility,
- access control, and
- commercial.

For more details on these domains, please refer to Withall (2007).

We extended the above-mentioned domains with the following domains relevant
for cloud computing, which we have identified throughout the project work. These
domains are:

- confidentiality,
- integrity,
- availability,
- authenticity,
- security management,
- non-repudiation,
- privacy,
- compliance, and
- transparency.

Structure of Requirement Patterns

We discuss the structure of security requirements patterns on the basis of the example
security requirement pattern given in Table 13.1. This pattern specifies the requirements to the integrity of the communication regarding cloud computing.

Table 13.1 Security Requirement Pattern "Integrity of cloud communication"

Security Requirement Pattern: I6 Integrity of cloud communication	
Section	**Meaning**
1. Basic details	• Index: I6 • Version number: 1.2 • Date of last change: 07.05.2013 • Specification language: English • Domain: Integrity • Related patterns: Aut13, SM10, SM11, SM12 • Anticipated frequency: in average 2 • Pattern classifications: Functional: No; Security: Yes
2. Applicability	Use the integrity of cloud communication security requirement pattern to specify for which communication connection the integrity must be preserved.
3. Discussion	An *integrity of cloud communication security requirement* specifies the two participants for whose communication the integrity has to be preserved. These security requirements are subdivided in the following categories: • **Category 1: Direct Stakeholder to cloud service:** Describe the need of integrity for the communication between direct stakeholders and cloud services. • **Category 2: Cloud services among each other:** Describe the need of integrity for the communication between cloud services.
4. Content	An *integrity of cloud communication security requirement* has to contain the two participants for whom the integrity of the communication has to be preserved. A cloud communication security requirement should contain: • **Cloud communication participants** The two participants in a cloud computing communication have to be – a direct stakeholder and a cloud service **or** – two cloud services.

5. Template(s)	**Summary**	**Definition**
	Integrity of the communication between <<direct stakeholder>> and <<cloud service>>	The integrity of the communication between <<direct stakeholder>> and <<cloud service>> shall be preserved.
	Integrity of the communication between <<cloud service>> and <<cloud service>>	The integrity of the communication between <<cloud service>> and <<cloud service>> shall be preserved.
6. Example(s)	**Summary**	**Definition**
	Integrity of the communication between administrators and IaaS	The integrity of the communication between administrators and IaaS shall be preserved.
	Integrity of the communication between end customers and SaaS	The integrity of the communication between end customers and SaaS shall be preserved.
	Integrity of the communication between IaaS and PaaS	The integrity of the communication between IaaS and PaaS shall be preserved.

Generally, a requirement pattern must have a name and a unique identifier. Our example (Table 13.1) considers a security requirement pattern named: "Integrity of cloud communication". It is identified through the unique index: "I6".

In Withall (2007), the structure of requirement patterns is defined by the sections *Basic Details*, *Applicability*, *Discussion*, *Content*, *Template(s)*, *Example(s)*, *Extra requirements*, *Considerations for development* and *Considerations for testing*.

These sections are discussed in the following. However, we omit the sections *Extra requirements*, *Considerations for development* and *Considerations for testing*, as we do not require them for our work.

Basic Details

The basic details provide basic information about a requirement pattern. One basic detail is the pattern manifestation. It represents meta-information about the requirement pattern such as version number, date of last change, or used specification language. If different variants or versions of a requirement pattern exist, more than one pattern manifestation is given. Because there exists only one version of our security requirement pattern example (see Table 13.1), it only has one pattern manifestation. It has the version number *1.2* and the last changes were made on *07.05.2013*. The specification language of our example is *English*. Another item of the basic details describes the domain to which a requirement pattern belongs. In our example, the security requirement pattern belongs to the domain *Integrity*.

Furthermore, related requirements patterns are stated. These are requirements patterns that are applied in the same context as the considered requirements pattern. If necessary, the kinds of the appropriate relationships can be described. Our example security requirements pattern is related to several of our other security requirement patterns, namely to *Physical protection/authentication of people (Aut13)*, *Destruction of cryptographic keys (SM10)*, *Securing integrity and confidentiality of cryptographic keys (SM11)*, and *Sufficient quality of cryptographic keys (SM12)*. The relations are depicted in Fig. 13.2. Example 13.1 shows the security requirement pattern *SM12*.[4] It states the requirement for the generation of cryptographic keys that are relevant in the context of the integrity of cloud communication.

The anticipated frequency represents the number of usages of the concerned requirement pattern in a typical requirement specification. We expect for our example requirement that it will be used twice in average. These estimates are based on previous experiences with SMEs.

A requirement pattern can be classified by a list of pattern classifications. The items of this list are represented by a name-value-tuple represented as *Name: Value* (Withall 2007). The different tuples are separated by semicolons (cf. Withall 2007). Our security requirement pattern example is assigned with *Security*. Therefore, the tuples for our example look as follows: Functional: No, Security: Yes.

Example 13.1 Mechanisms for the generation of cryptographic keys shall ensure sufficient quality of the keys in ≪*all cloud elements*≫.

The security requirement pattern *SM12*

---

[4]Note that SM stands for security management.

Applicability

Usually, the application of a requirement pattern can succeed only in one clear situation (cf. Withall 2007). This situation is described in the applicability section. If necessary, this section can also contain descriptions of situations where the requirement pattern should not be applied (cf. Withall 2007). Our security requirement pattern example (see Table 13.1) is used when the integrity of cloud communications has to be preserved.

Discussion

The discussion section provides context information that is helpful for the specification of the requirement. It can support the creation of requirements by describing a creation process, list topics that shall be considered or caution against pitfalls (cf. Withall 2007). The discussion in our example (see Table 13.1) mentions that it can be instantiated for the following two categories of cloud communication:

Category 1 Cloud communications between direct stakeholders and cloud services.

Category 2 Cloud communications between cloud services.

Content

This section lists items of information that a requirement of this type must convey or can optionally convey, respectively. An information item is a tuple of an item name and an informal text. Optional information items are marked by the keyword *optional* that is enclosed in parentheses. In our example (see Table 13.1), the content is described by one information tuple that contains the values "a direct stakeholder and a cloud service", and "two cloud services". It represents information regarding the participants in a cloud communication in the context of our security requirement pattern example.

Templates

Templates act as a starting point for describing requirements. They provide text passages that are characteristic for a requirement of a certain type. These text passages can contain placeholders that can be filled with appropriate information regarding the relevant actual situation. Each template corresponds to an information item of the content. In our example, each template considers either "two cloud services" or "a direct stakeholder and a cloud service". Here, a requirement pattern does not have to contain templates for all information items. In Withall (2007) the rule of thumb is mentioned that a requirement pattern shall only contain templates that are used in at least 20 % of the specified requirement patterns. It is not mandatory that all templates are used for the specification of a requirement. The modifying or disregarding of the text passages during the instantiation of the requirement pattern is also allowed.

A template consists of a *summary* and a *definition*. The definition represents the specification of a part of the requirement. A definition is identified by the corresponding summary. It typically consists of a text regarding the pattern name and a text that allows the distinction from the other templates of the requirement pattern. The summary gives a short hint about the content of the description.

The content and description can contain the above-mentioned placeholders. A placeholder is marked by doubled-angled brackets and an italic notation.

Templates can contain optional parts. These are text passages that can or cannot be used regarding the situation the requirement is specified for. Optional parts allow for the creation of alternative requirements for typical situations. As mentioned above, text passages could be disregarded, albeit they are no optional parts. Optional parts are marked by square brackets.

Our security pattern example specifies two templates (see Table 13.1). Here, each template relates to a category regarding the individual participants in cloud communications, for which the integrity of the communications has to be preserved. The templates contain the placeholders ≪direct stakeholder≫ and ≪cloud service≫ as well as just ≪cloud service≫, respectively. During the application of the ClouDAT framework, a cloud computing scenario is specified. This cloud computing scenario is based on the structural metamodel of a cloud mentioned in Sect. 13.2. The placeholders in our security requirement patterns can be replaced by the appropriate information from the specified cloud scenario taken from an instance of our cloud system analysis pattern (Chap. 10). Regarding the templates in our example, information in terms of the representation of direct stakeholders and cloud services in a corresponding cloud computing scenario can be inserted into the templates.

Examples

A requirement pattern should contain at least one requirement as an example of an instantiation of the requirement pattern. In our *Integrity of cloud communication* security requirement pattern the first two examples are instantiations of the first template. Their content belongs to Category 1, meaning that they specify requirements for the cloud communication between direct stakeholders and a cloud service (Sect. 13.5). The first example considers the communication between direct stakeholders, namely administrators, and the cloud service IaaS. In the second example, the communication between end customers of the cloud and the cloud service SaaS is considered. The third example is an instantiation of the second template. Its content belongs to Category 2. Accordingly, it specifies requirements for the communication between cloud services. In this case, the communication between cloud services IaaS and PaaS is referenced. The PaaS service has to contact the IaaS service to get access to cloud resources like databases.

In addition to creating a catalog of security requirement patterns, we mapped our security requirement patterns to appropriate security measures in form of the controls given in Annex A of ISO 27001 (ISO/IEC 2005). This mapping was conducted in the following four steps:

1. Determining particular assets, security goals and important terms (or keywords) regarding individual security requirement patterns from our catalog. By means of this additional information, the set of possibly relevant security requirement patterns regarding an ISO 27001 control can be narrowed down.
2. Screening the ISO 27001 Annex A and assign the relevant ISO 27001 controls to the individual security requirement patterns. This step contains the following substeps:

   a. Identifying a set of security requirement patterns for every ISO 27001 control with the help of the above-mentioned additional information.

   b. Detailed analysis of the meaning of each security requirement patterns determined in (a) in relation to the terms and meanings of the ISO 27001 controls.

   c. Capturing of relevant mappings.

3. Iterating over our security requirement patterns and executing full text searches in the ISO 27001 Annex A. Here, the search items are the according additional information (assets, security goals, keywords) from the appropriate security requirement pattern.

4. Mutual review among the project participants of the created mapping and discussion of open issues.

In Fig. 13.2, we illustrate the mapping of ISO 27001 controls to our example security requirement pattern given in Table 13.1. Furthermore, we provide the full text of

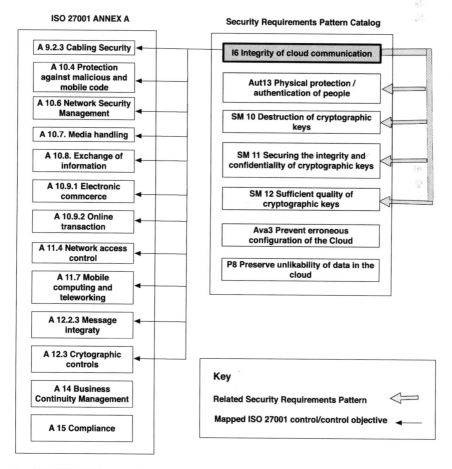

Fig. 13.2 ISO 27001-control-mapping for "integrity of cloud communication" security requirement pattern

the mapped ISO 27001 control *A.10.6.1* in Example 13.2. The control refers to the protection of information in transit over networks. We determined that this protection includes the protection of the integrity of the information.

Example 13.2 Networks shall be adequately managed and controlled, in order to be protected from threats, and to maintain security for the systems and applications using the network, including information in transit.

ISO 27001 control *A.10.6.1*

13.5 Discussion and Analysis

We analyzed our security requirements pattern catalog against scenarios developed in the ClouDAT project and discussed this catalog with security consultants, who have already used our catalog partially in industrial projects. This resulted in a consolidated set of security requirements patterns.

During the discussion, we found out that in ISO 27001 projects the selected controls have to be documented and reasoned in the so-called *Statement of Applicability*. Especially, the effort for this task is significantly reduced by the presented catalog and tool support. The security consultants also mentioned that this structured procedure

- helps to elicit security requirements in more detail,
- supports the identification of threats to assets,
- helps not to forget relevant security concerns, and
- supports the identification and classification of functional as well as security requirements.

We analyzed how many security requirements were selected and which were excluded, as well. Based on our experience in the field of cloud security, we reconstructed 73 % of the initial SRD documents, based on only the context description of these documents. The remaining requirements were very specific to the documents. In addition, using our catalog, we could elicit an average of 17 % of additional security requirements for the SRD documents.

In discussions with evaluators of the ISO 27001 standards, the following concerns toward our pattern catalog were raised:

- Security requirements patterns might hinder security reasoning, because security analysts will randomly select patterns.
- The reading of the entire pattern catalog is time consuming.
- The structure of the requirement patterns has to be learned beforehand, and
- Our method does not support the do-check-act phases of the ISO 27001 standard. We support exclusively the plan phase of the ISO 27001 standard.

13.6 Tool Support

In this section, we present our tool support, which is an extension of the Cloud Pattern tool shown in Sect. 10.4.5. First, we consider the Cloud System Analysis Pattern Tool (CSAP Tool). This tool supports the instantiation of the CSAP. It also provides modeling support that allows one to extend the CSAP with additional instantiable CSAP elements and the corresponding relations between them and other CSAP elements. Because this procedure is not part of our method, it will not be considered further. The architecture of our tool is shown in Fig. 13.3. The notation used to specify the pattern is based on UML, i.e., stick figures represent roles, boxes represent concepts or entities of the real world, and named lines represent relations (associations) equipped with cardinalities. Our tool is based on the Eclipse platform (Eclipse Foundation 2011a) as well as its plug-ins Eclipse Modeling Framework (EMF) (Eclipse Foundation 2012a) and the Graphical Editing Framework (GEF) (Eclipse Foundation 2012b). We further use the Graphical Modeling Framework (GMF) (Eclipse Foundation 2011b) to generate graphical editors.

Our CSAP metamodel (Sect. 10.4.5) serves as the basis for generating the appropriate (ecore) model using EMF. This (ecore) model enables the generation of

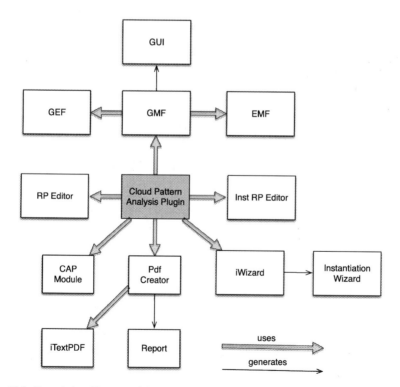

Fig. 13.3 Extended architecture of the cloud pattern tool

components representing CSAP information within the CSAP Tool. The cloud system analysis pattern GUI is generated by GMF using our CSAP metamodel. The CSAP Tool uses the Eclipse interface IWizard to create a wizard to support the instantiation of a cloud pattern. The wizard provides a graphical interface that asks the user for the necessary information to instantiate stakeholders, cloud elements, and assets. It asks, for example, for the name and owner of an asset. In addition, the wizard supports instantiating several instances of one instantiable CSAP element. For example, the wizard can instantiate four indirect stakeholders at once.

Furthermore, we equipped the wizard with validation capabilities. An example for an already implemented validation condition is to check whether all fields of a stakeholder in the corresponding template have entries. It is also possible to generate a report, called CSAP report. It contains the graphical representation of the model as well as the texts provided in the stakeholder template. We use the iTextPDF interface of Eclipse to generate the pdf files for the report. For the management of security requirement patterns, we provide a Requirements Pattern Editor (RP Editor). Its implementation is also based on the Eclipse platform (Eclipse Foundation 2011a) as well as the aforementioned plug-ins. The RP Editor provides functionality for displaying, creating, modifying, and deleting security requirement patterns.

Figure 13.4 shows the modification of a predefined security requirement pattern. The requirement with the index A4 (Note that in the tool the index is called Prefix) is

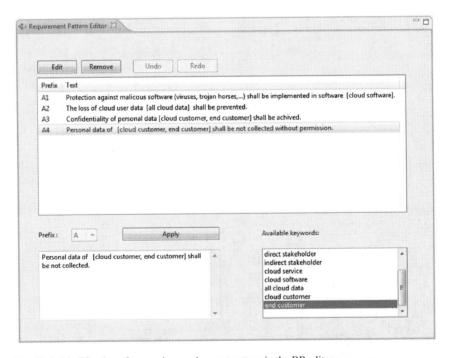

Fig. 13.4 Modification of a security requirement pattern in the RP editor

changed from the text "Personal data of [cloud customer, end customer] shall not be collected without permission" to "Personal data of [cloud customer, end customer] shall not be collected". The changed text is shown on the lower left corner of the figure.

For creating and modifying security requirement patterns, the RP Editor provides keywords for referencing CSAP elements. These keywords represent the instance types of the corresponding CSAP elements. Considering a newly created or modified security requirement pattern, it has to be ensured that its structure is valid. Based on the RP Editor it provides an appropriate validation function, which is executed before adding a newly created or modified security requirement pattern. The management of security requirements is provided by another editor, the so-called *Instantiated Requirement Pattern Editor (InstRP)*. The functionality of the InstRP Editor comprises instantiating security requirement patterns as well as displaying, modifying, and deleting security requirements. The implementation of the InstRP Editor is based on the same technologies as the RP Editor. During the instantiation of security requirement patterns, the names of the referenced CSAP elements are inserted into the corresponding generic text passages automatically. For this procedure, the InstRP Editor provides an appropriate wizard that allows the selection of the relevant CSAP element names.

The representation of predefined security requirement patterns in the InstRP Editor is shown in Fig. 13.5. Potential cloud customers have the possibility to select the relevant predefined security requirement patterns that shall be instantiated. In

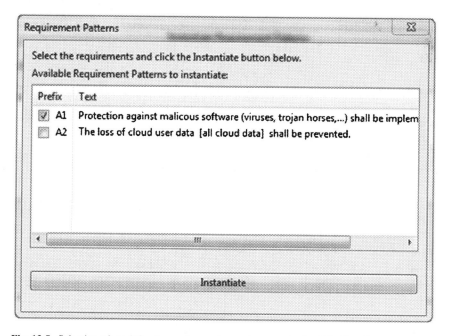

Fig. 13.5 Selection of predefined security requirement patterns in InstRP editor for instantiation

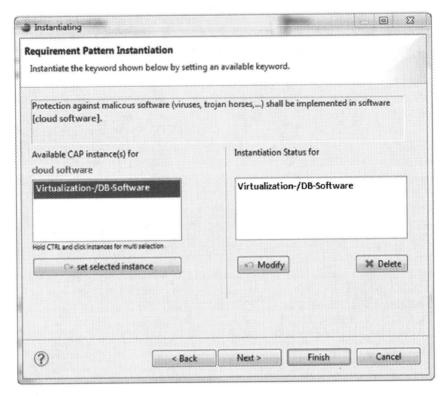

Fig. 13.6 Instantiation of a security requirement pattern in the InstRP editor

Fig. 13.5, for example, only the security requirement pattern from Example 1 is selected. Figure 13.6 shows the instantiation of the above-selected security requirement pattern in the InstRP Editor. Here, the InstRP Editor shows the security requirement pattern that shall be instantiated and the term (or keyword) "cloud software" that shall be replaced by the appropriate information of the corresponding CSAP instance. In our instantiation the cloud pattern, depicted in Fig. 12.7, is considered. According to this figure, the only selectable CSAP information is Virtualization-/DB-Software. The resulting security requirement is given in Fig. 13.7.

The set of security requirements has to be consistent to the corresponding cloud pattern instance. This consistency has to be ensured by an appropriate validation function, meaning the instance type of a referenced cloud pattern element has to be captured in the representation of a security requirement. Based on the name and the instance type, the validation function can compare the references in the security requirements against the elements in the corresponding cloud pattern instance.

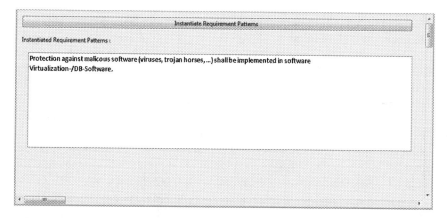

Fig. 13.7 Representation of security requirements in the InstRP editor

13.7 Discussions with Practitioners

We discussed the procedure presented in Chap. 12 with the industrial partners from the NESSoS project.[5] The practitioners valued the idea of reusing analysis results, e.g., the assets identified using the pattern and the idea of consistency checks of the instantiations of the pattern. They also appreciated the possibility to extend the cloud pattern. The practitioners liked that these features have the potential to reduce the time for establishing a security standard and lead to consistent documents, as well.

The practitioners encouraged us to continue on our work and considered the following issues, which they are struggling with:

- Scope changes shall be possible at any point during the process. During the establishment of the standard, it is usually discussed repeatedly what is in or out of the scope. These discussions can occur until the very last step of the process, and a quick change of scope and all related artifacts would be beneficial. Specific consistency checks for this purpose would be a worthy aim.
- Risk management is essential to the establishment of the standard (Sect. 5.6). In practice, according to the experience of the industrial partners, employees accept risks far too quickly. This leads to an acceptance of risks that should not be, considering an objective perspective. The practitioners suggested to make the risk management method exchangeable, so every user can apply the risk management method, which addresses the mentioned problem best. Risk management is already a single step in PACTS and has defined inputs and outputs. We therefore assume that it should be possible to exchange CORAS as the risk management procedure.
- The industrial partners stated that the risk management study regarding clouds from the European Network and Information Security Agency (2009) is highly relevant for our method and should be included. We considered the study in our followup

---

[5]The NESSoS project: http://www.nessos-project.eu.

work of PACTS, which is described in this chapter. In particular, Sect. 13.3 on considers the ENISA study.

- The industrial partners suggested to have a template-based guide through the entire method. At the moment, we provide tool support for parts of PACTS such as the scope definition, but we do not have a tool support yet that covers the entire method. When we create the tool support, we will consider this suggestion.
- The management commitment is a document in our method. The industrial partners suggested that our method should also provide assistance in dealing with the management if they do not want to uphold their statements given in the management commitment. We would like to consider this issue, but this should be done in collaboration with experts in the organizational management science.
- The industrial partners suggested that they would like to see a different definition of cloud computing as the basis for PACTS.
  They wanted to consider a particular definition of cloud computing, which states that it is impossible to determine the location of stakeholder's data in the cloud. The reason for this definition should be a particular technical implementation of clouds. We declined this request, because we showed in Sect. 12.12 that the German Federal Data Protection Act (BDSG) restricts a cloud from transferring personal data outside of the EU.

We presented our PACTS method in an early stage to the industrial partners of the ClouDAT[6] project, as well. The ClouDAT project aims at developing a framework and an open-source tool (also called ClouDAT) (see Sect. 13.2 on for details). The ClouDAT tool will provide functionalities for analyzing and documenting cloud computing systems with regard to security. The tool shall be able to check consistency and generate standard documentation compliant to security standards. In particular, the tool shall serve the needs of small companies in their certification efforts.

The industrial partners of ClouDAT are enthusiastic about the promise of the PACTS method, and one of them even joined in the effort of improving the PACTS method and completing the journal publication that is the basis for this chapter. Furthermore, the partners of the ClouDAT project choose to base their framework and tool support on the PACTS method.

The ClouDAT partners analyzed and applied the PACTS method to their cloud. They focused in their analysis on the applicability of the method specifically for small and medium enterprises. Their analysis resulted in the following suggestions, which the partners already partially implemented:

- The partners added a specific template for the ISMS policy. An ISMS policy requires a high level scenario description. The partners discuss replacing the UML use case diagrams with existing business processes instead. The processes can provide all the required information and do not require any effort for drawing new diagrams. This relies on the assumptions that business processes exist and are modeled. An open issue is, of course, how to integrate the different graphical notations or even textual descriptions of business processes in the PACTS method.

---

[6]The ClouDAT project: http://ti.uni-due.de/ti/clouddat/de/.

- The ClouDAT partners are not familiar with the CORAS method and they decided to replace it with a different method for risk management. The partners devised their own template-based method for risk management based on the ISO 27005 standard. The assumption is that removing CORAS from the PACTS method and replacing it with a series of templates shall reduce the effort for risk management, because the templates do not require users to learn any graphical notation.
- The usage of the policy pattern to select appropriate security controls seemed also difficult to the partners. At the moment, different options are analyzed, among them are the possibility to create textual security requirements patterns and define a mapping of these to ISO 27001 security controls. This idea is further explored in Chap. 13.
- The ClouDAT project focuses on supporting German small and medium enterprises. In Germany, the Federal Department of Information Security (BSI) has published a series of security standards (BSI 2011), similar to ISO 27001. The partners aim to extend PACTS to support also these methods.

Our decision with the practitioners has led to an intensive collaboration with the ClouDAT industrial partners, which led to the work presented in this chapter and a significant improvement of our research results. The collaboration continues and is likely to produce further results in the future.

13.8 Summary

In this chapter, we presented the results of the validation of our context-pattern approach and in particular the cloud pattern (Chap. 10) and the PACTS method (Chap. 12). One part of the validation was to present our works to practitioners and discuss the usefulness and applicability of our work in industry. The results of the discussion are presented in Sect. 13.7, which were generally favorable and highlight numerous possibilities for improvement of our work, which we will pursue in the future.

The second part of the validation is executed by introducing our previously mentioned work to the researchers of the ClouDAT[7] project and jointly created a new framework specifically to support the needs of the ClouDAT project, the so-called *ClouDAT framework*. These needs are in particular to provide support for Small and Medium Enterprises (SME)s for cloud-specific ISO 27001 certifications. In particular, we presented in this chapter a novel catalog of security requirements patterns for the cloud computing domain.

Our catalog is currently composed of 78 security requirements patterns.

Our pattern catalog offers the following main contributions:

- Starting from a context description of a cloud scenario, our pattern catalog can be used to elicit security requirements with little effort.

---

[7]The ClouDAT project: http://ti.uni-due.de/ti/clouddat/de/.

- The instantiation is tool-supported and uses the context description as an input. Hence, the instantiation of the patterns refers to particular parts of a cloud.
- Using a pattern catalog decreases the risk of underspecifying security requirements for a given cloud scenario. In addition, the catalog can also be used to reason why particular patterns are not selected. This documentation can be used for audits to ensure that correct decisions were taken.
- We provide a mapping of our security requirements patterns to ISO 27001 controls, which helps the user not only to define his/her needs, but also to specify how protection mechanisms should look like.

We showed a structured method for mining security requirements patterns for the cloud computing domain. The method is based on existing risk and threat catalogs and our cloud system analysis pattern.

References

Beckers, K., Côté, I., Goeke, L., Güler, S., & Heisel, M. (2013). Structured pattern-based security requirements elicitation for clouds. In *Proceedings of the international conference on availability, reliability and security (ARES)—7th international workshop on secure software engineering (SecSE 2013)* (pp. 465–474). IEEE Computer Society.
Beckers, K., Côté, I., & Goeke, L. (2014a). A catalog of security requirements patterns for the domain of cloud computing systems. In *Proceedings of the 29th Symposium on Applied Computing* (pp. 337–342). ACM.
Beckers, K., Côté, I., Goeke, L., Güler, S., & Heisel, M. (2014b). A structured method for security requirements elicitation concerning the cloud computing domain. *International Journal of Secure Software Engineering (IJSSE)*, 5(2), 20–43.
BITKOM. (2009). Cloud-computing—Evolution in der technik, revolution im business.
BSI. (2010). IT-Grundschutzkataloge. Bonn, Germany: Bundesamt für Sicherheit in der Informationstechnik (BSI)—Federal Office for Information Security Germany. (http://www.bsi.bund. de).
BSI. (2011, August). BSI Grundschutz Homepage. Bonn, Germany: Bundesamt für Sicherheit in der Informationstechnik (BSI)—Federal Office for Information Security Germany. (http://www. bsi.bund.de/DE/Themen/ITGrundschutz/itgrundschutznode.html).
Cloud Security Alliance (CSA). (2010). Top threats to cloud computing v1.0. (https:// cloudsecurityalliance.org/topthreats/csathreats.v1.0.pdf).
Eclipse Foundation. (2011a). Eclipse—An open development platform [Computer software manual]. (http://www.eclipse.org/).
Eclipse Foundation. (2011b). Eclipse graphical modeling framework (GMF). (http://www.eclipse. org/modeling/gmf/).
Eclipse Foundation. (2012a). Eclipse modeling framework project (EMF). (http://www.eclipse.org/ modeling/emf/).
Eclipse Foundation. (2012b). Graphical editing framework project (GEF). (http://www.eclipse.org/ gef/).
Essoh, A. D. (2010). Cloud computing und sicherheit–geht denn das? Bonn, Germany: Bundesamt für Sicherheit in der Informationstechnik (BSI)—Federal Office for Information Security Germany. (http://www.bsi.bund.de/cae/servlet/contentblob/808266/publicationFile/46724/ 07_essoh_bsi.pdf).
Eurocloud. (2010). Eurocloud prüfkatalog.

European Network and Information Security Agency (ENISA). (2009). Cloud computing—Benefits, risks and recommendations for information security.

Fabian, B., Gürses, S., Heisel, M., Santen, T., & Schmidt, H. (2010). A comparison of security requirements engineering methods. *Requirements Engineering—Special Issue on Security Requirements Engineering*, **15**(1), 7–40.

Gartner. (2008). Assessing the security risks of cloud computing. (http://www.gartner.com/id=685308).

ISO/IEC. (2005). Information technology—Security techniques—Information security management systems—Requirements (ISO/IEC 27001). Geneva, Switzerland: International Organization for Standardization (ISO) and International Electrotechnical Commission (IEC).

ISO/IEC. (2013). Information technology—Security techniques—Information security management systems—Requirements (ISO/IEC 27001). Geneva, Switzerland: International Organization for Standardization (ISO) and International Electrotechnical Commission (IEC).

Liu, W., He, K.-Q., Zhang, K., & Wang, J. (2008). Combining domain-driven approach with requirement assets for networked software requirements elicitation. In *Proceedings of the 2008 IEEE International Conference on Semantic Computing* (pp. 354–361). IEEE Computer Society.

Mather, T., Kumaraswamy, S., & Latif, S. (2009). *Cloud security and privacy*. O'Reilly.

Mell, P., & Grance, T. (2009). Effectively and securely using the cloud-computing paradigm. NIST. (presentation at NIST).

Palomares, C., Quer, C., Franch, X., Renault, S., & Guerlain, C. (2013). A catalogue of functional software requirement patterns for the domain of content management systems. In *Proceedings of the 28th annual acm symposium on applied computing* (pp. 1260–1265). ACM.

Streitberger, W., & Ruppel, A. (2009). Cloud-Computing Sicherheit—Schutzziele. Taxonomie. Marktübersicht, Fraunhofer Institute for Secure Information Technology (SIT). Technical report. Darmstadt, Germany: Fraunhofer Institute for Secure Information Technology (SIT).

Withall, S. (2007). Software requirement patterns. Microsoft Press.

Chapter 14
Conclusion

Abstract We have shown in the contents of this book how to support the establishment of security standards with security requirements engineering methods and patterns. This final chapter of our book summarizes our research results and our key findings when applying our work and discussing the results with practitioners in the field. We reflect on their feedback and discuss how possible improvements of our work may look like. In addition, we discuss and critically review how successful our work is at answering the research questions stated in the first chapter of this book. The answers to these research questions show under which circumstances our work is applicable and what areas require further investigations. Furthermore, we provide ideas for future directions of research of the work presented in this book. Our work has the potential to impact the way standardization efforts are conducted today, and we aim to investigate how we can support them even further in the future.

14.1 Overview

We summarize the results of this book and our key findings in Sect. 14.2. Afterward in Sect. 14.3, we discuss and critically review how successful our work is at answering the research questions stated in Chap. 1. Finally, in Sect. 14.4, we propose future directions of research for the work presented in this book.

14.2 Key Findings

We state the key findings of this book and summarize how these findings contribute to the state of the art of engineering secure systems in compliance with security standards. Based on these findings and our discussions with practitioners, we revisit our results and propose modifications to address the outcome of our discussions.

© Springer International Publishing Switzerland 2015
K. Beckers, *Pattern and Security Requirements*,
DOI 10.1007/978-3-319-16664-3_14

14.2.1 Security-Requirements-Engineering-Based Establishment of Security Standards

The establishment of security standards demands the refinement of all ambiguities and abstract descriptions in security standards and to fulfill the documentation demands of the standards. We propose to create methods for security standard establishment based on Security Requirements Engineering methods (SREs). SREs have been used for over a decade (Fabian et al. 2010, p. 9) in software engineering to elicit requirements and describe the environment of the system-to-be, refine these requirements from abstract concepts to a level of detail on which it is possible to derive architectures, and in some cases, even formalize security requirements to a degree that makes it possible to conduct mathematic proofs that the architecture is a sound and complete refinement of the requirement. SREs demand to document their application in models and texts, as well. Moreover, many SREs contain structured methods that describe in steps how to achieve refinement and documentation. These features are essential for a successful establishment.

We provided relations between two influential security standards, namely ISO 27001 and Common Criteria (Chap. 1), and the conceptual framework of SRE (Chap. 5). The framework has relations to numerous SRE methods. In combination with the relations from the framework to the standards, we concluded which SRE method supports what standard best. We showed three examples of extensions for SREs tailored to the establishment of security standards in Chaps. 6–8. We discuss the detailed results of each example in Sect. 14.3. In addition, we focus our review in this section on the overall results of our contribution to use SRE for security standard establishment.

Moreover, we showed three examples of extensions of individual SREs each tailored to support the establishment of a single security standard: *Si*-ISMS*, *ISMS-CORAS*, and *UML4PF-CC*. We used the relations to the conceptual framework from standards and SREs to identify suitable SRE candidates and extended those successfully to support standards. We showed the results to different practitioners and the industrial partners of the NESSoS project, and we also received reviews of our work from the forums in which we published our work.

Based on the feedback from practitioners and researchers in this area, we reached the following conclusions:

- The documentation part of security standards is currently based on texts and tables. SRE methods are often based on graphical models and introducing models to the certification process seemed desirable to practitioners, because in their opinion models ease the effort of understanding the system documentations. Moreover, security reasoning of different parties benefits from the use of models, because models ease to gain a common understanding of a system and its security issues.
- SREs provide structured methods for tasks such as the identification of the stakeholders of a system, its assets, threats, and countermeasures, as well. These methods provide defined starting points of the security analysis and also define when the security analysis is finished. This is helpful for practitioners particularly if they

are not extensively experienced in the field. In addition, if the engineers agree on an SRE method for establishing and certifying a security standard, they agree on the steps of the method. This relieves them of discussing or even coming up with a structured procedure to establish a security standard. Hence, the engineers can pool their efforts if the security analysis is conducted properly according to the chosen SRE method.

- The security analysis according to standards depends on the chosen level of abstraction. The problem is that for the task of security standard establishment only a limited amount of resources are available with regard to the working time of the engineers who conduct the establishment. Choosing an appropriate abstraction level helps to use these limited resources reasonably. If the abstraction is too high, important threats might not be considered, and if the abstraction is too detailed, the time resources are not sufficient. SREs have rules and examples on how detailed the system model (meaning its abstraction level) should be and help engineers to make the right decision.

- Security analysis has the goal to protect a system against attackers from the inside and outside of the system. Understanding how to model the attackers and elicit the threats these attackers can cause is an essential goal. SREs provide support in understanding how attackers work. In addition, security standards demand description of attackers in terms of resources they have, e.g., time and equipment. Combining SRE with security standards' demands holds the potential to create a very detailed attacker modeling and a structured elicitation of security threats.

- Once attackers are properly modeled and the threats they cause are elicited, countermeasures for these threats have to be devised and described in sufficient detail. The combination of SRE and security standards provides a useful contribution. The reason is that SREs offer a structured way to refine security issues. These refinements in turn help to select the right countermeasure. Security standards offer best practices like categories of countermeasures that prevailed in the past, and the selection of these countermeasures should be based on these catalogs of countermeasures the standards provide.

- SREs demand a structured and model-based description of a system and in some cases also of its environment. Based on these models, specific methods help not to forget relevant domain knowledge such as assumptions, facts, and existing policies and countermeasures. In security analysis it is often only slightly considered to describe detailed domain knowledge, which is relevant for the analysis. Changes to the domain knowledge have significant impact on the outcome of a security analysis if a software component was assumed to be secure and is often not protected by any countermeasure. Checking periodically if these assumptions hold is fundamental to the evaluation of a security analysis. Moreover, security standards demand the documentation of assumptions and other domain knowledge, and SREs help to improve the situation via structured methods.

Nevertheless, considering the practitioners and the industrial partners of the NES-SoS project, we also concluded some concerns toward our contributions. The practitioners mentioned the following limitations of our work, as well:

- The notations of the SREs have to be learned beforehand, and their methods have to be practiced in order to be understood in enough detail. In some cases, tool support exists, e.g., for UML4PF and Security Analysis for the Cloud Pattern, but it is not matured to an industry level of usability. These issues impose serious barriers for industrial application of SRE and have to be addressed in the future.
- The modeling is time-consuming in comparison to writing texts in order to establish security standards. For industrial application, the return on investment of models has to be proven in empirical studies. We strongly believe that models significantly support security reasoning in the resulting quality of the analysis, because of all the reasons stated earlier, and we will work on empirical proof toward that fact in the future.
- The complexity of models is sometimes distracting. In some cases, models can contain numerous elements and span over several screens, or the amount of text in, e.g., UML class diagrams can be confusing. Ideally, models have only very few elements and provide a lot of overview of a system, but the lack of complexity can result in missing threats during the security analysis. A useful compromise is to introduce views to a model, which limits the number of elements in the model that are shown to the user. Thus, automatic analysis of the model can analyze the entire model, or engineers can analyze the model in parts. However, not all SREs support the concept of views yet. This situation has to be improved in order to eliminate the concern of practitioners regarding the complexity of models.
- Our methods do not support the entire process of Common Criteria or ISO 27001 establishment. The complexity of both standards demands further extensions of the methods we presented in Chaps. 6–8. We developed these methods to a point in which we could show how SREs can support both standards significantly. We chose not to concentrate on one standard and one SRE, because we aimed to show that our contribution is valid for multiple SREs and standards. We aim to extend our methods in the future to support the entire standards.

14.2.2 Knowledge Transfer of Our Results to the Establishment of Safety Standards

We did not restrict the idea of applying SRE methods to standards to the knowledge area of security. We showed that the scope of the application of requirements engineering methods extends to safety. We did not use a security requirements engineering method, but a safety requirements engineering method. We enhanced a safety requirements engineering method to support the establishment of a safety standard by applying the approach taken in Chap. 8 to the ISO 26262 standard for safety of road vehicles. We conducted this research with a significant industrial partner and SME consultants, who are domain experts in safety. Safety is similar to security in the sense that a system should be protected from harm. In security, the harm is caused by the threats of an attacker, while in safety the cause is a hazard that is part of the system. In Chap. 8, we proposed to define parameters of how to conduct a model-based threat analysis including formal checks to validate the model. We also proposed a similar method for hazard analysis in safety in Chap. 9. Thus, we showed the applicability of our concept of using requirements engineering methods for the safety knowledge area. The practitioners of that area found the outcome useful and appreciated the possibility of automatic validation of models instead of the current usage of text-based approaches. They recommended to develop tool support that can transform the information from Microsoft (MS) Word or MS Excel to requirements engineering models and back. We will work on this tool support in the future. Furthermore, we applied our concept of using requirements engineering to establish security standards successfully to safety, but the application to further knowledge areas has yet to be proven.

14.2.3 Structured Elicitation of the Environment with Context Patterns

Requirements define what properties and functionality a software should have. It is impossible to assess the quality of a software without requirements. Moreover, writing requirements is only possible if the domain knowledge of the system-to-be and its environment is known and considered thoroughly. We address this problem by describing common structures and stakeholders for several different domains in so-called *context patterns*. We show several example patterns and tool support in Chap. 10. The idea of describing the system in a reusable structure has been proposed in requirements engineering, but very few examples have been shown how to address this. In addition, methods for eliciting domain knowledge in a structured way are missing. It is often implicitly assumed that domain knowledge is somehow present. We show a meta-model for context patterns in Chap. 11. The meta-model shows a common structure of our context patterns and supports software and security engineers in describing their own context patterns. Based on the insights gained from the meta-model, we also defined relations between our existing context patterns. Thus, we showed the fundamentals of a pattern language for context patterns.

Context patterns can be used for the structured collection of domain knowledge. This domain knowledge can be in turn used as input for a given software engineering process. Moreover, we can create methods for security analysis for specific contexts using our patterns. We showed the PACTS method (Chap. 12) as an example for cloud security analysis in compliance to ISO 27001. We discussed our context patterns contribution with the pattern community at EuroPlop 2013 and also with the industrial members of the NESSoS and the ClouDAT[1] projects. Our pattern-based approach toward structured domain knowledge elicitation seemed novel to the pattern experts and practitioners. We create graphical models for specific contexts, e.g., clouds that can be instantiated. The graphical models are accompanied with

---

[1]The ClouDAT project: http://ti.uni-due.de/ti/clouddat/de/.

templates for additional information about elements in the graphical model and also with methods that describe in steps how to instantiate our models and templates and analyze the collected information. We discussed existing related work to our context patterns (Chap. 1) and illustrated that they fill the existing research gap of specific patterns for the elicitation of domain knowledge. We discussed with numerous people that domain knowledge elicitation is a problem, but the only related work that was mentioned multiple times in our discussions is Fowler's work regarding analysis patterns. However, his work focuses on specific analysis pattern for economic enterprises and lacks the generality of our approach. We provide a pattern language that supports the description of any kind of domain knowledge.

We had a discussion with practitioners on our usage of context patterns to establish the ISO 27001 standard (Chap. 12 and in particular Sect. 13.7). In our discussions based on the practitioners' opinions, we concluded that the idea of reusing analysis results, e.g., the identified threats based on a context pattern and the consistency checks of the instantiation of context patterns, are novel and useful contributions. These features have the potential to reduce the time for establishing a security standard and produce consistent documents. In the discussions with the practitioners, we reached the conclusion that a support mechanism should guide users through the instantiation of templates and help with analyzing the content of the instantiated templates. We concluded based on their opinion that tool support should contain wizards for this purpose that guide through the instantiation process and fill templates in the background with meta information. The wizard should help users going back and forth in the method and visualize the effects changes have, e.g., if elements are inside or outside of the scope.

The industrial partners of the ClouDAT project liked the context-pattern-based approach of establishing security standards. In collaboration, we developed an extension of the approach that considers textual security requirements patterns. These can be instantiated using elements of context patterns (Chap. 13). The feedback from the usage of the cloud pattern in the project is that the pattern can be used with little effort and is easily extensible. For example, the documented cloud in the project has the specific characteristic that stakeholders can administrate parts of it in a so-called *Cloud Controller*. This element was added to the pattern with little effort. Some concerns toward the cloud are that the pattern and all the templates presented in the PACTS method (Chap. 12) might raise the difficulty of applying the pattern. Current efforts try to make the approach simpler for application in small business scenarios.

14.3 Answers to Our Research Questions

In Chap. 1, we claimed that our PEERESS framework (Chap. 3) enables the use of patterns and existing security requirements engineering methods to support the security analysis and documentation demands of security standards. In Sect. 14.2, we illustrated our works and the overall feedback we received from practitioners and industrial partners of the NESSoS and the ClouDAT projects. The overall positive feedback confirms our claim, even though we still have to enhance our work with the numerous suggested improvements. We refined our main research question into several more detailed research questions (Chap. 1), and we aim to answer these in the following. In order to avoid an overlap with Sect. 14.2, we focus only on relevant details for the particular question.

RQ 1   Which concepts of security requirements engineering methods can be used and improved such that they support the establishment of security standards?

Answer   Chapter 5 of our work showed that in our experience security standards require consideration of a majority of the concepts presented in the conceptual framework of Fabian et al. (2010).[2] We could not identify a concept that is not relevant in our examples with the ISO 27001 and the Common Criteria standards. This statement may vary for other security standards, which we will consider in the future. Unfortunately, SREs often do not contain all of the concepts of the framework. We used the framework successfully to identify which concepts we had to extend in order to enable security standard establishment.

Overall, the conceptual framework helped us to successfully identify SREs that contain many of the concepts needed to establish a particular security standard. The Si*-ISMS, ISMS-CORAS, and UML4PF-CC methods shown in Chaps. 6–8 are proof-of-concept applications of our approach. The main difficulty we encountered was to align the artifacts produced by SREs with the documentation demands of standards. This leads to an analysis of these demands, which suggested that significant support for these tasks can be provided. For example, practitioners and reviewers of scientific conferences liked our transformation of information in UML models to texts and tables compliant with Common Criteria (Chap. 8).

RQ 2   Are the identified techniques to extend security requirements engineering methods for security standard isolated to the security knowledge area, or can the techniques also be applied to other knowledge areas such as safety?

Answer   We collaborated with experts and practitioners in the safety knowledge area to show that our approach is not limited to the security knowledge area. The UML4PF-ISO26262 method presented in Chap. 9 is proof of this claim. Note that due to the difference of the knowledge areas, we used a safety requirements engineering method to support the establishment of the ISO 26262 safety standard. We are not aware of any obstacle that may prevent that also further knowledge areas benefit from our approach.

---

[2]Note that this conceptual framework is in turn part of our PEERESS framework.

RQ 3   How can security analysis contexts be described in a uniform and reusable way in alignment with the documentation demands of security standards?

Answer   We analyzed the work of the pattern community to identify a way to structure domain knowledge for different domains such as clouds. Our solution is a specific type of pattern called *context pattern*. We described several patterns in Chap. 10 and showed how typical security tasks like asset identification can be performed using context patterns. We discussed these patterns with the partners of the NESSoS project, and the partners highly valued the simplicity of the context patterns and their application for security tasks. The partners requested a closer alignment of the context patterns with security standards. We also proposed a pattern language for context patterns in Chap. 11. The language helps engineers to describe their own context patterns.

RQ 4   Can security standard establishment be based on context patterns that support security requirements engineering methods?

Answer   In Chap. 12 we presented the PACTS method, which is based on our cloud pattern. PACTS shows how the procedure for establishing an ISMS in the ISO 27001 profits from the usage of a context pattern. We conducted several security analysis tasks based on the pattern and derived several artifacts such as a list of assets that can be reused. We discussed the method with the partners of NESSoS and ClouDAT. The partners of ClouDAT decided to adapt PACTS as a core part of their security framework (Chap. 13).

14.4 Directions for Future Research

In this section, we present several ideas for the continuation of our research in the future.

14.4.1 Ontology-Based Support for Identifying Knowledge Objects to Support Security Standard Establishment

The common body of knowledge (CBK)[3] (Schwittek et al. 2012; Beckers et al. 2012; Beckers and Heisel 2013) of NESSoS is an ontology that contains knowledge objects (methods, tools, notations, etc.) for secure systems engineering. The CBK is intended to support one of the main goals of the NESSoS NoE, namely to cre-

---

[3]The CBK homepage: http://www.nessos-project.eu/cbk.

ate a long-lasting research community on engineering secure software services and systems and to bring together researchers and practitioners from security engineering, service computing, and software engineering. Hence, the CBK contains numerous knowledge objects and will most likely contain more in the future.

Knowledge objects are stored in the CBK using a template with fields such as context, problem, solution, and consequences. These templates are inspired by the patterns of Gamma et al. (1994). The CBK contains also numerous search functionalities for identifying relevant knowledge objects, e.g., searching for specific words in the problem description of a knowledge object.

In the future, we are planning to contribute a method to define specific search words to identify relevant knowledge objects to support the establishment of a certain security standard. The method has the potential to identify relevant knowledge objects for security standards with little effort.

14.4.2 Investigating the Relations Between SRE and Security Testing

FI systems are so complex that it is unlikely to avoid all exploitable vulnerabilities. One way to find these vulnerabilities is penetration testing, which is a practical technique in support of finding vulnerabilities of a running system, often through the insertion of input data. However, the number of possible combinations is so enormous that it is not feasible to comprehensively test all of them. Techniques have to devise input data, which are the most likely to find a vulnerability. We will survey existing techniques for that purpose. In addition, we will look for techniques that also consider and refine security requirements up to a point where it is possible to describe input data that can find vulnerabilities. We will base our work on the testing of Voice-over-IP terminal devices and servers that we conducted in the past as presented in Seedorf et al. (2008, 2009).

Furthermore, we will discuss our work with practitioners in the knowledge area of security testing. We will work on feedback cycles between security testing and security requirements engineering with practitioners. On the one hand, security testing practitioners will help us in deriving domain-specific security requirements patterns. These patterns capture types of vulnerabilities encountered in the domain. We assume it will be justified if three known findings of a vulnerability exist to describe a pattern. For vulnerabilities that can cause significant harm to an asset, even less than three cases might be sufficient. On the other hand, for each security requirement at least one test should be created. We will also create patterns for test cases that are linked to the security requirements, which are derived from the experience of security testers.

Moreover, we envision to prevent vulnerabilities during the design and implementation of software engineering. We envision that our test case patterns and security requirements patterns can also be mapped to exercises and lectures that teach software engineers how to prevent these vulnerabilities in the first place.

14.4.3 Empirical Studies

As future work, we want to conduct an extensive study with industrial patterns to ensure the applicability of our work to a wide variety of scenarios. We would like to engage in case studies, in which students and practitioners use our work and establish security standards. We are aiming at analyzing the number of person hours required to apply our work in comparison to the application of existing methods to establish security standards. We also aim to analyze the difference of the quality of the resulting documentation. An analysis of the kind and number of identified assets, threats, etc. may provide insights into which parts of our work require further improvement.

14.5 Summary

In this book, we have shown how to address the ambiguity problem of security standards using security requirements engineering methods and context patterns. We have presented how to build methods to support security standard establishment by enhancing security requirements engineering methods and customizing context patterns. All of these are accompanied by illustrative examples using current scenarios such as smart grid or cloud computing. These methods and the knowledge of how to build further ones should lead to a better security standard adoption in organizations, because they contain means and tools to ease the efforts of security analysis and fulfilling documentation demands. This is a key factor for a broader acceptance and use of security standards. In particular, small and medium-sized businesses have to be supported in this effort, due to their limited available resources.

Increased adoption of security standards in the future depends on the creation and application of easy-to-use methods and tools that provide increased automation of tasks concerning the establishment of security standards. For example, the automated document generation from models shown in Chap. 8 of this book is an important step toward this goal. We believe these and similar efforts will pave the way to a widespread security standard adoption. In particular, the knowledge of how to design similar methods contained in this book shall help to bring us a small step closer toward this goal.

References

Beckers, K., & Heisel, M. (2013). A usability evaluation of the NESSoS common body of knowledge. In *Proceedings of the International Conference on Availability, Reliability and Security (ARES) 2nd International Workshop on Security Ontologies and Taxonomies (SecOnT 2013)* (pp. 559–568). IEEE Computer Society.

Beckers, K., Eicker, S., Faßbender, S., Heisel, M., Schmidt, H., & Schwittek, W. (2012). Ontology-based identification of research gaps and immature research areas. In *Proceedings of the International Cross Domain Conference and Workshop (CD-ARES 2012)* (pp. 1–16). Springer.

Fabian, B., Gürses, S., Heisel, M., Santen, T., & Schmidt, H. (2010). A comparison of security requirements engineering methods. *Requirements Engineering   Special Issue on Security Requirements Engineering, 15*(1), 7–40.

Gamma, E., Helm, R., Johnson, R., & Vlissides, J. (1994). *Design patterns: Elements of reusable object-oriented software*. Reading: Addison-Wesley.

Seedorf, J., Beckers, K., & Huici, F. (2008). Testing dialog-verification of sip phones with single-message denial-of-service attacks. In *Global E-Security* (pp. 61–64). Springer.

Seedorf, J., Beckers, K., & Huici, F. (2009, April). Single-message denial-of-service attacks against voice-over-internet protocol terminals. *International Journal of Electronic Security and Digital Forensics, 2*, 29–34.

Schwittek, W., Schmidt, H., Beckers, K., Eicker, S., Faßbender, S., & Heisel, M. (2012). A common body of knowledge for engineering secure software and services. In *Proceedings of the International Conference on Availability, Reliability and Security (ARES)   1st International Workshop on Security Ontologies and Taxonomies (SecOnT 2012)* (pp. 499–506). IEEE Computer Society.

Appendix A
OCL-Expressions for Validation and Security Reasoning

This appendix contains the currently available OCL-expressions for the validation conditions.

The expressions are used to automatically check the consistency of the model and support security reasoning (see *UML4PF CC-Extension* in Chap. 8).

A.1 Common Criteria—Security Reasoning

```
1  let attackers : Set(Class) =
2
3  let stereotype : String =
4    'Attacker'
5  in
6  Class.allInstances()->select(
7  let first : Set(Stereotype) =
8    getAppliedStereotypes()->asSet()
9  in
10 first->union(first->closure(general.oclAsType(Stereotype)))
11 .name->includes(stereotype))
12 in
13
14 let assets : Set(Class) =
15
16 let stereotype : String =
17   'Asset'
18 in
19 Class.allInstances()->select(
20 let first : Set(Stereotype) =
21   getAppliedStereotypes()->asSet()
22 in
23 first->union(first->closure(general.oclAsType(Stereotype)))
24 .name->includes(stereotype))
25 in
26
27 Class.allInstances()->-(assets->union(attackers))
```

Listing A.1 AE01REA List all classes that are not assets or secondary assets or attackers

© Springer International Publishing Switzerland 2015
K. Beckers, *Pattern and Security Requirements*,
DOI 10.1007/978-3-319-16664-3

The expression collects all classes with the stereotype ≪Attacker≫ (lines 1–13), with the stereotype ≪Attacker≫ (lines 15–24), and it subtracts all attackers, assets from the set of all classes (line 27).

```
1   let assets : Set(Class) =
2
3   let stereotype : String =
4     'Asset'
5   in
6
7   Class.allInstances()->select(
8
9   let first : Set(Stereotype) =
10    getAppliedStereotypes()->asSet()
11
12  in
13  first->union(first->closure(general.oclAsType(Stereotype)))
14  .name->includes(stereotype)} in
15
16  let st: Stereotype = assets.getAppliedStereotypes()->select(name =
        'Asset')->asSequence()->first()
17    in
18
19  assets->select(getValue( st ,'needForProtection') = null)
```

Listing A.2 AE02REA List all assets that have no need for protection property

The expression collects all assets (lines 1–14) and checks if the 'needForProtection' attribute is empty (line 19).

```
1   let stereotype1 : String =
2     'ConnectionDomain'
3   in
4
5   let connectionDomains : Set (Class) =
6
7   Class.allInstances()->select(
8         let first : Set(Stereotype) =
9                 getAppliedStereotypes()->asSet()
10        in
11        first->union(first->closure(general.oclAsType(Stereotype))).name
                ->includes(stereotype1))->asSet()
12  in
13
14  --Association.allInstances().endType
15  Class.allInstances()->select(getAppliedStereotypes().name
        ->includes('ConnectionDomain'))->select( cd |
16  Association.allInstances()->select( a |
        a.endType.getAppliedStereotypes().name->includes('Asset')
17  and a.endType.getAppliedStereotypes().name->includes('ConnectionDomain')
        ).endType
18  ->select( getAppliedStereotypes().name->includes('ConnectionDomain')
        )->asSet()->excludes(cd))
```

Listing A.3 AE03REA List all connection domains that do not transmit assets

The expression collects all classes with the stereotype ≪ConnectionDomain≫ (lines 3–14), the expression checks further if associations not connect to assets and if associations that have a stereotype connection are connected to an asset (lines 17–20).

```
1     List all assets that have no description property
2  let assets : Set(Class) =
3
4  let stereotype : String =
5    'Asset'
6  in
7  Class.allInstances()->select(
8  let first : Set(Stereotype) =
9    getAppliedStereotypes()->asSet()
10 in
11 first->union(first->closure(general.oclAsType(Stereotype)))
12 .name->includes(stereotype))
13 in
14
15 let st: Stereotype = assets.getAppliedStereotypes()->select(name =
       'Asset')->asSequence()->first()
16   in
17
18 assets->select(getValue( st ,'description' ) = null)
```

Listing A.4 AE04REA List all assets that have no description property

The expression collects all classes with the stereotype ≪Asset≫ (lines 1–14), defines a reference to a stereotype (line 16), and collects all assets that do not have set the 'description' attribute (line 19).

```
1  let
2      stereotype2 : String =
3      'connection'
4  in
5
6  let
7
8  Connections : Set (Association) =
9
10 Association.allInstances()->select(
11 let
12   first : Set(Stereotype) =
13   getAppliedStereotypes()->asSet()
14 in
15
16 first->union(first->closure(general.oclAsType(Stereotype)))
17 .name->includes(stereotype2)) in
18
19 Connections->select(not
       endType.getAppliedStereotypes().name->includes('Asset'))->size() = 0
```

Listing A.5 AE05REA List all connections that do not transmit assets

The expression collects all associations that have the stereotype ≪connection≫ (lines 2–19), and that do not relate to a class with the stereotype ≪Asset≫ (line 21).

```
1  let domains : Set(Class) =
2
3  let stereotype : String =
4      'Domain'
5  in
6
7  Class.allInstances()->select(
8          let first : Set(Stereotype) =
9                  getAppliedStereotypes()->asSet()
10         in
11         first->union(first->closure(general.oclAsType(Stereotype))).name
12                 ->includes(stereotype))->asSet()
13 in
14 let targets : Set(Class) =
15 Dependency.allInstances()->select(source.getAppliedStereotypes().name
           ->includes('Assumption')
16 and domains->includes(target.oclAsType(Class)->asSequence()->first() )
           ).target.oclAsType(Class)->asSet()
17 in
18
19 domains->-(targets)
```

Listing A.6 AS01REA List all domains that have no assumptions

The expression collects all domains (lines 3–14), and all assumptions and their dependencies (lines 16–19). The set of all domains is subtracted by the set of all targets of dependency from an assumption (line 21).

```
1  let attackers : Set(Class) =
2
3  let stereotype : String =
4      'Attacker'
5  in
6
7  Class.allInstances()->select(
8  let first : Set(Stereotype) =
9      getAppliedStereotypes()->asSet()
10 in
11 first->union(first->closure(general.oclAsType(Stereotype))).name
           ->includes(stereotype).oclAsType(Class)->asSet()
12 in
13
14 let
15 threatDependency : Set(Dependency) =
16 Dependency.allInstances()->intersection(attackers.clientDependency)
17 in
18
19 attackers->asSequence()->select(att |
20 (att.clientDependency.getAppliedStereotypes()
21 .name->includes('controlThreat')
22 and not
23 att.clientDependency.getAppliedStereotypes()
24 .name->includes('observeThreat'))
25 or
26 (att.clientDependency.getAppliedStereotypes()
27 .name->includes('observeThreat')
28 and not
29 att.clientDependency.getAppliedStereotypes()
30 .name->includes('controlThreat')))
```

Listing A.7 AT01REA List all attackers that have only observe threats or only controls threats

The expression collects all attackers (lines 4–15), and all dependencies that originate from an attacker (lines 17–20), afterwards we check if an attacker causes only control threats or observe threats (lines 22–30).

```
1    -- List all assets that have no control Threats
2
3    let assets : Set(Class) =
4
5    let stereotype1 : String =
6      'Asset'
7    in
8
9    Class.allInstances()->select(
10   let first : Set(Stereotype) =
11     getAppliedStereotypes()->asSet()
12   in
13   first->union(first->closure(general.oclAsType(Stereotype))).name
           ->includes(stereotype1)).oclAsType(Class)->asSet()
14   in
15
16   let stereotype3 : String =
17     'controlThreat'
18   in
19
20   let
21   controlThreats : Set (Dependency) =
22   Dependency.allInstances()->select(
23   let first : Set(Stereotype) =
24     getAppliedStereotypes()->asSet()
25   in
26   first->union(first->closure(general.oclAsType(Stereotype))).name
           ->includes(stereotype3))
27   in
28
29   let AssetsWithControlThreats : Set (Class) =
30   controlThreats->select(
31   target.getAppliedStereotypes().name->includes('Assets')
32   ).target.oclAsType(Class)->asSet()
33   in
34
35   assets->-(AssetsWithControlThreats)
```

Listing A.8 CT01REA List all assets that have no control threats

The expression collects all assets (lines 1–12), and all dependencies with a
≪controlThreat≫ stereotype (lines 14–25). All assets that have a control threat
are collected in a set (lines 27–31). The expression subtracts these assets from the
set all assets.

```
1   - List all domains that have no facts or assumptions
2   let domains : Set(Class) =
3
4   let stereotype : String =
5     'Domain'
6   in
7
8   Class.allInstances()->select(
9         let first : Set(Stereotype) =
10                  getAppliedStereotypes()->asSet()
11        in
12        first->union(first->closure(general.oclAsType(Stereotype))).name
                  ->includes(stereotype))->asSet()
13  in
14
15  let targets : Set(Class) =
16  Dependency.allInstances()->select((source.getAppliedStereotypes().name
          ->includes('Assumption') or
          source.getAppliedStereotypes().name->includes('Fact'))
17  and domains->includes(target.oclAsType(Class)->asSequence()->first() )
          ).target.oclAsType(Class)->asSet()
18  in
19
20  domains->-(targets)
```

Listing A.9 DA01REA List all domains that have no facts or assumptions.

The expression collects all domains (lines 1–12) and all domains that have facts or assumptions (lines 14–17). These domains are subtracted from all domains (line 19).

```
1   let domains : Set(Class) =
2
3   let stereotype : String =
4     'Domain'
5   in
6
7   Class.allInstances()->select(
8         let first : Set(Stereotype) =
9                  getAppliedStereotypes()->asSet()
10        in
11        first->union(first->closure(general.oclAsType(Stereotype))).name
                  ->includes(stereotype))->asSet()
12  in
13
14  let targets : Set(Class) =
15  Dependency.allInstances()->select(source.getAppliedStereotypes().name
          ->includes('Fact')
16  and domains->includes(target.oclAsType(Class)->asSequence()->first() )
          ).target.oclAsType(Class)->asSet()
17  in
18
19  domains->-(targets)
```

Listing A.10 FA01REA List all domains that have no facts

The expression collects all domains (lines 3–15) and all domains that have facts (lines 16–19). These domains are subtracted from all domains (line 21).

```
1    let stereotype1 : String =
2       'ConnectionDomain'
3    in
4
5    let connectionDomains : Set (Class) =
6
7    Class.allInstances()->select(
8            let first : Set(Stereotype) =
9                    getAppliedStereotypes()->asSet()
10           in
11           first->union(first->closure(general.oclAsType(Stereotype))).name
                    ->includes(stereotype1))->asSet()
12   in
13
14   let
15      stereotype2 : String =
16      'connection'
17   in
18
19   let
20   Connections : Set (Association) =
21   Association.allInstances()->select(
22   let
23     first : Set(Stereotype) =
24     getAppliedStereotypes()->asSet()
25   in
26
27   first->union(first->closure(general.oclAsType(Stereotype))).name
              ->includes(stereotype2))
28   in
29
30   let
31      stereotype : String =
32      'threat'
33   in
34
35   let
36   Threats : Set (Dependency) =
37
38   Dependency.allInstances()->select(
39   let
40     first : Set(Stereotype) =
41     getAppliedStereotypes()->asSet()
42   in
43
44   first->union(first->closure(general.oclAsType(Stereotype))).name
              ->includes(stereotype))
45   in
46
47   let haveNetworkAttacker : Set(Class) =
48   Threats->select(
          target.getAppliedStereotypes().name->includes('ConnectionDomain') and
          source.getAppliedStereotypes().name
49   ->includes('NetworkAttacker') ).target.oclAsType(Class)->asSet()
50   in
51
52   Threats->select(
          target.getAppliedStereotypes().name->includes('ConnectionDomain')
53   and not source.getAppliedStereotypes().name->includes('NetworkAttacker') )
54   .target->select(getAppliedStereotypes().name->includes('ConnectionDomain'))
          .oclAsType(Class)->asSet()
55   ->reject( d| haveNetworkAttacker->includes(d) )->asSet()
56   ->union( connectionDomains->select(cd | not Threats.target->includes(cd) ) )
```

Listing A.11 NA01REA List all connection domains that are not threatened by a network attacker

The expression collects all classes that are connection domains (lines 3–14) and all associations that are connections (lines 16–29). It also collects all threats (lines 35–51) and collects all domains threatened by a network attacker (lines 53–57). The threats from network attackers are traced to connection domains and the domains are filtered that are not throated by a network attacker (lines 58–62).

```
1  let
2       stereotype2 : String =
3       'connection'
4  in
5
6  let
7
8  Connections : Set (Association) =
9
10 Association.allInstances()->select(
11 let
12   first : Set(Stereotype) =
13   getAppliedStereotypes()->asSet()
14 in
15
16 first->union(first->closure(general.oclAsType(Stereotype)))
17 .name->includes(stereotype2)) in
18
19
20 let
21     stereotype : String =
22     'threat'
23 in
24
25 let
26
27 Threats : Set (Dependency) =
28
29 Dependency.allInstances()->select(
30 let
31   first : Set(Stereotype) =
32   getAppliedStereotypes()->asSet()
33 in
34
35 first->union(first->closure(general.oclAsType(Stereotype)))
36 .name->includes(stereotype)) in
37
38 let haveNetworkAttacker : Set(Association) =
39 Threats.target->intersection(Connections).oclAsType(Association)->asSet()
40 in
41
42  Connections->-{haveNetworkAttacker}
```

Listing A.12 NA02REA List all connections that are not threatened by a network attacker

The expression collects all connections (lines 3–19), collects all threats (lines 22–28), and selects all connections that are threatened by a network attacker (lines 40–42). From all connections the ones that are threatened by a network attacker are subtracted (line 44).

```
1    -- List all connection domains that are not threatened by a network attacker
2    and do not have an assumption
3
4    let stereotype1 : String =
5      'ConnectionDomain'
6    in
7
8    let connectionDomains : Set (Class) =
9
10   Class.allInstances()->select(
11          let first : Set(Stereotype) =
12                  getAppliedStereotypes()->asSet()
13          in
14          first->union(first->closure(general.oclAsType(Stereotype))).name
                  ->includes(stereotype1))->asSet()
15   in
16
17   let
18      stereotype2 : String =
19      'connection'
20   in
21
22   let
23
24   Connections : Set (Association) =
25
26   Association.allInstances()->select(
27   let
28     first : Set(Stereotype) =
29     getAppliedStereotypes()->asSet()
30   in
31
32   first->union(first->closure(general.oclAsType(Stereotype))).name
           ->includes(stereotype2))
33   in
34
35
36   let
37      stereotype : String =
38      'threat'
39   in
40
41   let
42
43   Threats : Set (Dependency) =
44
45   Dependency.allInstances()->select(
46   let
47     first : Set(Stereotype) =
48     getAppliedStereotypes()->asSet()
49   in
50
51   first->union(first->closure(general.oclAsType(Stereotype))).name
           ->includes(stereotype))
52   in
53
54   let
55      stereotype3 : String =
56      'Assumption'
57   in
58
59   let
60
61   Assumptions : Set (Class) =
62
63   Class.allInstances()->select(
64   let
65     first : Set(Stereotype) =
66     getAppliedStereotypes()->asSet()
```

```
67  in
68
69  first->union(first->closure(general.oclAsType(Stereotype))).name
70  ->includes(stereotype3))
71  in
72
73  let haveNetworkAttacker : Set(Class) =
74  Threats->select( target.getAppliedStereotypes().name
          ->includes('ConnectionDomain') and source.getAppliedStereotypes().name
75  ->includes('NetworkAttacker') ).target.oclAsType(Class)->asSet()
76  in
77
78  let dependencyConnection : Set (Dependency) =
79
80  Dependency.allInstances()->select(
81  source.getAppliedStereotypes().name->includes('Assumption')
82  and
83  target.getAppliedStereotypes().name->includes('ConnectionDomain')
84  )
85
86  in
87
88  let connectionWithAssumption : Set (Class) =
89  dependencyConnection.target.oclAsType(Class)->asSet()
90  in
91
92  let connectionWithoutAssumption : Set (Class) =
93  connectionDomains->-(connectionWithAssumption).oclAsType(Class)->asSet()
94  in
95
96
97  Threats->select(
          target.getAppliedStereotypes().name->includes('ConnectionDomain')
98  and not source.getAppliedStereotypes().name->includes('NetworkAttacker') )
99  .target->select(getAppliedStereotypes().name->includes('ConnectionDomain'))
          .oclAsType(Class)->asSet()
100 ->reject( d| haveNetworkAttacker->includes(d) )->asSet()
101 ->union( connectionDomains->select(cd | not Threats.target->includes(cd) ) )
102 ->union(connectionWithoutAssumption)
```

Listing A.13 NA03REA List all connection domains that are not threatened by a network attacker

The expression collects all classes with the stereotype ≪ConnectionDomain≫ (lines 1–12) and all associations that are connections (lines 14–30). Afterwards all threats are collected (lines 33–54) and all assumptions (lines 51–68). The expression further determines which connection domains and connections are threatened by a network attacker or have assumptions (lines 75–91). Finally all connections are determined that are not threatened by a network attacker and do not have assumptions (lines 94–99).

```
1   -- List all connections that are not threatened by a network attacker
2   and do not have an assumption
3
4
5   let
6       stereotype2 : String =
7       'connection'
8   in
9
10  let
11
12  Connections : Set (Association) =
13
14  Association.allInstances()->select(
15  let
16    first : Set(Stereotype) =
17    getAppliedStereotypes()->asSet()
18  in
19
20  first->union(first->closure(general.oclAsType(Stereotype))).name
        ->includes(stereotype2))
21  in
22
23
24  let
25      stereotype : String =
26      'threat'
27  in
28
29  let
30
31  Threats : Set (Dependency) =
32
33  Dependency.allInstances()->select(
34  let
35    first : Set(Stereotype) =
36    getAppliedStereotypes()->asSet()
37  in
38
39  first->union(first->closure(general.oclAsType(Stereotype))).name
        ->includes(stereotype))
40  in
41
42  let
43      stereotype3 : String =
44      'Assumption'
45  in
46
47  let
48
49  Assumptions : Set (Class) =
50
51  Class.allInstances()->select(
52  let
53    first : Set(Stereotype) =
54    getAppliedStereotypes()->asSet()
55  in
56
57  first->union(first->closure(general.oclAsType(Stereotype))).name
58  ->includes(stereotype3))
59  in
60
61  let haveNetworkAttacker : Set(Association) =
62  Threats.target->intersection(Connections) .oclAsType(Association)->asSet()
63  in
64
65    Connections->-(haveNetworkAttacker).oclAsType(Association)->asSet()
66
```

```
67   ->-(Assumptions.clientDependency.target->intersection(Connections)
        .oclAsType(Association)->asSet() )
68
69   ->select( a | Assumptions.clientDependency.target
        ->intersection(Connections)->excludes(a) )
```

Listing A.14 NA04REA List all connections that are not threatened by a network attacker

The expression collects all associations that are connections (lines 1–21). It collects all threats (lines 24–40) and all assumptions (lines 42–59). Afterwards the expression collects all connections that have a network attacker (lines 61–63) and subtracts these from all connections (line 65). The expression also subtracts the connections that have assumptions (lines 67–69).

```
1    -- List all assets that have no observe Threats
2
3    let assets : Set(Class) =
4
5    let stereotype1 : String =
6      'Asset'
7    in
8
9    Class.allInstances()->select(
10   let first : Set(Stereotype) =
11     getAppliedStereotypes()->asSet()
12   in
13   first->union(first->closure(general.oclAsType(Stereotype))).name
           ->includes(stereotype1)).oclAsType(Class)->asSet()
14   in
15
16   let stereotype3 : String =
17     'observeThreat'
18   in
19
20   let
21   observeThreats : Set (Dependency) =
22   Dependency.allInstances()->select(
23   let first : Set(Stereotype) =
24     getAppliedStereotypes()->asSet()
25   in
26   first->union(first->closure(general.oclAsType(Stereotype))).name
           ->includes(stereotype3))
27   in
28
29   assets->-(observeThreats.target.oclAsType(Class)->asSet())
```

Listing A.15 OT01REA List all assets that have no observe Threats

The expression collects all classes with the stereotype ≪Asset≫ (lines 3–18) and all dependencies with the stereotype ≪observeThreat≫ (lines 16–27). The assets that are threatened are subtracted from the set of all assets (line 29).

```
1   -- List all biddable domains that are not threatened by a physical attacker
2
3   let stereotype1 : String =
4     'BiddableDomain'
5   in
6
7   let biddables : Set (Class) =
8
9   Class.allInstances()->select(
10          let first : Set(Stereotype) =
11                getAppliedStereotypes()->asSet()
12          in
13          first->union(first->closure(general.oclAsType(Stereotype))).name
                    ->includes(stereotype1))->asSet()
14  in
15
16  let stereotype : String =
17    'threat'
18  in
19
20  let
21
22  Threats : Set (Dependency) =
23
24  Dependency.allInstances()->select(
25  let first : Set(Stereotype) =
26    getAppliedStereotypes()->asSet()
27  in
28  first->union(first->closure(general.oclAsType(Stereotype))).name
            ->includes(stereotype))
29  in
30
31  let
32  threatenedClasses : Set (Class) =
33  Threats.target.oclAsType(Class)->asSet()
34  in
35
36  let havePhysicalAttacker : Set(Class) =
37
38  Threats->select( target.getAppliedStereotypes().name
          ->includes('BiddableDomain') and source.getAppliedStereotypes().name
          ->includes('PhysicalAttacker') ).target.oclAsType(Class)->asSet()
39
40  in
41
42  biddables->-(havePhysicalAttacker)
```

Listing A.16 PA01REA List all biddable domains that are not threatened by a physical attacker

The expression collects all classes with the stereotype ≪BiddableDomain≫ (lines
3–14) and all dependencies that are threats (lines 16–29). All biddable domains that
are threatened by a physical attacker are also collected (lines 35–40). These are
subtracted from the set of all biddable domains (line 42).

```
1   -- List all causal domains that are not threatened by a physical attacker and
         do not have assumptions

2
3   let stereotype1 : String =
4     'CausalDomain'
5   in
6
7   let causales : Set (Class) =
8
9   Class.allInstances()->select(
10          let first : Set(Stereotype) =
11                  getAppliedStereotypes()->asSet()
12          in
13          first->union(first->closure(general.oclAsType(Stereotype))).name
                     ->includes(stereotype1))->asSet()
14  in
15
16  let stereotype : String =
17    'threat'
18  in
19
20  let
21
22  Threats : Set (Dependency) =
23
24  Dependency.allInstances()->select(
25  let first : Set(Stereotype) =
26    getAppliedStereotypes()->asSet()
27  in
28  first->union(first->closure(general.oclAsType(Stereotype))).name
             ->includes(stereotype))
29  in
30
31  let
32  threatenedClasses : Set (Class) =
33  Threats.target.oclAsType(Class)->asSet()
34  in
35
36  let havePhysicalAttacker : Set(Class) =
37
38  Threats->select( target.getAppliedStereotypes().name
            ->includes('CausalDomain') and source.getAppliedStereotypes().name
39  ->includes('PhysicalAttacker') ).target.oclAsType(Class) ->asSet()
40
41  in
42
43  let dependencyConnection : Set (Dependency) =
44
45  Dependency.allInstances()->select(
46  source.getAppliedStereotypes().name->includes('Assumption')
47  and
48  target.getAppliedStereotypes().name->includes('CausalDomain')
49  )
50
51  in
52
53  let causalWithAssumption : Set (Class) =
54  dependencyConnection.target.oclAsType(Class)->asSet()
55  in
56
57  causales->-(havePhysicalAttacker)->-(causalWithAssumption)
```

Listing A.17 PA02REA List all biddable domains that are not threatened by a physical attacker and do not have assumptions

The expression collects all classes with the stereotype ≪CausalDomain≫ (lines 3–14) and all dependencies that are threats (lines 20–29). All causal domains that are threatened by a physical attacker are also collected (lines 31–41). All causal domains that have an assumption are also collected (lines 43–55). The causal domains that are threatened by a physical attacker and the ones that have assumptions are subtracted from the set of all causal domains.

```
1  let stereotype1 : String =
2    'CausalDomain'
3  in
4
5  let causales : Set (Class) =
6
7  Class.allInstances()->select(
8        let first : Set(Stereotype) =
9              getAppliedStereotypes()->asSet()
10       in
11       first->union(first->closure(general.oclAsType(Stereotype))).name
                  ->includes(stereotype1))->asSet()
12 in
13
14 let stereotype : String =
15    'threat'
16 in
17
18 let
19
20 Threats : Set (Dependency) =
21
22 Dependency.allInstances()->select(
23 let first : Set(Stereotype) =
24   getAppliedStereotypes()->asSet()
25 in
26 first->union(first->closure(general.oclAsType(Stereotype))).name
           ->includes(stereotype))
27 in
28
29 let threatendCausales : Set (Class) =
30
31 Threats->select(
32 (target.getAppliedStereotypes().name->includes('CausalDomain') and
          source.getAppliedStereotypes().name->includes('SoftwareAttacker')
33 )
34 ).target.oclAsType(Class)->asSet()
35
36 in
37
38 causales->-(threatendCausales)->asSet()
```

Listing A.18 SA01REA List all causal domains that are not threatened by a software attacker

The expression collects all classes with the stereotype «CausalDomain» (lines 3–14) and all dependencies that are threats (lines 20–29). All causal domains that are threatened by a physical attacker are also collected (lines 31–41). All causal domain that have an assumption are also collected (lines 43–55). The causal domains that are threatened by a physical attacker and the ones that have assumptions are subtracted from the set of all causal domains.

```
1   let stereotype1 : String =
2     'CausalDomain'
3   in
4
5   let causales : Set (Class) =
6
7   Class.allInstances()->select(
8           let first : Set(Stereotype) =
9                   getAppliedStereotypes()->asSet()
10          in
11          first->union(first->closure(general.oclAsType(Stereotype))).name
12              ->includes(stereotype1))->asSet()
12  in
13
14  let stereotype : String =
15    'threat'
16  in
17
18  let
19
20  Threats : Set (Dependency) =
21
22  Dependency.allInstances()->select(
23  let first : Set(Stereotype) =
24    getAppliedStereotypes()->asSet()
25  in
26  first->union(first->closure(general.oclAsType(Stereotype))).name
27      ->includes(stereotype))
27  in
28
29  let threatendCausales : Set (Class) =
30
31  Threats->select(
32  (target.getAppliedStereotypes().name->includes('CausalDomain') and
33      source.getAppliedStereotypes().name->includes('SoftwareAttacker')
33  )
34  ).target.oclAsType(Class)->asSet()
35
36  in
37
38  let haveAssumption : Set(Class) =
39
40  Dependency.allInstances()->select(source.getAppliedStereotypes().name
41      ->includes('Assumption')
41  and target.getAppliedStereotypes().name->includes('CausalDomain'))
42      .target.oclAsType(Class)->asSet()
43  in
44
45  causales->-(threatendCausales)->asSet()->-(haveAssumption)
```

Listing A.19 SA02REA List all causal domains that are not threatened by a software attacker and that does not have an assumption

The expression collects all classes with the stereotype ≪CausalDomain≫ (lines 3–14) and all dependencies that are threats (lines 20–27). All causal domains that are threatened by a physical attacker are also collected (lines 29–36). All causal domains that have an assumption are also collected (lines 38–43). The causal domains that are threatened by a physical attacker and the ones that have assumptions are subtracted from the set of all causal domains (line 45).

```
1    -- List all biddable domains that are not threatened by a social engineering
         attacker and are not attackers

2
3    let stereotype1 : String =
4      'BiddableDomain'
5    in
6
7    let biddables : Set (Class) =
8
9    Class.allInstances()->select(
10          let first : Set(Stereotype) =
11                getAppliedStereotypes()->asSet()
12          in
13          first->union(first->closure(general.oclAsType(Stereotype))).name
                ->includes(stereotype1))->asSet()
14   in
15
16   let stereotype2 : String =
17     'Attacker'
18   in
19
20   let attackers : Set (Class) =
21
22   Class.allInstances()->select(
23          let first : Set(Stereotype) =
24                getAppliedStereotypes()->asSet()
25          in
26          first->union(first->closure(general.oclAsType(Stereotype))).name
                ->includes(stereotype2))->asSet()
27   in
28
29   let stereotype : String =
30     'threat'
31   in
32
33   let
34
35   Threats : Set (Dependency) =
36
37   Dependency.allInstances()->select(
38   let first : Set(Stereotype) =
39     getAppliedStereotypes()->asSet()
40   in
41   first->union(first->closure(general.oclAsType(Stereotype))).name
         ->includes(stereotype))
42   in
43
44   let threatenedBiddables : Set (Class) =
45
46   Threats->select( target.getAppliedStereotypes().name
          ->includes('BiddableDomain') and  source.getAppliedStereotypes().name
          ->includes('SocialEngineeringAttacker')
47   ).target.oclAsType(Class)->asSet()
48   in
49
50   biddables->-(threatenedBiddables)->-(attackers)
```

Listing A.20 SE01REA List all biddable domains

The expression collects all classes with the stereotype ≪Biddable Domain≫ (lines
3–14) and all classes that are attackers (lines 16–27) and all dependencies that are
threats (lines 29–42). In addition, the expression collects all biddable domains that
are threatened by social engineering attackers (lines 44–48). These are subtracted
from the set of all biddable domains (line 50).

```
1    -- List all biddable domains that are not threatened by a social engineering
        attacker and that do not have an assumption specified.
2
3  let stereotype1 : String =
4    'BiddableDomain'
5  in
6
7  let biddables : Set (Class) =
8
9  Class.allInstances()->select(
10         let first : Set(Stereotype) =
11              getAppliedStereotypes()->asSet()
12         in
13         first->union(first->closure(general.oclAsType(Stereotype))).name
                ->includes(stereotype1))->asSet()
14 in
15
16 let stereotype2 : String =
17   'Attacker'
18 in
19
20 let attackers : Set (Class) =
21
22 Class.allInstances()->select(
23         let first : Set(Stereotype) =
24              getAppliedStereotypes()->asSet()
25         in
26         first->union(first->closure(general.oclAsType(Stereotype))).name
                ->includes(stereotype2))->asSet()
27 in
28
29 let stereotype : String =
30   'threat'
31 in
32
33 let
34
35 Threats : Set (Dependency) =
36
37 Dependency.allInstances()->select(
38 let first : Set(Stereotype) =
39   getAppliedStereotypes()->asSet()
40 in
41 first->union(first->closure(general.oclAsType(Stereotype))).name
        ->includes(stereotype))
42 in
43
44 let threatenedBiddables : Set (Class) =
45
46 Threats->select( target.getAppliedStereotypes().name
        ->includes('BiddableDomain') and  source.getAppliedStereotypes().name
        ->includes('SocialEngineeringAttacker')
47 ).target.oclAsType(Class)->asSet()
48 in
49
50 let stereotype3 : String =
51   'refersTo'
52 in
53
54 let
55
56 refersTo : Set (Dependency) =
57
58 Dependency.allInstances()->select(
59 let first : Set(Stereotype) =
60   getAppliedStereotypes()->asSet()
61 in
62 first->union(first->closure(general.oclAsType(Stereotype))) .name->includes(
63 stereotype3))
```

```
64 in
65
66
67 let haveAssumptions : Set (Class) =
68 refersTo->select( target.getAppliedStereotypes().name
       ->includes('BiddableDomain') and
       source.getAppliedStereotypes().name->includes('Assumption')
69 ).target.oclAsType(Class)->asSet()
70 in
71
72
73
74 biddables->-(threatenedBiddables)->-(attackers)->-(haveAssumptions)
```

Listing A.21 SE02REA List all biddable domains that are not threatened by a social engineering attacker

The expression collects all classes with the stereotype ≪Biddable Domain≫ (lines 3–14), all classes that are attackers (lines 16–27), and all dependencies that are threats (lines 33–42). In addition, all threatened biddable domains are collected (lines 44–48), and all dependencies that have a ≪refersTo≫ stereotype (lines 54–64). The expression also collects all biddable domains that have assumptions (lines 67–70). From the set of all biddable domains, the ones are subtracted that are threatened by a social engineering attacker and the ones that have assumptions (line 74).

```
1 let stereotype : String =
2   'SecondaryAsset'
3 in
4
5 Class.allInstances()->select(
6       let first : Set(Stereotype) =
7             getAppliedStereotypes()->asSet()
8       in
9       first->union(first->closure(general.oclAsType(Stereotype)))
             .name->includes(stereotype))->asSet()
```

Listing A.22 ST01REA List all secondary assets

The expression collects all classes that have the stereotype ≪SecondaryAsset≫.

```
1    -- List all assets that are not threatened
2
3    let assets : Set(Class) =
4
5    let stereotype1 : String =
6      'Asset'
7    in
8
9    Class.allInstances()->select(
10   let first : Set(Stereotype) =
11     getAppliedStereotypes()->asSet()
12   in
13   first->union(first->closure(general.oclAsType(Stereotype))).name
            ->includes(stereotype1)).oclAsType(Class)->asSet()
14   in
15
16   let stereotype4 : String =
17     'Threat'
18   in
19
20   let
21   threats : Set (Dependency) =
22   Dependency.allInstances()->select(
23   let first : Set(Stereotype) =
24     getAppliedStereotypes()->asSet()
25   in
26   first->union(first->closure(general.oclAsType(Stereotype))).name
            ->includes(stereotype4))
27   in
28
29   let AssetsWithThreats : Set (Class) =
30   threats->select(
31   target.getAppliedStereotypes().name->includes('Assets')
32   ).target.oclAsType(Class)->asSet()
33   in
34
35   assets->-(AssetsWithThreats)
```

Listing A.23 TH01REA List all assets that are not threatened

The expression collects all assets (lines 3–14) and all dependencies that are threats (lines 16–27). In addition, the expression collects all assets that are threatened (lines 29–33). These are subtracted from the set of all assets (line 35).

A.2 Common Criteria—Model Validation

```
1   Class.allInstances()->select(getAppliedStereotypes().name ->includes('Fact'))
2   ->forAll(f | f.clientDependency->select(getAppliedStereotypes().name
        ->includes('refersTo'))->size() > 0)
```

Listing A.24 FA01CON Fact—Refers to at least one domain

The expression collects all facts and checks if these have a dependency with a refersTo stereotype (line 1).

```
1  Class.allInstances()->select(getAppliedStereotypes().name
       ->includes('Assumption'))
2  ->forAll(f | f.clientDependency->select(getAppliedStereotypes().name
       ->includes('refersTo'))->size() > 0)
```

Listing A.25 AS01CON Assumption—Refers to at least one domain

The expression collects all assumptions (line 1) and checks that the number of dependencies is greater than 0 (line 2).

```
1  --assets has a connection to the ToE
2  let toes : Set(Class) =
3    let stereotype : String =
4      'TOE'
5    in
6    Class.allInstances()->select(
7    let first : Set(Stereotype) =
8      getAppliedStereotypes()->asSet()
9    in
10   first->union(first->closure(general.oclAsType(Stereotype))).name
         ->includes(stereotype))->asSet()
11 in
12 let assets1 : Set(Class) =
13   let stereotype : String =
14     'Asset'
15   in
16   Class.allInstances()->select(
17   let first : Set(Stereotype) =
18     getAppliedStereotypes()->asSet()
19   in
20   first->union(first->closure(general.oclAsType(Stereotype))).name
         ->includes(stereotype))->asSet()
21 in
22 let assets2 : Set(Class) =
23   let stereotype : String =
24     'Asset'
25   in
26   Property.allInstances()->select(
27   let first : Set(Stereotype) =
28     getAppliedStereotypes()->asSet()
29   in
30   first->union(first->closure(general.oclAsType(Stereotype))).name
         ->includes(stereotype)).class->asSet()
31 in
32 let assets : Set(Class) =
33   assets1->union(assets2)
34 in
35 assets->select(a | Set{a}->closure(c |
       Association.allInstances()->select(endType->includes(c))
       .endType.oclAsType(Class)->asSet())->intersection(toes)->isEmpty())
       ->asSet()->isEmpty()
36
37 and
38
39 --assets are not attackers
40 assets1->select( a | a.getAppliedStereotypes().name->includes('Attacker')
       )->isEmpty()
```

Listing A.26 AE01CON Asset—Has a relation to the ToE domain (e.g., composition) and is not an attacker

The OCL expression collects all classes with the ToEs (line 10) and Asset (line 20) stereotypes and also all properties with an Asset stereotype (line 30). The Assets have to have a relation to a ToE (line 35) and the assets are not allowed to be attackers, as well (line 40).

```
1   ---Asset is a domain
2   let stereotype : String =
3     'Asset'
4   in
5   Class.allInstances()->select(
6   let first : Set(Stereotype) =
7     getAppliedStereotypes()->asSet()
8   in
9   first->union(first->closure(general.oclAsType(Stereotype))).name
           ->includes(stereotype)).getAppliedStereotypes().name->includes('Domain')
10
11  or
12
13  (
14  ---Asset is part of a domain
15  Class.allInstances()->select(getAppliedStereotypes().name
           ->includes('SecondaryAsset')).getAssociations()->asSequence()
16  ->forAll(a : Association
           a.getEndTypes().getAppliedStereotypes().name->includes('SecondaryAsset')
           and a.getEndTypes().getAppliedStereotypes().name->includes('Asset') )
17
18  and
19
20  ---Assets have at least one association
21  Class.allInstances()->select(getAppliedStereotypes().name
           ->includes('Asset'))->forAll( ass |  ass.getAssociations()->size() > 0
22  ))
```

Listing A.27 AE02CON Asset—Is a domain or part of a domain ... only assets

The expression collects all domains (lines 1–10) and all assets of the model (lines 12–21), and also all assets that have a relation to a domain (line 25). We remove all assets that have associations (line 28).

```
1   ---Asset is a domain
2   let stereotype : String =
3     'Asset'
4   in
5   Class.allInstances()->select(
6   let first : Set(Stereotype) =
7     getAppliedStereotypes()->asSet()
8   in
9
10  ---Assets have at least one association
11  Class.allInstances()->select(getAppliedStereotypes()
12  .name->includes('Asset'))
13  ->forAll( ass |  ass.getAssociations()->size() > 0 )
14  )
```

Listing A.28 AE03CON Asset—Asset has at least one association

The expression collects all assets (lines 1–8) and checks if these have at least one association (lines 11–14).

```
1    --an secondary asset is not an attacker
2    let s : Set(Class) =
3    Class.allInstances()->select(getAppliedStereotypes()
4    .name->includes('SecondaryAsset')
5    and getAppliedStereotypes().name->includes('Attacker'))
6    in
7    s->isEmpty()
```

Listing A.29 ST01CON Secondary Asset—Is not an attacker

The expression checks if a secondary asset is not also an attacker (lines 2–7).

```
1    --an secondary asset Has a relation to an asset
2    Class.allInstances()->select(getAppliedStereotypes().name
          ->includes('SecondaryAsset')).getAssociations()->asSequence()
3    ->forAll(a : Association| a.getEndTypes().getAppliedStereotypes().name-
          >includes('SecondaryAsset') and
          a.getEndTypes().getAppliedStereotypes().name ->includes('Asset') )
```

Listing A.30 ST01CON Secondary Asset—Has a relation to an asset

The expression checks if the every secondary asset has an association to an asset (line 3).

```
1    let stereotype : String =
2       'Attacker'
3    in
4
5    let
6
7    attackers : Set (Class) =
8
9    Class.allInstances()->select(
10   let first : Set(Stereotype) =
11      getAppliedStereotypes()->asSet()
12   in
13   first->union(first->closure(general.oclAsType(Stereotype)))
            .name->includes(stereotype))
14   in
15
16      --attackers have at least one dependency
17   attackers->forAll( a | a->asSequence()->first().clientDependency->size()
18   >= 1 )
19
20   and
21
22      --at least one of these dependencies is a threat
23   attackers.clientDependency->forAll(getAppliedStereotypes()
24   .name ->includes('controlThreat') or getAppliedStereotypes().name
25   ->includes('observeThreat') or getAppliedStereotypes()
26   .name ->includes('Threat') or
            getAppliedStereotypes().name->includes('threat'))
27
28   and
29
30      --an attacker is not an asset
31   let s : Set(Class) =
32   Class.allInstances()->select(getAppliedStereotypes()
33   .name->includes('Attacker') and
            getAppliedStereotypes().name->includes('Asset'))
34   in
35   s->isEmpty()
```

Listing A.31 AT01CON Attacker—Present at least one threat and is not an asset

The expression collects all attackers (lines 1–15), collects attackers that have no
dependencies (lines 15–18), and for the attackers that have dependencies checks if
these are threats (lines 21–26). The expression also checks that the attacker is not an
asset (lines 29–34).

```
1   --assumption: attacker expression already run
2
3   let stereotype : String =
4      'NetworkAttacker'
5   in
6
7   let
8
9   networkattackers : Set (Class) =
10
11  Class.allInstances()->select(
12  let first : Set(Stereotype) =
13     getAppliedStereotypes()->asSet()
14  in
15  first->union(first->closure(general.oclAsType(Stereotype))).name
           ->includes(stereotype))
16  in
17  --check if all clientDependency are threats
18  networkattackers.clientDependency->forAll(getAppliedStereotypes().name
           ->includes('controlThreat') or getAppliedStereotypes().name
19  ->includes('observeThreat') or
           getAppliedStereotypes().name->includes('Threat') or
           getAppliedStereotypes().name->includes('threat'))
20
21  and
22
23  --check if the threats threaten only networks
24  networkattackers.clientDependency.target->forAll(
           getAppliedStereotypes().name->includes('ConnectionDomain') or
           getAppliedStereotypes().name->includes('wan') or
           getAppliedStereotypes().name->includes('lmn') or
           getAppliedStereotypes().name->includes('han') or
           getAppliedStereotypes().name->includes('wireless') or
           getAppliedStereotypes().name->includes('network_connection'))
```

Listing A.32 NA01CON Network Attacker—Threatens only connection domains, connections,
or subtypes related to networks

This expression collects all network attackers (lines 1–15), checks if all dependencies
are threats (lines 18–20), and also checks if only networks are threatened (line 24).

```
 1  let stereotype : String =
 2      'PhysicalAttacker'
 3  in
 4
 5  let
 6
 7  physicalAttackers : Set (Class) =
 8
 9  Class.allInstances()->select(
10  let first : Set(Stereotype) =
11      getAppliedStereotypes()->asSet()
12  in
13  first->union(first->closure(general.oclAsType(Stereotype))).name
            ->includes(stereotype))
14  in
15
16  --checken ob clientDependency ein threat ist
17
18  physicalAttackers->forAll( a |
19      a->asSequence()->first().clientDependency->size() >= 1 )
20  and
21
22  (physicalAttackers.clientDependency->forAll(getAppliedStereotypes().name
            ->includes('controlThreat') or getAppliedStereotypes().name
23  ->includes('observeThreat') or getAppliedStereotypes().name
            ->includes('Threat') or getAppliedStereotypes().name->includes('threat'))
24
25  or
26
27  physicalAttackers.clientDependency.target->forAll(getAppliedStereotypes()
            .name ->includes('Domain') or
            getAppliedStereotypes().name->includes('physical'))
28  )
```

Listing A.33 PA01CON Physical Attacker—Threatens a domain or physical connection

The expression collects all physical attackers (line 1–22), checks if these cause threats (line 25–45) and that the physical attackers only threaten physical domains or connections (line 48–62).

```
1   let stereotype : String =
2     'SoftwareAttacker'
3   in
4
5   let
6
7   softwareAttackers : Set (Class) =
8
9   Class.allInstances()->select(
10  let first : Set(Stereotype) =
11    getAppliedStereotypes()->asSet()
12  in
13  first->union(first->closure(general.oclAsType(Stereotype))).name
          ->includes(stereotype))
14  in
15
16
17  softwareAttackers->forAll( a | a->asSequence()->first()
          .clientDependency->size() >= 1 )
18
19  and
20
21  (softwareAttackers.clientDependency->forAll(getAppliedStereotypes().name
          ->includes('controlThreat') or getAppliedStereotypes().name
22  ->includes('observeThreat') or getAppliedStereotypes().name
          ->includes('Threat') or
23          getAppliedStereotypes().name->includes('threat'))
24
25  or
26
27  softwareAttackers.clientDependency.target->forAll(getAppliedStereotypes()
          .name ->includes('CausalDomain'))
28  )
```

Listing A.34 SA01CON Software Attacker—Threatens only causal domains

The expression collects all software attackers (lines 7–14), checks if these have dependencies (line 17), checks if these are all threats (lines 21–22), and if the threats only threaten only causal domains (line 26).

```
 1  let stereotype : String =
 2    'SocialEngineeringAttacker'
 3  in
 4
 5  let
 6
 7  socialEngineeringAttackers : Set (Class) =
 8
 9  Class.allInstances()->select(
10  let first : Set(Stereotype) =
11    getAppliedStereotypes()->asSet()
12  in
13  first->union(first->closure(general.oclAsType(Stereotype))).name
            ->includes(stereotype))
14  in
15
16
17  socialEngineeringAttackers->forAll{ a | a->asSequence()->first()
          .clientDependency->size() >= 1 }
18
19  and
20
21  (socialEngineeringAttackers.clientDependency->forAll(getAppliedStereotypes()
          .name->includes('controlThreat') or getAppliedStereotypes().name
22  ->includes('observeThreat') or getAppliedStereotypes().name
          ->includes('Threat') or
23          getAppliedStereotypes().name->includes('threat'))
24
25  or
26
27  socialEngineeringAttackers.clientDependency.target
          ->forAll(getAppliedStereotypes() .name->includes('BiddableDomain'))
28  )
```

Listing A.35 SE01CON Social Engineering Attacker—Threatens only biddable domains

The expression collects all social engineering attackers (lines 7–14), if these have dependencies (line 17), if the dependencies are threats, and if these threaten only biddable domains (line 26).

```
 1  let stereotype : String =
 2    'threat'
 3  in
 4
 5  let
 6
 7  Threats : Set (Dependency) =
 8
 9  Dependency.allInstances()->select(
10  let first : Set(Stereotype) =
11    getAppliedStereotypes()->asSet()
12  in
13  first->union(first->closure(general.oclAsType(Stereotype)))
          .name->includes(stereotype))
14  in
15
16  Threats.target->forAll(getAppliedStereotypes().name->includes('asset') or
          getAppliedStereotypes().name->includes('Asset'))
```

Listing A.36 TH01CON Threat—Threatens only assets

The expression collects all threats (lines 7–14) and checks if these threats threaten all assets (line 16).

```
1    let stereotype : String =
2      'Attacker'
3    in
4
5    let
6    attackers : Set (Class) =
7    Class.allInstances()->select(
8    let first : Set(Stereotype) =
9      getAppliedStereotypes()->asSet()
10   in
11   first->union(first->closure(general.oclAsType(Stereotype)))
            .name->includes(stereotype))
12   in
13
14   let allStereotypes : Set(Stereotype) =
15   attackers.getAppliedStereotypes().oclAsType(Stereotype)->asSet()
16   in
17
18
19   let stereotype1 : String =
20     'observeThreat'
21   in
22
23   let
24   observeThreats : Set (Dependency) =
25   Dependency.allInstances()->select(
26   let first : Set(Stereotype) =
27     getAppliedStereotypes()->asSet()
28   in
29   first->union(first->closure(general.oclAsType(Stereotype)))
            .name->includes(stereotype1))
30   in
31
32   observeThreats->asSequence()->collect(ob | ob.source)->forAll(att|
33
34   let st: Stereotype = att.getAppliedStereotypes()
            ->intersection(allStereotypes)
35     ->asSequence()->first()
36     in
37
38     let ty: Set(EnumerationLiteral) = att.getValue (st,'windowOfOppertunity')
            .oclAsType(EnumerationLiteral)->asSet()
39
40     in
41       not ty.name->includes('None **')
42
43   )
```

Listing A.37 OT01CON observeThreat—Window of opportunity of the attacker is greater than 0

The expression collects all observe threats (lines 1–11), and checks if any of
the attackers causing the threats has a windows of opportunity that is none (lines
14–25).

```
1  let stereotype : String =
2    'Attacker'
3  in
4
5  let
6  attackers : Set (Class) =
7  Class.allInstances()->select(
8  let first : Set(Stereotype) =
9    getAppliedStereotypes()->asSet()
10 in
11 first->union(first->closure(general.oclAsType(Stereotype)))
          .name->includes(stereotype))
12 in
13
14 let allStereotypes : Set(Stereotype) =
15 attackers.getAppliedStereotypes().oclAsType(Stereotype)->asSet()
16 in
17
18
19 let stereotype1 : String =
20   'controlThreat'
21 in
22
23 let
24 observeThreats : Set (Dependency) =
25 Dependency.allInstances()->select(
26 let first : Set(Stereotype) =
27   getAppliedStereotypes()->asSet()
28 in
29 first->union(first->closure(general.oclAsType(Stereotype)))
          .name->includes(stereotype1))
30 in
31
32 observeThreats->asSequence()->collect(ob | ob.source)->reject(att|
33
34 let st: Stereotype =
35 att.getAppliedStereotypes()->intersection(allStereotypes)
36    ->asSequence()->first()
37   in
38
39   let ty: Set(EnumerationLiteral) = att.getValue (st,'windowOfOppertunity')
           .oclAsType(EnumerationLiteral)->asSet()
40
41   in
42     not ty.name->includes('None **')
43
44 )
```

Listing A.38 CT01CON controlThreat—Window of opportunity of the attacker is greater than 0

The expression collects all control threats (lines 1–11), and checks if any of the attackers causing the threats has a windows of opportunity that is none (lines 14–25).

Appendix B
Comparing ISO 27001 and ISO 31000

Although the ISO 27001 standard and the CORAS approach are the most important references for ISMS-CORAS, we have also based the approach on ISO 31000. Due to some differences between the three, we make a comparison between them in this appendix. We show the differences of several terms of risk management in these two standards, and how these are defined in CORAS. Afterwards, we compare relevant terms and sections of the standards. The aim of our work is to create a method that supports the ISO 27001 standard. Hence, our aim is to identify which sections in ISO 31000 are similar to ISO 27001 sections, and which ISO 27001 sections that do not have equivalents in ISO 31000.

B.1 Terminology Comparison: Risk Assessment

In ISO 31000 (ISO 2009, Sect. 2.1) risk is defined as the effect of uncertainty on objectives. This is a quite general definition, but five notes are added to elaborate on the term. (1) An effect is a deviation from the expected—positive and/or negative. (2) Objectives can have different aspects (such as financial, health and safety, and environmental goals) and can apply at different levels (such as strategic, organization-wide, project, product, and process). (3) Risk is often characterized by reference to potential events and consequences, or a combination of these. (4) Risk is often expressed in terms of a combination of the consequences of an event (including changes in circumstances) and the associated likelihood of occurrence. (5) Uncertainty is the state, even partial, of deficiency of information related to, understanding or knowledge of an event, its consequence, or likelihood. The CORAS approach is mainly based on the definition of the fourth note, i.e., the likelihood of an unwanted incident (an event) and its consequence for a specific asset. ISO 27001 does not give an explicit definition of the term risk in isolation; the standard only contains definitions of related terms, such as risk treatment and risk acceptance. Notice however that its terminology refers heavily to ISO/IEC Guide 73 where risk is defined as in ISO 31000.

© Springer International Publishing Switzerland 2015
K. Beckers, *Pattern and Security Requirements*,
DOI 10.1007/978-3-319-16664-3

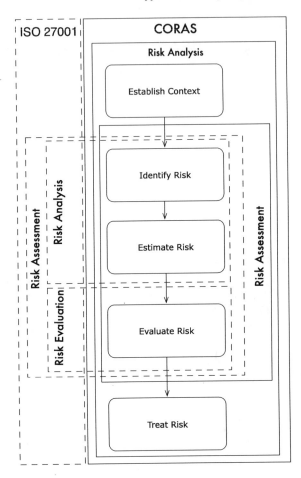

Fig. B.1 Risk terms in CORAS and ISO 27001

We will discuss the similarities and differences of further relevant terms in the following.

Figure B.1 gives an overview of the terms that are used to define the risk assessment process in ISO 27001 and CORAS. In ISO 27001 *risk assessment* includes *risk analysis* and *risk evaluation*, and risk analysis in turn includes *risk identification* and *risk estimation*.

CORAS is based on the ISO 31000, both of which use slightly different definitions of the terms risk analysis and risk assessment. ISO 31000 defines risk assessment as the "overall process of risk identification, risk analysis and risk evaluation" (ISO 2009, p. 4). Risk analysis is a "process to comprehend the nature of risk and to determine the level of risk". Two notes further elaborate on the term by stating that risk analysis provides the basis for risk evaluation and decisions about risk treatment, and that risk analysis includes risk estimation (ISO 2009, p. 5). ISO 31000 further states that the risk levels are derived from the likelihoods and consequences.

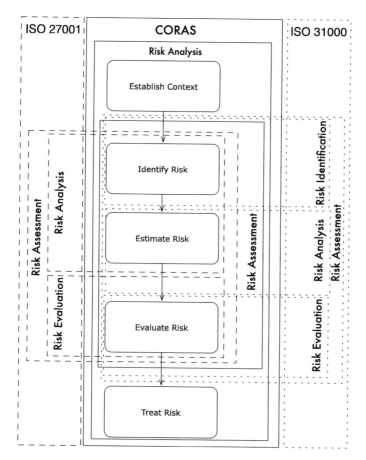

Fig. B.2 Risk terms in CORAS, ISO 27001 and ISO 31000

The term risk analysis is defined differently in CORAS, where it is defined as a process that includes phases to *establish context*, *identify risk*, *estimate risk*, *evaluate risk*, and *treat risk*. Whereas the term *risk assessment* includes *risk analysis* in the ISO 27001 standard, this is not the case in CORAS. However, on both accounts *risk assessment* involves the activities *identify risk*, *estimate risk*, and *evaluate risk*.

To sum up CORAS, ISO 27001, and ISO 31000 all define risk assessment to include risk identification, risk estimation, and risk evaluation. However, the definition of risk analysis differs in all three. In the following we use the terminology according to the definitions in CORAS, but we state explicitly the mapping to the ISO 27001 terminology.

ISMS-CORAS is an approach to conduct and document a security risk assessment, and this activity as demanded by ISO 27001 risk assessment can be achieved by the CORAS steps risk identification, risk estimation, and risk evaluation. In addition, the context establishment part of the CORAS process must be included since these produce essential inputs for the subsequent risk assessment. The Legal CORAS extension satisfies also the condition that the risk assessment methodology shall be able to consider legal and regulatory requirements (ISO 27001 Sect. 4.2.1 c. 1).

B.2 General Term Comparison

In order to elicit the requirements that CORAS has to fulfill in order to support the ISO 27001 standard, we have to analyze the differences between the ISO 31000 standard (which CORAS already supports) and the ISO 27001 standard. In Tables B.1,

Table B.1 ISO 27001 and ISO 31000 comparison of the Chapters (1/3)

ISO 27001	ISO 31000
3. Terms and definitions	2. Terms and definitions
asset	risk
availability	risk management
confidentiality	risk management framework
information security	risk management policy
information security event	risk attitude
information security incident	risk management plan
information security management system ISMS	risk owner
integrity	risk management process
residual risk	establishing the context
risk acceptance	external context
risk analysis	internal context
risk assessment	communication and consultation
risk evaluation	stakeholder
risk management	risk assessment
risk treatment	risk identification
statement of applicability	risk source
	event
	consequence
	likelihood
	risk profile
	risk analysis
	risk criteria
	level of risk
	risk evaluation
	risk treatment
	control
	residual risk
	monitoring
	review

Table B.2 ISO 27001 and ISO 31000 comparison of the Chapters (2/3)

ISO 27001	ISO 31000
4. Information security management system	**4. Framework**
4.1 General requirements	4.1 General
4.2 Establishing and managing the ISMS	4.2 Mandate and commitment
4.2.1 Establishing and managing the ISMS	4.3 Design of framework for managing risk
4.2.1 a Define scope and boundaries	4.3.1 Understanding of the organization and its context
4.2.1 b Define ISMS policy	
4.2.1 c Define risk assessment	4.3.2 Establishing risk management policy
4.2.1 d Identify the risk	4.3.3 Accountability
4.2.1 e Analyse and evaluate risk	4.3.4 Integration into organizational processes
4.2.1 f Identify risk treatment	4.3.5 Resources
4.2.1 g Select controls	4.3.6 Establishing internal communication and
4.2.1 h,i Obtain management approval	reporting mechanisms
4.2.1 j Prepare a statement of applicability	4.3.7 Establishing external communication and reporting mechanisms
4.2.2 Implement and operate the ISMS	
4.2.3 Monitor and review the ISMS	4.4 Implementing risk management
4.2.4 Maintain and improve the ISMS	4.4.1 Implementing the framework for managing risk
4.3 Documentation requirements	
4.3.1 General	4.4.2 Implementing the risk management process
4.3.2 Control of documents	4.5 Monitoring and review of the framework
4.3.3 Control of records	4.6 Continual improvement of the framework

B.2 and B.3 we show the document outlines of the standards, where the former is structured according to the defined terms. We focus on the terms of ISO 27001 and not ISO 31000, because our aim is to support the former. The following terms are defined in ISO 27001 only:

- asset
- availability
- confidentiality
- information security
- information security event
- information security incident
- information security management system ISMS
- integrity
- statement of applicability

Table B.3 ISO 27001 and ISO 31000 comparison of the Chapters (3/3)

ISO 27001	ISO 31000
5. Management responsibility	**5. Process**
5.1 Management commitment	5.1 General
5.2 Resource management	5.2 Communication and consultation
	5.3 Establishing the context
	5.3.1 General
	5.3.2 Establishing the external context
	5.3.3 Establishing the internal context
	5.3.4 Establishing the context of the risk management process
	5.3.5 Defining risk criteria
	5.4 Risk assessment
	5.4.1 General
	5.4.2 Risk identification
	5.4.3 Risk analysis
	5.4.4 Risk evaluation
	5.5 Risk treatment
	5.5.1 General
	5.5.2 Selection of risk treatment options
	5.5.3 Preparing and implementing risk treatment plans
	5.6 Monitoring and review
	5.7 Recording the risk management process

B.3 Section Comparison

The ISO 27001 standard describes the requirements for an Information Security Management System (ISMS). The subsections of *Section 4.2.1 Establishing and managing the ISMS* describe the required steps to built an ISMS. *Sect. 4.2.1 a* requires a context and scope description of the ISMS. In the ISO 31000 standard *Sect. 4.3.1* demands also an description of the organisation and its context, and ISO 31000 *Sect. 5.3* demands a description of the risk management context.

ISO 27001 *Sect. 4.2.1 b* requires a definition of an *ISMS policy*. ISO 31000 requires a *risk management policy*. An ISMS policy (ISO/IEC 2005b, Sect. 4.2.1):

1. includes a framework for setting objectives and establishes an overall sense of direction and principles for action with regard to information security;
2. takes into account business and legal or regulatory requirements, and contractual security obligations;
3. aligns with the organization's strategic risk management context in which the establishment and maintenance of the ISMS will take place;
4. establishes criteria against which risk will be evaluated (4.2.1 c); and
5. has been approved by management.

ISO 31000 does not demand an ISMS policy, but a risk management policy instead. A risk management policy according to ISO 31000 is "statement of the overall intentions and direction of an organization related to risk management" (ISO 2009, Sect. 2.4). Risk management is "coordinated activities to direct and control an organization with regard to risk" (ISO 2009, Sect. 2.2).

Hence, both standards demand policy statements concerning risk. However, the ISMS policy demands further directions for information security that also consider business and legal regulations. In order to extend CORAS to support ISMS policy design, we have extend the approach for these tasks. CORAS has already extensions for legal regulations (Lund et al. 2010), thus the focus of the extension is on information security.

The following demands exist in both standards:

- ISO 27001 *Sect. 4.2.1 c* requires a risk assessment, this is also required for ISO 31000 Sect. *5.4*.
- ISO 27001 *Sect. 4.2.1 d* requires risk identification, which is also demand in ISO 31000 *Sect. 5.4.2*.
- ISO 27001 *Sect. 4.2.1 e* requires risk analysis and evaluation, which is demanded in ISO 31000 *Sects. 5.4.3* and *5.4.4*.
- ISO 27001 *Sect. 4.2.1 f* requires risk treatment identification, which is demanded in ISO 31000 *Sect. 5.5*.
- ISO 27001 *Sect. 4.2.1 h,i* demands management approval, which is similar to ISO 31000 *Sect. 4.2*.

ISO 27001 *Sect. 4.2.1 j* demands a statement of applicability. A statement of applicability is a (ISO/IEC 2005b, Sect. 3.16) "documented statement describing the control objectives and controls that are relevant and applicable to the organization's ISMS. NOTE: Control objectives and controls are based on the results and conclusions of the risk assessment and risk treatment processes, legal or regulatory requirements, contractual obligations and the organization's business requirements for information security". This is similar to ISO 31000 *Sect. 5.5.2 Preparing and implementing risk treatment plans.*

ISO 27001 *Sect. 4.2.1 g* demands to select controls, which is similar to ISO 31000 *Sect. 5.5*. The term control is defined in the standard ISO 27002, which refines the ISO 27001. According to ISO 27002 (ISO/IEC 2005a, Sect. 2.2) a Control is a "means of managing risk, including policies, procedures, guidelines, practices or organizational structures, which can be of administrative, technical, management, or legal nature. NOTE Control is also used as a synonym for safeguard or countermeasure".

The ISO 31000 (ISO 2009, Sect. 2.26) defines a control as a "measure that is modifying risk (2.1). NOTE 1 Controls include any process, policy, device, practice, or other actions which modify risk. NOTE 2 Controls may not always exert the intended or assumed modifying effect."

In both standards controls modify risks and can include policies, procedures, guidelines, practices. In the ISO 27001 a control can also explicitly be an organisational measure, e.g., an administrative or legal action. These are also implicitly in the ISO 31000 definition, because the can be measures that modify risk.

Appendix C
Comparing Annex A of ISO 27001 and ISO 27001:2013

We show a mapping of ISO 27001 controls to ISO 27001:2013 controls in Table C.1 and a mapping of all the subcontrols in Table C.2. All fields in the ISO 27001:2013 column marked with an "-"mean that this controls does not have corresponding control in ISO 27001:2013. Subcontrols in ISO 27001:2013 that have no corresponding controls in ISO 27001 are listed in Table C.3.

Table C.1 Comparison of Annex A controls from ISO 27001 to ISO 27001:2013

ISO 27001 Controls	Corresponding ISO 27001:2013 Control(s)
A.5 Security policy	A.5 Information security policies
A.6 Organization of information security	A.6 Organization of information security
A.7 Asset management	A.8 Asset management
A.8 Human resources security	A.7 Human resource security
A.9 Physical and environmental security	A.11 Physical and environmental security
A.10 Communications and operations management	A.12 Operations security, A.13 Communications security
A.11 Access Control	A.9 Access control
A.12 Information systems acquisition development	A.14 System acquisition, development and maintenance
A.13 Information security incident management	A.16 Information security incident management
A.14 Business continuity management	A.17 Information security aspects of business continuity management
A.15 Compliance	A.18 Compliance

© Springer International Publishing Switzerland 2015
K. Beckers, *Pattern and Security Requirements*,
DOI 10.1007/978-3-319-16664-3

Table C.2 Comparision of Annex A subcontrols from ISO 27001 to ISO 27001:2013

ISO 27001 Controls	Corresponding ISO 27001:2013 Control(s)
A.5 Security policy	*A.5 Information security policies*
A.5.1 Information security policy	A.5.1 Management direction for information security
A.5.1.1 Information security policy document	A.5.1.1 Policies for information security
A.5.1.2 Review of the information security policy	A.5.1.2 Review of the policies for information security
A.6 Organization of information security	*A.6 Organization of information security*
A.6.1 Internal organization	A.6.1 Internal Organization
A.6.1.1 Management commitment to information security	-
A.6.1.2 Information security coordination	-
A.6.1.3 Allocation of information security responsibilities	A.6.1.1 Information security roles and responsibilities
A.6.1.4 Authorization process for information processing facilities	-
A.6.1.5 Confidentiality agreements	-
A.6.1.6 Contact with authorities	A.6.1.3 Contact with authorities
A.6.1.7 Contact with special interest groups	A.6.1.4 Contact with special interest groups
A.6.1.8 Independent review of information security	-
A.6.2 External parties	-
A.6.2.1 Identification of risks related to external parties	-
A.6.2.2 Addressing security when dealing with customers	-
A.6.2.3 Addressing security in third party agreements	-
A.7 Asset management	*A.8 Asset management*
A.7.1 Responsibility for assets	A.8.1 Responsibilities for assets
A.7.1.1 Inventory of assets	A.8.1.1 Inventory of assets
A.7.1.2 Ownership of assets	A.8.1.2 Ownership of assets
A.7.1.3 Acceptable use of assets	A.8.1.3 Acceptable use of assets
A.7.2 Information classification	A.8.2 Information classification
A.7.2.1 Classification guidelines	A.8.2.1 Classification of information
A.7.2.2 Information labelling and handling	A.8.2.2 Labelling of information
A.8 Human resources security	*A.7 Human resource security*
A.8.1 Prior to employment	A.7.1 Prior to employment
A.8.1.1 Roles and responsibilities	-
A.8.1.2 Screening	A.7.1.1 Screening
A.8.1.3 Terms and conditions of employment	A.7.1.2 Terms and conditions of employment
A.8.2 During employment	A.7.2 During employment

(continued)

Table C.2 (continued)

ISO 27001 Controls	Corresponding ISO 27001:2013 Control(s)
A.8.2.1 Management responsibilities	A.7.2.1 Management responsibilities
A.8.2.2 Information security awareness, education and training	A.7.2.2 Information security awareness, education and training
A.8.2.3 Disciplinary process	A.7.2.3 Disciplinary process
A.8.3 Termination or change of employment	A.7.3 Termination and change of employment
A.8.3.1 Termination responsibilities	A.7.3.1 Termination or change of employment responsibilities
A.8.3.2 Return of assets	-
A.8.3.3 removal of access rights	-
A.9 Physical and environmental security	*A.11 Physical and environmental security*
A.9.1 Secure areas	A.11.1 Secure areas
A.9.1.1 Physical security perimeter	A.11.1.1 Physical security perimeter
A.9.1.2 Physical entry controls	A.11.1.2 Physical entry controls
A.9.1.3 Securing offices, rooms and facilities	A.11.1.3 Securing offices, rooms and facilities
A.9.1.4 Protecting against external and environmental threats	A.11.1.4 Protecting against external and environmental threats
A.9.1.5 Working in secure areas	A.11.1.5 Working in secure areas
A.9.1.6 Public access, delivery and loading areas	A.11.1.6 Delivery and loading areas
A.9.2 Equipment security	A.11.2 Equipment
A.9.2.1 Equipment siting and protection	A.11.2.1 Equipment siting and protection
A.9.2.2 Supporting utilities	A.11.2.2 Supporting utilities
A.9.2.3 Cabling security	A.11.2.3 Cabling security
A.9.2.4 Equipment maintenance	A.11.2.4 Equipment maintenance
A.9.2.5 Security of equipment off-premises	A.11.2.6 Security of equipment and assets off-premises
A.9.2.6 Secure disposal or re-use of equipment	A.11.2.7 Secure disposal or reuse of equipment
A.10 Communications and operations management	*A.12 Operations security*
A.10.1 Operational procedures and responsibilities	A.12.1 Operational procedures and responsibilities
A.10.1.1 Documented operating procedures	A.12.1.1 Documented operating procedures
A.10.1.2 Change management	A.12.1.2 Change management
A.10.1.3 Segregation of duties	A.6.1.2 Segregation of duties
A.10.1.4 Separation of development, test and operational duties	A.12.1.4 Separation of development, testing and operational environments
A.10.2 Third party service delivery management	-
A.10.2.1 Service delivery	-
A.10.2.2 Monitoring and review of third party services	-

(continued)

Table C.2 (continued)

ISO 27001 Controls	Corresponding ISO 27001:2013 Control(s)
A.10.2.3 Managing changes to third party services	-
A.10.3 System planning and acceptance	-
A.10.3.1 Capacity management	A.12.1.3 Capacity management
A.10.3.2 System acceptance	A.14.2.9 System acceptance testing
A.10.4 Protection against malicious and mobile code	A.12.2 Protection from malware
A.10.4.1 Controls against malicious code	A.12.2.1 Controls against malware
A.10.4.2 Controls against mobile code	-
A.10.5 Back-up	A.12.3 Backup
A.10.5.1 Information back-up	A.12.3.1 Information backup
A.10.6 Network security management	A.13.1 Network security management
A.10.6.1 Network controls	A.13.1.1 Network controls
A.10.6.2 Security of network services	A.13.1.2 Security of network services
A.10.7 Media handling	A.8.3 Media handling
A.10.7.1 Management of removable media	A.8.3.1 Management of removable media
A.10.7.2 Disposal of media	A.8.3.2 Disposal of media
A.10.7.3 Information handling procedures	-
A.10.7.4 Security of system documentation	-
A.10.8 Exchange of information	A.13.2 Information transfer
A.10.8.1 Information exchange policies and procedures	A.13.2.1 Information transfer policies and procedures
A.10.8.2 Exchange agreements	A.13.2.2 Agreements on information transfer
A.10.8.3 Physical media in transit	A.8.3.3 Physical media transfer
A.10.8.4 Electronic messaging	A.13.2.3 Electronic messaging
A.10.8.5 Business information systems	-
A.10.9 Electronic commerce services	-
A.10.9.1 Electronic commerce	-
A.10.9.2 On-line transaction	-
A.10.9.3 Publicly available information	-
A.10.10 Monitoring	A.12.4 Logging and monitoring
A.10.10.1 Audit logging	A.12.4.1 Event logging
A.10.10.2 Monitoring system use	-
A.10.10.3 Protection of log information	A.12.4.2 Protection of log information
A.10.10.4 Administrator and operator logs	A.12.4.3 Administrator and operator logs
A.10.10.5 Fault logging	-
A.10.10.6 Clock synchronization	A.12.4.4 Clock synchronization
A.11 Access Control	A.9 Access control
A.11.1 Business requirement for access control	A.9.1 Business requirement of access control
A.11.1.1 Access control policy	A.9.1.1 Access control policy

(continued)

Table C.2 (continued)

ISO 27001 Controls	Corresponding ISO 27001:2013 Control(s)
A.11.2 User access management	A.9.2 User access management
A.11.2.1 User registration	A.9.2.1 User registration and de-registration
A.11.2.2 Privilege management	A.9.2.3 Management of privileged access rights
A.11.2.3 User password management	-
A.11.2.4 Review of user access rights	A.9.2.5 Review of user access rights
A.11.3 User responsibilities	A.9.3 User responsibilities
A.11.3.1 Password use	A.9.3.1 Use of secret authentication information
A.11.3.2 Unattended user equipment	A.11.2.8 Unattended user equipment
A.11.3.3 Clear desk and clear screen policy	A.11.2.9 Clear desk and clear screen policy
A.11.4 Network access control	-
A.11.4.1 Policy on use of network	-
A.11.4.2 User authentication for external connections	-
A.11.4.3 Equipment identification in networks	-
A.11.4.4 Remote diagnostic and configuration port protection	-
A.11.4.5 Segregation in networks	A.13.1 Network security management
A.11.4.6 Network connection control	-
A.11.4.7 Network routing control	-
A.11.5 Operating system access control	A.9.4 System and application access control
A.11.5.1 Secure log-on procedures	A.9.4.2 Secure log-on procedures
A.11.5.2 User identification and authentication	-
A.11.5.3 Password management system	A.9.4.3 Password management system
A.11.5.4 use of system utilities	A.9.4.4 Use of privileged utility programs
A.11.5.5 Session time-out	-
A.11.5.6 Limitation of connection time	-
A.11.6 Application and information access control	A.9.4 System and application access control
A.11.6.1 Information access restriction	A.9.4.1 Information access restriction
A.11.6.2 Sensitive system isolation	-
A.11.7 Mobile computing and teleworking	A.6.2 Mobile devices and teleworking
A.11.7.1 Mobile computing and communication	A.6.2.1 Mobile device policy
A.11.7.2 Teleworking	A.6.2.2 Teleworking
A.12 Information systems acquisition development	*A.14 System acquisition, development and maintenance*
A.12.1 Security requirements of information systems	A.14.1 Security requirements of information systems
A.12.1.1 Security requirements analysis and specification	A.14.1.1 Information security requirements analysis and specification

(continued)

Table C.2 (continued)

ISO 27001 Controls	Corresponding ISO 27001:2013 Control(s)
A.12.2 Correct processing in applications	–
A.12.2.1 Input data validation	–
A.12.2.2 Control of internal processing	–
A.12.2.3 Message integrity	–
A.12.2.4 Output data validation	–
A.12.3 Cryptographic controls	A.10.1 Cryptographic controls
A.12.3.1 Policy on the use of cryptographic controls	A.10.1.1 Policy on the use of cryptographic controls
A.12.3.2 Key management	A.10.1.2 Key Management
A.12.4 Security of system files	–
A.12.4.1 Control of operational software	A.12.5 Control of operational software
A.12.4.2 Protection of system test data	A.14.3.1 Protection of test data
A.12.4.3 Access control to program source code	A.9.4.5 Access control to program source code
A.12.5 Security in development and support processes	A.14.2 Security in development and support processes
A.12.5.1 Change control procedures	A.14.2.2 System change control procedures
A.12.5.2 Technical review of applications after operating system changes	A.14.2.3 Technical review of applications after operating platform changes
A.12.5.3 Restriction on changes to software packages	A.14.2.4 Restriction on changes to software packages
A.12.5.4 Information leakage	–
A.12.5.5 Outsourced software development	A.14.2.7 Outsourced development
A.12.6 Technical vulnerability management	A.12.6 Technical vulnerability management
A.12.6.1 Control of technical vulnerabilities	A.12.6.1 Management of technical vulnerabilities
A.13 Information security incident management	*A.16 Information security incident management*
A.13.1 Reporting information security events and weaknesses	A.16.1 Management of information security incidents and improvements
A.13.1.1 Reporting information security events	A.16.1.2 Reporting information security events
A.13.1.2 Reporting security weaknesses	A.16.1.3 Reporting information security weaknesses
A.13.2 Management of information security incidents and improvements	A.16.1 Management of information security incidents and improvements
A.13.2.1 Responsibilities and procedures	A.16.1.1 Responsibilities and procedures
A.13.2.2 Learning from information security incidents	A.16.1.6 Learning from information security incidents
A.13.2.3 Collection of evidence	A.16.1.7 Collection of evidence

(continued)

Table C.2 (continued)

ISO 27001 Controls	Corresponding ISO 27001:2013 Control(s)
A.14 Business continuity management	*A.17 Information security aspects of business continuity management*
A.14.1 Information security aspects of business continuity management	A.17.1 Information security continuity
A.14.1.1 Including information security in the business continuity management process	A.17.1.1 Planning information security continuity
A.14.1.2 Business continuity and risk assessment	-
A.14.1.3 Developing and implementing continuity plans including information security	-
A.14.1.4 Business continuity planning framework	-
A.14.1.5 Testing, maintaining and reassessing business continuity plans	-
A.15 Compliance	*A.18 Compliance*
A.15.1 Compliance with legal requirements	A.18.1 Compliance with legal and contractual requirements
A.15.1.1 Identification of applicable legislation	A.18.1.1 Identification of applicable legislation and contractual requirements
A.15.1.2 Intellectual property rights (IPR)	A.18.1.2 Intellectual property rights
A.15.1.3 Protection of organizational records	A.18.1.3 Protection of records
A.15.1.4 Data protection and privacy of personal information	A.18.1.4 Privacy and protection of personally identifiable information
A.15.1.5 Prevention of misuse of information processing facilities	-
A.15.1.6 Regulation of cryptographic controls	A.18.1.5 Regulation of cryptographic controls
A.15.2 Compliance with security policies and standards, and technical compliance	A.18.2 Information security reviews
A.15.2.1 Compliance with security policies and standards	A.18.2.2 Compliance with security policies and standards
A.15.2.2 Technical compliance checking	A.18.2.3 Technical compliance review
A.15.3 Information systems audit considerations	-
A.15.3.1 Information systems audit controls	-
A.15.3.2 Protection of information systems audit tools	-

Table C.3 Subcontrols in Annex A that are exclusive to ISO 27001:2013

A.6 Organization of information security
A.6.1.1 Information security roles and responsibilities
A.6.1.5 Information security in project management
A.8 Asset management
A.8.1.4 Return of assets
A.8.3 Media handling
A.9 Access control
A.9.2.2 User access provisioning
A.9.2.4 Management of secret authentication information of users
A.9.3.1 Use of secret authentication information
A.12 Operations security
A.12.6.2 Restriction on software installation
A.13 Communications security
A.13.2.4 Confidentiality or non-disclosure agreements
A.14 System acquisition, development and maintenance
A.14.1.2 Securing application services on public networks
A.14.1.3 Protecting application services transactions
A.14.2.1 Secure development policy
A.14.2.5 Secure system engineering principles
A.14.2.6 Secure development environment
A.14.2.8 System security testing
A.15 Supplier relationships
A.15.1 information security in supplier relationships
A.15.1.1 Information security policy for supplier relationships
A.15.1.2 Addressing security within supplier agreements
A.15.1.3 Information and communication technology supply chain
A.15.2 Supplier service delivery management
A.15.2.1 Monitoring and review of supplier services
A.15.2.2 Managing changes to supplier services
A.16 Information security incident management
A.16.1.4 Assessment of and decision on information security events
A.16.1.5 Response to information security incidents
A.17 Information security aspects of business continuity management
A.17.1.2 Implementing information security continuity
A.17.1.3 Verify, review and evaluate information security continuity
A.17.2 Redundancies
A.17.2.1 Availability of information processing facilities
A.18 Compliance
A.18.2.1 Independent review of information security

Appendix D
Template for Security Standards

We show our templates for describing security standards according to our conceptual model shown in Chap. 4 in Tables D.1, D.2, and D.3. We have elicited a series of questions for each building block, which shall help to fill in the required information. In addition, we stated which common terms (see Sect. 4.3) are relevant for each part of the template.

Table D.1 Security analysis context and preparation part of the template for security standard description

Security analysis context and preparation
Environment description
• Which essential parts of the environment have to be described?
• How do relations between these parts have to be described?
• What is the required abstraction level of the description?
Relevant common terms: *machine, environment*
Stakeholder description
• How are stakeholders defined?
• Which relation to the machine is required to be a stakeholder?
• Are there restrictions on stakeholders, e.g., do they have to be humans?
Relevant common terms: *stakeholder*
Asset identification
• How are assets identified?
• Which relation does a stakeholder have to an asset?
• Are assets categorized?
Relevant common terms: *asset*
Risk level definition
• What kinds of risk levels are defined?
• What is the required abstraction for these risk levels?
• How do the risk levels relate to assets and stakeholders?

© Springer International Publishing Switzerland 2015
K. Beckers, *Pattern and Security Requirements*,
DOI 10.1007/978-3-319-16664-3

Table D.2 Security analysis process part of the template for security standard description

Security analysis process

Security property description

- Do specific security goals have to be considered for assets, e.g., confidentiality?

- Which further security properties are used and how are they defined?

- What kind of methodology is required to elicit security goals?

Relevant common terms: *security goal, availability, confidentiality, integrity*

Control assessment

- How are existing security controls identified?

- Is it mandatory to described the threats that existing controls mitigates?

- Is it required to describe which assets an existing control protects?

Relevant common terms: *security control*

Vulnerability and threat analysis

- What kind of attacker model does the standard consider?

- Which activities does the standard demand for threat and vulnerability analysis?

- When is the threat and vulnerability analysis complete?

Relevant common terms: *attacker, vulnerability, threat*

Risk determination

- How is risk defined e.g. as a product of likelihoods and consequences?

- Is a process for risk management defined?

- Is a qualitative or quantitative risk determination required?

Table D.3 Security analysis product part of the template for security standard description

Security analysis product

Security assessment

- How are controls selected?

- Does a categorization exist for controls, e.g., types of threats the controls protect against?

- Do relations between controls have to be considered, e.g., one control has a working access control as a precondition?

Relevant common terms: *security requirements, policies*

Security measures

- What criteria are used to determine that a control is relevant to mitigate a particular threat?

- Is there a demand to describe the improved protection these controls provide?

- How is the reasoning done that the selected controls are sufficient and no further controls are required?

Relevant common terms: *security functions, policies*

Risk acceptance

- How are acceptable risk levels defined?

- Which kind of assessment determines that a security control reduces the risk to an acceptable risk?

- What kind of review is required to ensure that the risk is acceptable?

Security and risk documentation

- What methods are used to document the results e.g. templates, check lists?

- What kind of documents are required for certification?

- Can documents from other certifications be re-used?

Printed in the United States
By Bookmasters